# ENERGY, ENTROPY AND ENGINES

# ENERGY, ENTROPY AND ENGINES

## AN INTRODUCTION TO THERMODYNAMICS

**Sanjeev Chandra**
*University of Toronto, Canada*

To my wife Smita, whose support has made everything possible, and my sons Rohan and Varun, always my favourite students.

# Contents

# Preface

This book is a text for teaching a one-semester, introductory engineering thermodynamics course. Its most important goal is to make students understand the meaning of fundamental concepts such as energy, entropy, equilibrium and reversibility, which form the foundation of engineering science. It uses simple, direct language and relies on physical rather than abstract, mathematical definitions. Every new concept is introduced starting from first principles, and only after explaining why it is necessary.

Thermodynamics is different from most other engineering courses, in that it expects students to grasp an entirely new concept, entropy, which they have never encountered before. Traditional thermodynamics texts resort to giving a purely mathematical definition of entropy, and students learn to use the property for solving problems without ever forming a physical picture of what it means. This book introduces entropy by combining macroscopic definitions with statistical descriptions based on the energy distribution of molecules. Readers are not expected to learn statistical mechanics but use analogies to acquire an intuitive grasp of the concept of entropy and understand why the second law of thermodynamics is a result of the laws of probability.

Chapters 1 to 3 are intended for students to read on their own, with only selected portions being discussed in lectures. Chapter 1 describes how thermodynamics grew out of attempts to understand and improve steam engines and puts the first and second laws in context. It is useful in motivating the study of thermodynamics and explaining why it is such a fundamental part of science and engineering. Most students will already be familiar with some of the material covered in Chapters 2 and 3, including Newton's laws, the definitions of kinetic and potential energy, molar quantities and the ideal gas equation and can review these sections independently.

Thermodynamics textbooks typically start, immediately after the introduction, by teaching how to read tabulated properties of saturated liquids and vapours. Students are immediately overwhelmed by terms such as internal energy and enthalpy before they understand how these properties are used, and they are left with the impression that thermodynamics is largely an exercise in reading tables and charts. In the first six chapters of this book, while students are still becoming familiar with the laws of thermodynamics, there is no discussion of phase change. Once the second law has been understood liquid–vapour mixtures are treated as systems in equilibrium that can be analysed using the laws of thermodynamics. Chapter 7 starts with a brief discussion of the chemical potential and the Clausius–Clapeyron equation. Covering this material takes only one or two lectures and makes it much easier to understand phase equilibrium and the significance of property tables. However, if instructors prefer not to include it, it is possible to omit the relevant sections (Sections 7.3–7.7) without any loss of continuity.

It should be possible to cover the entire book in a one-semester introductory course that teaches the fundamentals of thermodynamics and their application in the analysis of heat engines and refrigerators. A slower paced course may leave out discussions of exergy (Section 6.14) and review only a selection of the engine and refrigeration cycles described in Chapters 9 and 10.

# About the Companion Website

This book is accompanied by a companion website:

**www.wiley.com/go/chandraSol16**
The website includes:

- Solutions for the Problems given at the end of each chapter.

# 1

# Introduction: A Brief History of Thermodynamics

---

**In this chapter you will:**

- Review the historical development of heat engines.
- Learn how thermodynamics grew out of efforts to improve the performance of heat engines.
- Gain an overview of concepts such as energy and entropy and the laws of thermodynamics.

---

## 1.1 What is Thermodynamics?

When earth's creatures were created, according to Greek legends, each received its own gift of speed or strength or courage. Some animals received wings to soar on, others claws to defend themselves, but finally, when it was the turn of humans, nothing remained. Prometheus saved mankind by stealing fire from the gods, making people far more powerful than any animal. Such myths – and similar stories exist in almost every society – trace the birth of human civilisation to the discovery of fire, which gave warmth, nourishment and the ability to craft objects out of stone and metal.

Fire alone would not have allowed humans to survive in the wilderness – they also needed tools. Life without sharp claws or fangs is possible if you can make knives and spearheads. Humans may not have the speed of a gazelle but they discovered wheels; levers and pulleys can lift heavier loads than any elephant. Tools improved slowly over time as wheelbarrows evolved into horse drawn carts and stones for grinding grain became windmills, but there were

---

*Energy, Entropy and Engines: An Introduction to Thermodynamics*, First Edition. Sanjeev Chandra.
© 2016 John Wiley & Sons, Ltd. Published 2016 by John Wiley & Sons, Ltd.
Companion website: www.wiley.com/go/chandraSol16

few gains made in the power used to drive them. Animals, water and wind were all harnessed to drive machines, but there is a limit to how effective any of these power sources are. Winds are unreliable, there are a finite number of sites with running water, and only a few horses can be hitched to a cart at one time. This lack of power sources limited how fast technology could evolve over most of human history. An ancient Egyptian, transported 30 centuries forward to medieval Europe, would have had little difficulty in recognising the machines used.

Then, a little over 300 years ago, fire was used as a power source for the first time. The first practical steam engine marked a turning point in human history, for it put enormous reserves of energy at our disposal. We are no longer restricted to capturing forces exerted by the elements or animals. Gases expand when heated and exert tremendous pressures that can be exploited to drive power plants, aircraft and automobiles. The only constraint on generating power is the amount of heat available, and the technology used to generate heat has advanced rapidly, whether it is by burning fuel, capturing solar radiation, or splitting atoms in nuclear reactions. Today, machines that use heat to produce work are everywhere.

As steam engines became more common, questions about them multiplied. What is the relation between heat and work? How much work can be obtained if a given amount of fuel is burned? Can the performance of engines be improved? Thermodynamics was the science that grew from efforts to answer these questions. The word itself is a combination of the Greek *therme*, meaning heat, and *dynamis*, meaning force, and thermodynamics is often defined as the science that studies the relationship between heat and work. Engineers struggling to understand how engines work formulated the principles of thermodynamics, but they have since been used in the study of all phenomena that involve changes in energy. Astrophysicists use the laws of thermodynamics to predict the fate of an exploding star, biologists apply them to the metabolism of animals and chemists rely on them to determine the products of chemical reactions.

## 1.2   Steam Engines

Using heat to produce motion is not a very novel achievement. As early as the first century a Greek inventor had designed a toy in which steam escaping from nozzles mounted on the surface of a metal sphere made it spin, but there seems to have been no practical application of this device. In subsequent centuries cannons became the most impressive illustrations of how objects could be transported by generating heat. But, no matter how spectacular the discharge of a cannon, it is difficult to harness it for any constructive purpose. For that we need a "heat engine", defined as a device that operates continuously, producing work as long as heat is supplied to it.

We can mark precisely the date when the first industrial heat engine was invented, for in 1698 the king of England was pleased to grant Thomas Savery a patent for a "fire engine" to be used "… for raising of water, and occasioning motion to all sorts of mill works …". Savery's machine did not resemble our typical image of a steam engine, for it had no furiously driving pistons or spinning flywheels. It consisted (see Figure 1.1) of a large chamber that was first filled with steam from a boiler, sealed and then sprayed with cold water to condense the steam in the vessel and create a partial vacuum that sucked water up from an underground mine. High-pressure steam was used to empty the chamber by pushing the water in it up to a

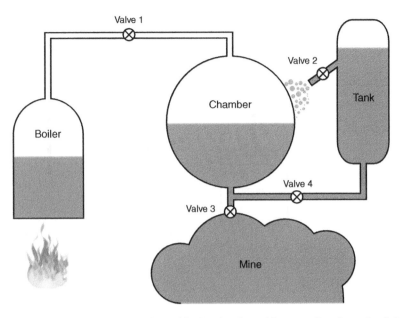

**Figure 1.1** Savery Engine. Opening valve 1 fills the chamber with steam. Opening valve 2 douses the chamber with cold water, condensing the steam and creating a partial vacuum. Opening valve 3 sucks water into the evacuated chamber from the flooded mine. Opening valve 4 while filling the chamber with steam pushes water into the tank.

higher level. Valves controlling the flow of steam and water were operated manually and a good operator could complete several cycles in a minute.

Savery intended to sell his pumps to English coal mines where flooding was a frequent occurrence, so that water had to be drained by hand or horse driven pumps. Sadly, his pumps proved to be rather leaky so that it was hard to hold a very good vacuum in the chamber. Savery claimed that his engine could raise water by about 80 feet (24.4 m), which would have required steam pressures of almost three atmospheres. Frequent explosions of poorly made boilers were so common that high-pressure steam was viewed with fear and not used again for more than a century when manufacturing techniques had greatly improved.

Savery was unable to produce a commercially successful engine, but he proved that steam could be used to drive machines. By the time Savery's 14-year patent expired in 1712 Thomas Newcomen was ready with his design for a new engine, which looks far more recognisable to us as a steam engine (Figure 1.2). Newcomen had a piston moving back and forth in a cylindrical chamber, one side of which was connected to a boiler producing steam. When the chamber was filled with steam the piston rose up. Spraying water into the chamber condensed the steam, producing a partial vacuum so that atmospheric pressure forced the piston down. A beam connected to the piston oscillated as the piston moved up and down, which could be used to drive a pump or some other machine. This was an "atmospheric engine", in which work was done by the atmosphere pushing a piston against a vacuum. Steam pressure was never much higher than one atmosphere, minimising the hazard of explosions.

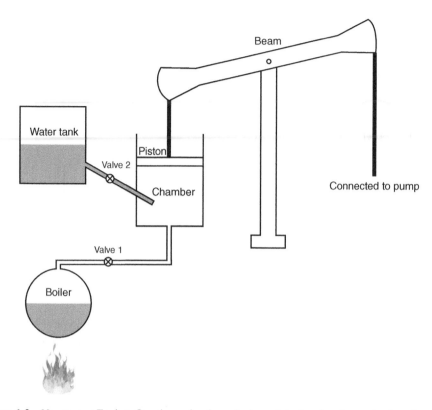

**Figure 1.2**  Newcomen Engine. Opening valve 1 sends steam into the chamber and raises the piston. Opening valve 2 sprays water into the chamber, condensing the steam so that atmospheric pressure forces the piston down. A beam connected to the piston oscillates up and down and drives the pump.

Newcomen's engine was an immense success and hundreds were built and sold. They were initially used to power pumps in coalmines but soon found new applications in textile mills and other factories. For over 50 years Newcomen's engines represented the most sophisticated technology available and sparked a remarkable technical and social transformation that changed human history. For the first time machines could work non-stop without depending on beasts of burden or being subject to the vagaries of weather. Any factory had access to as much power as it needed, no matter where it was located. The steam engine gave birth to the industrial revolution and created the modern world.

Newcomen's engines were a tremendous accomplishment but consumed enormous amounts of coal to generate steam, most of which was wasted. At the start of each cycle, when steam entered the cylinder, much of its energy went into heating the walls of the cylinder, only to have to cool them down again when the steam was condensed with a water spray. While engines were confined to coalmines this was not of great concern since fuel was practically free, but when they began to be used in factories far from fuel supplies operating costs became a serious problem.

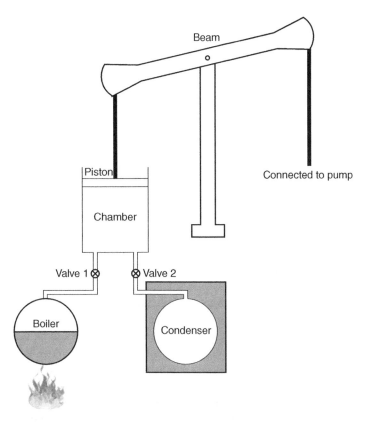

**Figure 1.3**  Watt's engine with an external condenser. Opening valve 1 allows steam from the boiler to fill the chamber and raises the piston. Opening valve 2 lets steam from the chamber escape into the condenser, creating a vacuum under the piston so that atmospheric pressure pushes the piston down.

James Watt, a young instrument maker at the University of Glasgow, proposed a solution. Watt had been assigned the job of fixing a model of a Newcomen engine used for laboratory demonstrations. He found that the model worked as designed, but consumed all the steam supplied simply to heat the cylinder wall. After much thought he solved the problem by adding an external chamber in which steam was condensed (Figure 1.3). In 1769 Watt obtained a patent for the external condenser, entitled a "new method for lessening the consumption of steam and fuel in fire engines". The new engines were so much more efficient than the older Newcomen engines that it became feasible to use them in many new industrial applications.

Many brilliant inventors have pioneered new technologies, only to see others reap the profits. Watt escaped this fate for he had the good fortune of entering into partnership with Matthew Boulton, an extremely shrewd businessman and manufacturer. For decades the firm of Boulton and Watt held the most important patents related to steam engines, enforced them vigorously and dominated the steam engine manufacturing industry. They produced atmospheric engines that used the pressure of the atmosphere to drive their pistons; steam was used only to produce a vacuum. Watt strongly opposed the use of high-pressure steam to drive engines, firmly convinced that they were too dangerous, and even tried to get the British parliament to pass a law banning high-pressure steam. Even without the law, his control of patents on the external condenser and several other technologies essential to steam engines ensured that he could block any new development that he did not agree with. Watt's fear of high-pressure engines was well founded in his experience of boiler explosions in the early days of steam, but manufacturing techniques were also improving rapidly. In 1776 John Wilkinson invented a new type of lathe that made it possible to bore cylinders up to 18 inches (46 cm) in diameter with great precision. Wilkinson had designed his machines to produce better cannon barrels, but they also proved eminently suitable for making engine cylinders with piston seals strong enough to withstand high pressures.

James Watt's patent on the external condenser finally expired in 1800 and the field of engine design was again open to new developments. That same year, Richard Trevithick built a high-pressure steam engine to operate a pump in a Cornish mine. In his engine, steam at more than twice atmospheric pressure pushed against the piston to deliver power and was then released into the air instead of being condensed. The sound of steam being vented led to the engines being popularly known as "puffers". Once it was demonstrated that high-pressure engines could be safely built and run, their advantages became immediately obvious. A relatively small engine could deliver more power than a much larger atmospheric engine and, even better, it had no need for a heavy condenser. Such a light engine could be used to power a vehicle and by 1804 Trevithick had built a locomotive that could move 25 tons (25.4 t) at a speed of 3.7 miles / h (6 km / h). This was a remarkable machine for no one had ever seen a self-propelled vehicle before, but it was not a commercial success since it frequently broke down and the iron rails available at that time could not withstand the weight of the locomotives for long. The now elderly James Watt launched virulent attacks, railing that Trevithick "deserved hanging for bringing into use the high-pressure engine".

George Stephenson finally solved the technical problems related to both engines and rails when he built his locomotive in 1813 and was operating a commercial train service by the time Trevithick died in poverty in 1833. Stephenson is now acclaimed as the inventor of the locomotive while Trevithick is rarely remembered, but Trevithick's firm belief that high-pressure engines could be safely operated and would prove more efficient was vindicated as operating pressures steadily increased.

It is frequently said that the great age of steam is over, and it is certainly true that there are not many steam locomotives running today. However, steam is used to produce electricity in power plants in every part of the world. In a modern steam power plant water is pumped into boilers where it is heated as it flows through tubes (Figure 1.4). Hot gases flow across the outer surfaces of the boiler tubes, heating the water until it emerges as high-pressure steam. The heat source is most often a burning fuel, but can also be a nuclear reaction or a renewable source such as sunlight. The steam is fed into a turbine, jetting out of nozzles at velocities greater than the speed of sound, and impinges on turbine blades projecting from a rotating shaft. The turbine shaft spins and when connected to an electric generator produces electricity. Low-pressure steam emerging from the turbine passes through a condenser, where it flows through tubes that are cooled by water running over them. Water condensing in the tubes is collected and pumped back into the boiler to complete the cycle.

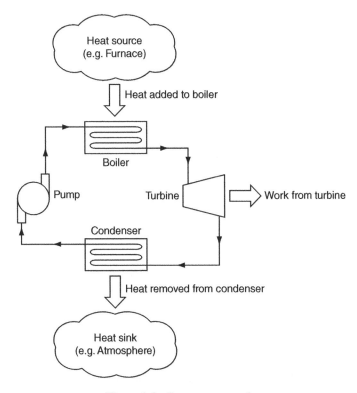

**Figure 1.4**   Steam power cycle.

## 1.3   Heat Engines

The first question a mine owner thinking of installing Newcomen's steam engine probably asked was "How much does this cost to run?" Methods of describing the performance of heat engines are almost as old as engines themselves. Operators using an engine for pumping water out of a mine defined the "duty" of an engine as the pounds of water raised one foot in height when one bushel (about 80 lbs; 36 kg) of coal was burned. Early Newcomen engines had a duty of 4.7 million ft.lb. / bushel which increased to 9.0 million ft.lb. / bushel with gradual improvements in design and construction.

Figures for duty were helpful in mines, but of no use when engines began to be used in textile mills for turning looms. Engines were typically replacing horses and for marketing purposes it made sense to compare the capacity of engines with the ability of a horse to lift a weight by a given height in a given length of time. One foot-pound (ft.lb.) was defined as the work required for raising a 1 lb weight by 1 ft. James Watt fixed one horsepower at 33 000 ft.lb. per minute. The unit still survives today, even though it would be a rare buyer who cares about comparing engine performance with that of a horse.

The horsepower of an engine tells us about its output – the rate at which it can do work – but nothing about the input, how much heat has to be supplied to it. We need a more general way of looking at heat engines. *A heat engine is any continuously operating device to which we*

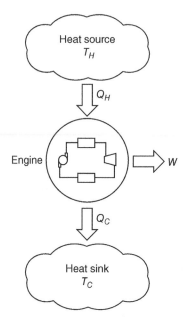

**Figure 1.5**   A schematic representation of a heat engine.

*input heat and from which we extract work.* This is a broad enough definition to encompass everything from Savery's engine to a modern nuclear power plant. We can represent a heat engine with a diagram such as Figure 1.5, in which heat ($Q_H$) from a high temperature ($T_H$) source (a coal furnace or nuclear reactor) is supplied to the engine. The engine does work ($W$) and loses heat ($Q_C$) to a low temperature ($T_C$) heat sink (the atmosphere or a river).

The question we would like to answer is: for a given heat input ($Q_H$) what is the maximum work output ($W$) that can be achieved? But how do we compare work with heat? What is the link between them? In the eighteenth century, there did not seem to be any obvious connection. In the era of James Watt, work was measured in ft.lbs. and heat in British Thermal Units (BTUs), defined as the amount of heat required to heat 1 lb of water by 1 °F (approx. 0.45 kg, 0.6 °C). Relating the two seems like comparing apples and oranges. But asking this question was the first step towards developing the science of thermodynamics.

## 1.4   Heat, Work and Energy

When Newcomen and Watt were designing their engines, the scientists of the day could offer them little assistance. There was still no clear understanding of what heat was. The prevailing theory, supported by the great French chemist Antoine Lavoisier, was that heat was a weightless, invisible fluid known as "caloric" that flowed from high temperatures to low temperatures. Most physical phenomena associated with heating an object could be explained with this theory. As caloric flowed into an object its temperature rose. Caloric seeping into the interstices between the atoms of an object being heated made it expand. Everybody had a maximum amount of caloric that it could contain, its "heat capacity". When this limit was reached in a liquid it was "saturated" with caloric, at which point it changed to vapour. Rubbing two objects together squeezed caloric out of them and produced heat. Caloric theory has disappeared from textbooks, but we still use phrases such as "heat flow", "heat capacity" and "saturated liquid", evidence of the dominant role that it once played in science.

A fundamentally new way of thinking about heat was developed by three men who were not academic scientists, but amateurs with little formal training. Their lack of academic credentials allowed them to reject established wisdom, but it proved very difficult for them to get their ideas accepted, especially since they were directly contradicting Lavoisier, the most famous scientist of the era.

The first challenge to caloric theory came from Benjamin Thomson, an American born in the town of Concord, New Hampshire, which had formerly been known as Rumford. A loyalist during the American Revolution, he acted as a spy for the British, sending dispatches in an invisible ink of his own formulation. After the British surrender he left for England in 1776, where he found fame and fortune by inventing a stove that was efficient and smoke-free. He then entered the service of the Duke of Bavaria, rising to the rank of major general in charge of the defence of Munich. When offered a title of nobility he selected the name of his old New England hometown and called himself Count Rumford, the name by which he is remembered today. Rumford was in charge of manufacturing brass cannons that were bored out with a huge lathe, when he noted the heat generated by cutting tools rubbing against metal. He devised an experiment in which a very dull boring bit was used on a cannon submerged in water, whose temperature rose until it boiled. Rumford concluded that if heat could be generated without limit it "cannot possibly be a material substance" since it was impossible for caloric to be produced endlessly.

Rumford's presentation of his experiments was met mostly by incomprehension and indifference. Supporters of the caloric theory speculated that perhaps fresh caloric was continuously uncovered as the drill removed layers of metal. Even worse, Rumford had no alternate theory of heat to offer. He speculated vaguely that heat is a form of motion – but motion of what? Rumford had nothing more specific to add. For lack of any better ideas caloric theory continued to hold sway and Rumford's work was virtually ignored.

Inspiration sometimes strikes in the most unexpected places. The insight that had eluded generations of the best scientists in the world came from an odd source – a provincial German doctor. When the 26 year old Robert Julius Mayer signed on as a doctor on board a Dutch ship bound for Indonesia in 1840, he was looking for nothing more than some adventure before settling down to his small town medical practice, but the observations he made during the voyage diverted his life onto an unexpected path. When the ship was in tropical waters he observed that the blood of his patients was bright red, a sign of high oxygen concentration in the blood. Mayer knew that the human body is heated by oxidation of carbon in food to produce heat and he concluded that in a hot climate less heat is required and therefore less oxygen is used. This was an interesting observation, but hardly revolutionary. It was the astounding leap of intuition that followed that led Mayer to a radically new way of thinking about heat and work.

Mayer reasoned that the same metabolic processes that generate heat in the human body also allow it to do work, so somehow heat and work must be related. Mayer was familiar with Lavoisier's statement of the law of conservation of mass: during chemical reactions the outward form of matter can change significantly – from liquid to vapour for example – but the total amount of mass remains the same. Perhaps there was a similar relation between heat and work – there was some underlying quantity that is conserved and cannot be destroyed, but appears in various forms. He was struggling for a term that would encompass both work and heat, what we now call energy, but the word did not have the same meaning at the time. Mayer called it "force". His 1842 paper concluded that this "force" was "indestructible and trans-formable", the earliest statement of a law of conservation of energy. He did no experiments but he gave an estimate for the mechanical equivalent of heat – the quantity of heat generated when a certain amount of work is done – purely from theoretical analysis of published data on temperature changes in gases as they are expanded or compressed.

Mayer's idea that heat and work were equivalent would eventually transform science, but it made little impact on the academic world when he first proposed it. Mayer's exposition was confused, he had little training in physics, and his mathematics was faulty. Most established researchers chose to simply ignore the paper. Deeply wounded by the rejection and facing family tragedies, Mayer despaired of receiving recognition and attempted suicide in 1850. He spent the next several years in institutions seeking help.

If Robert Mayer was an unlikely scientist, James Prescott Joule had been raised from birth for that role. The heir of a wealthy English brewer, he had been tutored at home by the famed chemist John Dalton. For the rest of his life Joule was to pursue both vocations, tending to the brewing business by day and pursuing his scientific pursuits as a gifted amateur. The two interests were not unrelated – brewing made him an expert at temperature measurements, indispensable in keeping vast brewing vats at controlled conditions. It also made him keenly aware of steam engines and the new technology of electric motors. In 1840, when he was just 22, his experiments led him to discover the relation between the heat released in a resistor and the current flowing through it, which is still known as Joule's law.

Unaware of Mayer's work on work and heat, Joule devised a simple experiment to measure the mechanical equivalent of heat. The apparatus he used was simple but very ingenious, con-sisting of a paddle wheel that rotated in a closed, well insulated container filled with water (Figure 1.6). A falling weight pulled a rope wrapped around the shaft of the paddle, making it rotate. The height of fall gave precisely the amount of mechanical work done in driving the paddle and a thermometer in the water measured the corresponding rise in temperature. After a series of tests, Joule calculated a value for the mechanical equivalent of heat that was within

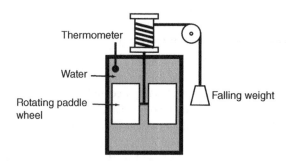

**Figure 1.6**   Joule's apparatus to determine the mechanical equivalent of work.

1% of modern measurements, an impressive feat of experimental research. But when Joule first presented his results in 1845 in Britain few found them noteworthy and his paper to the Royal Society was initially rejected.

Joule's work may have excited little comment initially, but he was not fated to face the total rejection that Mayer had. Attending his presentation was a young professor named William Thomson, who was later to become one of the most famous scientists in Britain and given the title of Lord Kelvin. Kelvin immediately grasped the importance of Joule's measurements and did much to publicise his work. Now it was time for the amateurs to make way for the professionals, and in 1847 a German professor, Hermann von Helmholtz, gave a talk to the Berlin Physical Society in which he used the phrase "conservation of energy" for the first time. In precise mathematical terms he showed that heat and work are both different forms of a fundamental property known as energy. He clearly stated the first law of thermodynamics, that energy can be transferred in the form of either work or heat, but it cannot be destroyed or created.

Helmholtz also made amends for an old injustice when he referred to the insights of both Mayer and Joule in his publications and recognised their pioneering work. By the time Mayer died in 1878 he had received many awards from the German government and the highest honour given by the British Royal Society. His greatest memorial, though, remains the principle of conservation of energy that is now one of the foundations of physics.

## 1.5   Energy and the First Law of Thermodynamics

Energy is a word that we are extremely familiar with in the modern world. We constantly monitor how much we use; we debate the price of it; we worry whether the world has enough. The concept of energy, though, is rather difficult to explain. In physics classes we define kinetic and potential energy and claim that these are related to chemical energy found in coal and electrical energy carried by copper wires. What makes all of these things the same?

A system is said to possess energy if it is capable of lifting a weight. When we raise a mass being pulled down by gravity we increase its potential energy, because if it is dropped it can lift another body to which it is tied by a rope passing over a pulley. A moving mass has kinetic energy: it is not difficult to devise a mechanism that uses its motion to lift another weight. A fuel contains energy in its chemical bonds, which can be released by burning it. Figure 1.7 shows a cylinder, one of whose walls is a movable piston, which contains a fuel gas. When the gas is ignited it heats and expands, lifting the piston. The motion of the piston can be used to do work – this is how a car engine works.

Energy can also be extracted in the form of heat. If a moving projectile hits a wall it comes to rest and all its kinetic energy is released as heat. Instead of letting the gas in Figure 1.7 expand and do work, we can lock the piston in place and let the gas cool. In both instances we have extracted energy from the cylinder, in the first case as work and in the second as heat. Work and heat are both measured using the same units, fittingly called joules.

The conservation of energy principle, known as the First Law of Thermodynamics, states that energy can neither be created nor destroyed. In a continuously operating engine the energy supplied must equal the energy leaving. Energy is added to the engine as heat ($Q_H$) and leaves

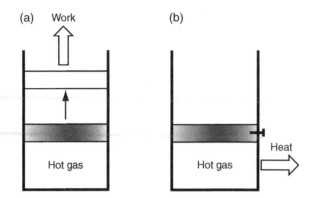

**Figure 1.7** Energy extracted from a cylinder containing hot gas: (a) work is done as the gas expands and the piston is raised, (b) heat is removed from the cylinder while the piston is locked in place.

it both as work (*W*) and waste heat ($Q_C$). We can write an energy balance:

$$Q_H = W + Q_C. \tag{1.1}$$

We can now return to our original question – how do we describe the performance of an engine? If work and heat are both forms of energy, with the same units, we can calculate their ratio and define the thermal efficiency of an engine:

$$\eta_{th} = \frac{W}{Q_H}, \tag{1.2}$$

or, substituting Equation (1.1) into Equation (1.2),

$$\eta_{th} = \frac{Q_H - Q_C}{Q_H} = 1 - \frac{Q_C}{Q_H}. \tag{1.3}$$

The thermal efficiency is a simple measure of the performance of an engine that tells us what fraction of the energy we supply as heat is recovered in the form of work

A calculation of the efficiency of a Newcomen Engine gives $\eta_{th} = 0.34\%$. Almost all the heat generated by burning coal was lost to the surroundings and went into heating up the chamber. James Watt, by providing an external condenser, increased efficiency more than 10-fold, to about 4%. This may not sound like a lot, but it was enough to power the industrial revolution and reshape the world. Further improvements increased engine efficiency to about 15% by the middle of the nineteenth century.

So how efficient are modern heat engines, the products of two centuries of research and development? The answer is surprising: not very. A typical steam power plant has an overall thermal efficiency of $\eta_{th} = 30\%$. Approximately two-thirds of the heat generated by burning an expensive fuel such as natural gas or oil is wasted, with only one-third converted to useful work.

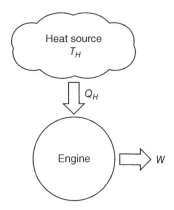

**Figure 1.8**   An engine with 100% efficiency.

Why are engine efficiencies so low? Could we increase them significantly through smarter engineering? It is quite easy to think of how we could get $\eta_{th} = 100\%$. If we build an engine in which there is no heat loss so $Q_C = 0$, like the one shown in Figure 1.8, then $\eta_{th} = 100\%$. Is it possible to develop an engine like this? If not, what is the highest efficiency that is achievable?

## 1.6   The Second Law of Thermodynamics

What is the maximum efficiency that a heat engine can have? This was a question raised almost 200 years ago by a young French engineer, Sadi Carnot. It was much more than an abstract puzzle for Carnot – whoever found the answer, he was convinced, would help his country achieve global supremacy.

Sadi Carnot was born in a family close to the centre of power in revolutionary France. His father, Lazare Carnot, was a brilliant engineer who was an expert on water driven turbines and also served as minister of war in Napoleon's cabinet. Sadi Carnot was tutored in science, mathematics and languages by his father and enrolled in the École Polytechnique in Paris where he was trained by some of the finest mathematicians and scientists in the world. He joined the army as a military engineer but Napoleon's defeat in 1815 left Carnot with little employment, giving him time to develop his interest in steam power. Resumption of trade with Britain gave him a chance to see the new engines developed by Watt and he realised that French technology had fallen far behind. Even worse, in his view, men with little scientific training had been trying to improve engines in a laborious process of trial-and-error. Carnot had in mind a far more ambitious undertaking: he would discover the theoretical principles that governed the operation of all engines and use these to design the most efficient engine possible.

For decades engineers had laboured to improve the performance of steam engines. Better lubrication, improved manufacturing tolerances, higher operating pressures had all helped to slowly edge efficiencies upwards. Carnot envisioned a much greater leap of the imagination. Let us take all engineering improvements as given, so that we have the perfect engine, frictionless, not leaking any steam and losing no heat to the surroundings. What would be the efficiency of this marvellous machine?

Carnot published the results of his study in 1824 in a book entitled *Reflections on the Motive Power of Fire*. In it he focused on the considerable loss of heat from an operating engine to its surroundings. Was there any way of completely eliminating this heat loss? Carnot's answer was no, there was no possibility of building a heat engine that did not entail heat rejection, and therefore an engine could never achieve 100% efficiency. But why was this? Carnot argued by analogy. Just as a water wheel needs water flowing continuously through it to operate, so an engine needs to have caloric flowing through it. If the outlet of a water wheel is blocked water will accumulate and soon stop the wheel from turning, so will caloric accumulate in an engine and stop it. Carnot was not satisfied with this explanation but, given the current status of knowledge about heat, it was the best that he could come up with. In spite of the flaws in reasoning his fundamental insight – that a heat engine cannot operate without losing heat – was correct and was later recognised as the second law of thermodynamics, which was actually discovered before the first.

Carnot's book was the only one that he ever published. In June 1832 a cholera epidemic swept through Paris and he fell ill and died at the age of only 36 years. Notes discovered in his papers, which were not understood until they were examined decades after his death, showed that he was preparing to abandon caloric theory and conduct experiments almost identical to those done by Joule 20 years later. Given a little more time Carnot would probably have discovered both the first and second laws of thermodynamics.

Carnot's reasoning may have been based on a false theory, but his conclusion was correct: an engine must lose heat to the surroundings to operate. Equally surprising was another deduction from Carnot's analysis – the efficiency of an ideal engine depended only on the temperatures of the heat source and heat sink it operated between. This implied that once the temperature of the boiler and the condenser were fixed, so was the engine efficiency. The working fluid in the engine, whether air or steam, made no difference. This was a truly remarkable result, but it had little impact at the time. Few practicing engineers read Carnot's book, viewing it as extremely abstract. Lay readers, for whom Carnot wrote the book, did not understand its significance.

By the middle of the nineteenth century the technology of steam engines was well developed but the science to explain them was still rudimentary. Caloric theory was on its way out and the idea of conservation of energy was beginning to be accepted. Carnot's book was out of print and almost forgotten. But all the pieces required to develop a comprehensive theory of thermodynamics were in place, waiting for someone to put them together.

Rudolf Clausius was a brilliant theoretical physicist who served as professor in most of the leading universities in Germany and Switzerland. In 1850 he published a paper that established the modern science of thermodynamics. He had before him the work of Helmholtz on the conservation of energy and, with a clear understanding of energy, he could revisit the work of Carnot and put it on a sound theoretical footing, explaining why a heat engine must continuously reject heat.

The first law tells us that transferring $Q_H$ joules of heat from a source at temperature $T_H$ to an engine increases its energy by an amount $Q_H$. Clausius postulated that heat transfer also makes another, previously undefined, property of the engine increase by an amount $Q_H / T_H$. He named this property "entropy", from the Greek word for transformation, deliberately choosing a name similar to "energy" to emphasise the relation between the two properties.

Adding or removing energy in the form of heat to any system always changes its entropy. When the engine loses $Q_C$ joules of heat to a sink at temperature $T_C$ its energy decreases by $Q_C$ and its entropy by $Q_C / T_C$.

According to the first law the net change in energy ($\Delta E$) of an engine must be zero after it has completed a heat addition and removal cycle. Rearranging Equation (1.1),

$$\Delta E = Q_H - W - Q_C = 0. \tag{1.4}$$

Clausius's work led to the development of the second law of thermodynamics which stated that over a cycle the net change in entropy ($\Delta S$) of the engine must also be zero. Stated mathematically,

$$\Delta S = \frac{Q_H}{T_H} - \frac{Q_C}{T_C} = 0. \tag{1.5}$$

Rearranging Equation (1.5) and substituting into Equation (1.3) gives

$$\eta_{th} = 1 - \frac{T_C}{T_H}. \tag{1.6}$$

The second law of thermodynamics leads to the same conclusion that Carnot had reached earlier – that the efficiency of an engine depends only on the temperatures of the heat source and sink.

But what is the physical meaning of the ratio $Q / T$? We often combine other seemingly unrelated properties: for example, mass times velocity gives momentum. But whereas it is easy to understand the physical significance of momentum – being hit by a flying ball gives a quick appreciation of it – classical thermodynamics offered no easy insight into the meaning of entropy. It is useful here to draw a parallel with energy, another concept that we use very frequently but that is hard to visualise. We lift a weight or heat up a gas and say that energy has increased, but we have no direct way of measuring this change. Instead we measure other properties such as the elevation of the weight or the temperature of the gas and use these to calculate changes in energy. Similarly, we can calculate changes in entropy of a gas by measuring its temperature and pressure and developing mathematical equations that tell us how entropy depends on these properties.

## 1.7  Entropy

But we are still left with that nagging question: what does entropy mean? It is a word that is used quite frequently in ordinary language, often as a synonym for "chaos" or "disorder". An Internet search for "entropy" yields links to art galleries and rock bands among others. What does any of this have to do with engines?

The nineteenth century saw, for the first time, the use of atomic theory in physics. The idea that matter was composed of atoms had been around since ancient times, but it was an idea that was difficult to prove or disprove, or to apply to any practical problem. Several scientists had proposed treating gases as a mass of molecules in motion, but nobody could conceive a method of applying the laws of mechanics to the unimaginably large number of molecules present in even a tiny volume of gas. The English physicist James Clerk Maxwell showed in 1866 how to solve what had appeared to be an insurmountable problem. When dealing with large populations it is difficult to predict the behaviour of any individual element, but probability theory can calculate average values very accurately. Assuming that the motion of each molecule in a gas is random, Maxwell showed how to use statistical theory to predict the average energy of molecules, while noting ruefully that probability was generally associated with "… gambling, dicing and wagering, and therefore highly immoral …". His results showed that the temperature of a gas is a measure of its molecular velocity: the hotter a gas, the faster its molecules travel. Summing the energy of all the molecules gives the total energy stored in the gas. Raising the temperature of a gas therefore increases its energy.

Maxwell's work linked energy, which had been considered a macroscopic property of a substance, to its molecular state. Perhaps entropy, which is also a macroscopic property, could also be interpreted in the same way?

Ludwig Boltzmann was a professor of physics at the University of Graz in Austria who crossed the last major hurdle in developing a science of statistical thermodynamics when he gave a molecular interpretation of entropy in 1871. He showed that the properties of a substance depend not only on how much energy it has, but also how that energy is distributed among its molecules. We can illustrate this idea by imagining a container filled with gas in which all molecules have exactly the same, low, value of energy. A few, very fast-moving molecules are introduced into one corner of the container, which is then sealed and left undisturbed. As the high-energy molecules collide with slower moving ones, some will accelerate and others decelerate until eventually there is an equilibrium distribution of velocities, with a few molecules moving slowly, a few very fast and the remainder with a range of velocities varying between the two extremes. Boltzmann related the entropy of the gas to the energy distribution of molecules. The gas in its initial state, with a few molecules possessing most of the energy, has low entropy. As energy is transferred from one molecule to another and spreads out in space entropy increases. Entropy is therefore a measure of how well the energy of a system is distributed among its molecules.

The molecular definition of entropy gives us new insight into this property. If a small amount of hot gas is injected into a large volume of cold gas and the mixture left isolated, energy diffuses out from the region where it was initially concentrated as molecules collide and exchange energy. Macroscopically we would observe this as a transfer of heat that produces an increase in entropy until the gas reaches equilibrium with the same temperature everywhere. However, the converse is never seen. If a gas is at equilibrium its molecules never rearrange themselves spontaneously so that all the energy in the system is confined to a small region. The entropy of an isolated system always increases until it reaches equilibrium and can never decrease. This is an observation with profound consequences, since it implies an asymmetry in nature: in an isolated system some processes only proceed in one direction and cannot be reversed.

Today the principles of thermodynamics are applied in every field of science and technology. Using the same fundamental theories mechanical engineers calculate the efficiency of energy conversion processes, chemical engineers analyse the products of chemical reactions, materials scientists predict the structure of alloys, biologists investigate the functioning of living cells and astronomers study the entire universe. The laws of thermodynamics that we will study in the following pages have been used in ways that pioneers such as Carnot, Joule, Maxwell and Boltzmann could never have imagined.

## Further Reading

1.  H. C. Von Baeyer (**1999**) Warmth Disperses and Time Passes: The History of Heat, Modern Library Paperbacks, London.
2.  M. Goldstein, I. F. Goldstein (**1995**) The Refrigerator and the Universe: Understanding the Laws of Energy, Harvard University Press, Harvard.
3.  P. Atkins (**2010**) The Laws of Thermodynamics: A Very Short Introduction, Oxford University Press, Oxford.
4.  D. S. L. Cardwell (**1971**) From Watt to Clausius – The Rise of Thermodynamics in the Early Industrial Age, Cornell University Press, Cornell.

# 2

# Concepts and Definitions

> **In this chapter you will:**
>
> - Review the fundamentals of Newtonian mechanics, including mechanical work, kinetic and potential energy.
> - Identify various types of thermodynamic systems such as open, closed and isolated systems.
> - Define thermodynamic properties.
> - Distinguish equilibrium from steady state.
> - Classify different types of equilibrium including mechanical, thermal and phase equilibrium.
> - Learn about thermodynamic states, processes and cycles.
> - Become familiar with problem solving methods in thermodynamics.

## 2.1 Fundamental Concepts from Newtonian Mechanics

How do you describe colours to someone who cannot see? Some ideas are so fundamental that the only way to grasp them is to use your own senses. All sciences are based on a small number of concepts that have to be understood intuitively before it is possible to conduct any meaningful discussion. In Newtonian mechanics there are three fundamental concepts that cannot be explained in terms of simpler ideas – we have to experience them ourselves to know what they are. These three concepts are: length, mass and time.

*Energy, Entropy and Engines: An Introduction to Thermodynamics*, First Edition. Sanjeev Chandra.
© 2016 John Wiley & Sons, Ltd. Published 2016 by John Wiley & Sons, Ltd.
Companion website: www.wiley.com/go/chandraSol16

## 2.1.1   Length

All physical objects occupy space. We specify the size of a body in any one dimension by measuring its *length* (L). Since length is a fundamental concept, it is measured using arbitrarily defined units. In the International System of units that we use (abbreviated as SI from its French title, *Système Internationale*) the unit of length is the *metre* (m). The metre was initially defined in 1799, when the distance between the pole and the equator of the Earth, measured along a line of longitude, was specified to be 10 million metres. A one-metre long platinum bar was made and kept in a vault in Paris and used as a standard. Since then more precise definitions have been adopted, based on the distance light travels in a given time interval, but the unit is still arbitrary.

Having established the concept of length, we can use it to define the *position* (x) of a body, which is its distance from the origin of a coordinate system that we have defined. The *displacement* (Δx) of a body in motion is the distance between its initial and final position. The concept of length can also be extended into two and three dimensions, to give us measurements of *area* (A = L²) and *volume* (V = L³).

## 2.1.2   Mass

All matter has *mass* (m). Mass is measured in *kilograms* (kg), which is again an arbitrary unit equivalent to the mass of a platinum–iridium cylinder kept by the International Bureau of Weights and Measures in Paris. It is very difficult to give a precise definition of mass: we detect that an object has mass by observing its interactions with the surroundings. For example, we know that any mass in the earth's gravitational field experiences a force, its weight, which we can measure.

## 2.1.3   Time

We measure the passage of *time* (t) in units of seconds (s). Whereas we can move backwards and forwards in space, we can only move in one direction in time. This is an important distinction, one that we will include in our development of thermodynamics when we discuss the second law of thermodynamics.

## 2.2   Derived Quantities: Velocity and Acceleration

We can combine these three fundamental quantities to get additional properties. *Velocity* (**V**) is defined as the rate of change of displacement:

$$\mathbf{V} = \lim_{\Delta t \to 0} \frac{\Delta x}{\Delta t} = \frac{dx}{dt}. \tag{2.1}$$

The units of velocity are those of length/time (L / t), or m / s. Similarly, the rate of change of velocity is defined to be the *acceleration* (a)

$$a = \lim_{\Delta t \to 0} \frac{\Delta \mathbf{V}}{\Delta t} = \frac{d\mathbf{V}}{dt} = \frac{d^2 x}{dt^2}. \tag{2.2}$$

The units of acceleration are m / s².

**Example 2.1**
**Problem:** The position of a body along the $x$ axis, in metres, is given by the equation $x = t^3 - 30t^2 + 5$, where $t$ is the time in seconds. Find its velocity and acceleration as a function of time.
**Find:** Velocity $V(t)$, and acceleration $a(t)$ as functions of time
**Known:** Position as a function of time, $x(t)$.
**Governing Equations:**

Velocity
$$V = \frac{dx}{dt}$$

Acceleration
$$a = \frac{d^2x}{dt^2}$$

**Solution:**

$$x = t^3 - 30t^2 + 5$$

$$V(t) = \frac{dx}{dt} = 3t^2 - 60t$$

$$a(t) = \frac{d^2x}{dt^2} = 6t - 60$$

**Answer:** The velocity as a function of time is $3t^2 - 60t$, and the acceleration as a function of time is $6t - 60$.                                                                                    ■

**Example 2.2**
**Problem:** Brakes are applied to a car travelling at 100 km / h so that its speed decreases to 30 km / h over a distance of 125 m. Find the acceleration, assuming it to be constant.
**Find:** Acceleration $a$.
**Known:** Initial velocity $V_1 = 100$ km / h, final velocity $V_2 = 30$ km / h, distance travelled during braking $\Delta x = 125$ m.
**Assumptions:** Acceleration is constant.
**Governing equations:**

Velocity
$$V = \frac{dx}{dt}$$

Acceleration
$$a = \frac{dV}{dt}$$

**Solution:**

$$a = \frac{dV}{dt},$$

Given $V = V_1$ when $t = 0$,

$$V = \int a\,dt = at + V_1 = V_2.$$

Therefore,

$$t = \frac{V_2 - V_1}{a}.$$

Given $x = 0$ when $t = 0$,

$$x = \int V dt = \frac{1}{2}at^2 + V_1 t = \Delta x.$$

We want to eliminate $t$ in this equation. Substituting for $t$ from above,

$$x = \frac{1}{2}a\left(\frac{V_2 - V_1}{a}\right)^2 + V_1\left(\frac{V_2 - V_1}{a}\right).$$

Simplifying,

$$a = \frac{V_2^2 - V_1^2}{2\Delta x},$$

$$V_1 = 100\frac{km}{h} = \frac{100\,km \times 1000\,m\,/\,km}{1\,h \times 3600\,s\,/\,h} = 27.78\,m\,/\,s,$$

$$V_2 = 30\frac{km}{h} = \frac{30\,km \times 1000\,m\,/\,km}{1\,h \times 3600\,s\,/\,h} = 8.333\,m\,/\,s,$$

$$a = \frac{\left(8.333\,m\,/\,s\right)^2 - \left(27.78\,m\,/\,s\right)^2}{2 \times 125\,m} = -2.809\,m\,/\,s^2.$$

**Answer:** The acceleration of the car is –2.8 m / s. Negative acceleration means that the car is slowing down. ∎

## 2.3   Postulates: Newton's Laws

If we want to accelerate a mass, we must exert a *force* ($F$) on it. Newton's second law gives the relationship between force, mass and acceleration:

$$F = ma. \tag{2.3}$$

Newton's first law is a special case of the second law, stating that when the net applied force is zero the acceleration is zero. Newton's laws are *postulates*, statements that we cannot prove from any more fundamental principle, but believe to be true based on observation of the world around us. The science of mechanics is based on Newton's laws and in later chapters we will state several more postulates that constitute the laws of thermodynamics.

If there are several forces acting on a mass we sum them up:

$$\sum_i F_i = ma. \tag{2.4}$$

Force has units of kg m / s². This is a unit that is used so often that it is convenient to give it a name, and it is called the *newton* and represented by the symbol N.

All masses experience a gravitational force when they are in the proximity of the earth, which is known as the *weight* ($F_w$) of the body. At the earth's surface all bodies fall with an acceleration $g = 9.81$ m / s². The weight of a body is therefore

$$F_w = mg. \tag{2.5}$$

The force of gravity decreases as we move further away from earth, and therefore its weight becomes less. In outer space, far from the gravitational attraction of any planet or other large body, objects experience no gravitational force and become weightless. However their mass does not change, since they possess the same amount of matter.

### Example 2.3

**Problem:** A 10 kg mass is lifted vertically upwards with an acceleration of 5 m / s². What is the force raising it?

**Find:** Force $F$ required to raise the mass.

**Known:** Mass of object $m = 10$ kg, upward acceleration $a = 5$ m / s².

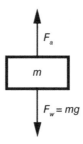

**Figure E2.3**   A free-body diagram of the mass.

**Diagram:** A free-body diagram of the mass is shown in Figure E2.3.

**Assumptions:** Acceleration due to gravity is at earth's surface $g = -9.81$ m/s²

### Governing Equation:

Sum of forces acting on an object                    $\sum_i F_i = ma$

**Solution:**

The force due to gravity acting in the negative downwards direction is

$$F_w = mg.$$

Substituting into the governing equation,

$$F_a + F_w = ma$$

$$F_a = ma - F_w = ma + mg = m(a + g)$$

$$F_a = 10\,\text{kg}\left(5\,\text{m/s}^2 + 9.81\,\text{m/s}^2\right) = 148.1\,\text{N}$$

**Answer:** The force required to raise the mass is 148.1 N.                                                        ■

## 2.4   Mechanical Work and Energy

If a force acts on a mass and moves it through a finite distance, it is said to have done work ($W$). When an object is moved (see Figure 2.1) from an initial position ($x_1$) to a final position ($x_2$) by a force ($F$) the amount of work done is

$$W = \int_{x_1}^{x_2} F\,dx. \tag{2.6}$$

The units of work are Nm and since this is a unit that is used very frequently we give it the name *joule* (J). If the force applied is constant, Equation (2.6) can be integrated to give:

$$W = F\int_{x_1}^{x_2} dx = F(x_2 - x_1) = F\Delta x. \tag{2.7}$$

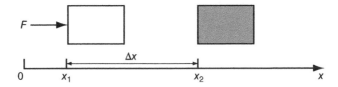

**Figure 2.1**   A mass is moved through a distance ($\Delta x$) by a force ($F$).

**Example 2.4**

**Problem:** The force acting on a 5 kg object varies with time as $F = (50 + 10t)$ N, where $t$ is the time in seconds. If the body starts from rest, how much work is done on it in 10 s?

**Find:** Work $W$ done on the object.

**Known:** Mass of body $m = 5$ kg, force applied $F(t) = 50 + 10t$, total time travelled $t_2 = 10$ s.

**Assumptions:** The body starts from rest, $\mathbf{V}(0) = 0$ m / s.

**Governing Equation:**

Mechanical work
$$W = \int_{x_1}^{x_2} F dx.$$

**Solution:**

The independent variable in the governing equation is the distance travelled, $x$. We would like it to be the elapsed time $t$,

$$W = \int_{x_1}^{x_2} F(t) dx = \int_0^{t_2} F \frac{dx}{dt} dt = \int_0^{t_2} F(t) \mathbf{V}(t) dt.$$

We need to express $\mathbf{V}$ as a function of $t$. We have been given $F$ as a function of $t$ and we know that

$$F(t) = ma = m \frac{d\mathbf{V}}{dt},$$

$$(5 \text{ kg}) \frac{d\mathbf{V}}{dt} = 50 + 10t \text{ N}.$$

Rearranging and integrating, assuming that $\mathbf{V}(0) = 0$,

$$\mathbf{V}(t) = \int (10 + 2t) dt = 10t + t^2.$$

Substituting expressions for both $\mathbf{V}$ and $F$,

$$W = \int_0^{t_2} F \mathbf{V} dt = \int_0^{t_2} (50 + 10t)(10t + t^2) dt = \int_0^{10} (500t + 150t^2 + 10t^3) dt,$$

$$= \left[ 250t^2 + 50t^3 + \frac{10}{4} t^4 \right]_0^{10} = 25\ 000\,\text{J} + 50\ 000\,\text{J} + 25\ 000\,\text{J},$$

$$= 100\ 000\,\text{J},$$

$$= 100 \text{ kJ}.$$

**Answer:** The work done on the object is 100 kJ.  ■

## 2.4.1   Potential Energy

In Figure 2.2 a body with mass $m$, originally at rest at a height $(z_1)$ above the ground, is raised slowly through a distance $(\Delta z)$ to a higher position $(z_2)$. How much work is required to do this? The force required to lift the mass is constant, equal to the weight $F = F_w = mg$. The work done is therefore

$$W = F\left(z_2 - z_1\right) = F\Delta z. \tag{2.8}$$

Doing work on the body alters its properties because, once it is raised to its new position, it is possible to use it to do work on another body. For example, we could connect the weight to another mass $(m')$ using a rope and pulley, as shown in Figure 2.3. If we allow mass $m$ to drop, it will raise mass $m'$, doing work on it. The property that gives a system the ability to raise a weight is known as *energy* $(E)$. Energy stored in a body by raising it in a gravitational field is known as *gravitational potential energy* $(PE)$ defined as

$$PE = mgz. \tag{2.9}$$

The change in gravitational potential energy is then

$$d\left(PE\right) = d\left(mgz\right). \tag{2.10}$$

If $m$ and $g$ are constant, the change in potential energy when the mass is raised through a height $(\Delta z)$ is

$$\Delta PE = mg\Delta z. \tag{2.11}$$

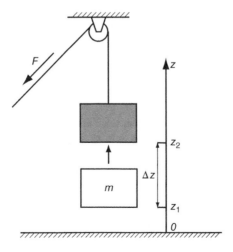

**Figure 2.2**   A mass $(m)$ is raised through a height $(\Delta z)$ by a force $(F)$.

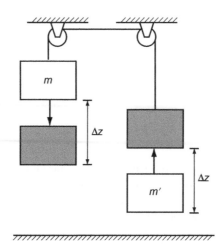

**Figure 2.3** A mass (*m*) falls through a height (Δ*z*), raising a second mass (*m'*) by the same distance. The potential energy stored in *m* is used to do work on *m'*.

Lowering the weight through a distance Δ*z* would reduce the gravitational potential energy by the same amount. The height of the body (*z*) is measured from some arbitrary datum plane – typically we choose it to be ground level, but this is not necessary, and we can choose any convenient level where we define the potential energy to be zero. The potential energy of any body raised above this level is positive, and of any body below it is negative. What is significant is the change in energy (Δ*PE*) as the body is moved, which does not depend on the choice of the datum level.

The dimensions of potential energy are the same as those of work, force times distance, and the units are also the same: joules or kilojoules.

Energy is frequently described, rather imprecisely, as the capacity to do work. This is not a real definition since, in a circular piece of reasoning, work is defined as the transfer of energy. We cannot give a more precise definition of energy since it is a fundamental concept. We add it to the list of three concepts that we already have (length, mass and time), which will be essential to developing our understanding of thermodynamics.

### Example 2.5

**Problem:** Acceleration due to gravity at Earth's surface is $g = 9.80665$ m / s$^2$ and decreases by $3.3 \times 10^{-6}$ m / s$^2$ for each metre of height above sea level. What is the potential energy of a body with 100 kg mass raised to an altitude of 1000 m: (a) assuming constant $g$, (b) accounting for the decrease in $g$ with height?

**Find:** Potential energy of a body raised to an altitude under (a) constant $g$ and (b) decreasing $g$ with altitude.

**Known:** On Earth's surface $g = 9.806\ 65$ m / s$^2$, as height $z$ increases $g$ decreases such that $g(z) = 9.806\ 65 - 3.3 \times 10^{-6} z$, mass of body $m = 100$ kg, the body is raised to height $z = \Delta z = h = 1000$ m.

**Assumptions:** For part (a) $g$ is constant, for part (b) $g$ decreases with height.

**Governing Equation:**

Change in potential energy
$$d(PE) = d(mgz)$$

**Solution:**

Assuming constant $g$,

$$d(PE) = d(mgz) \rightarrow \Delta PE = mg\Delta z$$

$$\Delta PE = mgh = 100\,\text{kg} \times 9.806\ 65\,\text{m}/\text{s}^2 \times 1000\,\text{m} = 980\ 665\,\text{J}$$

(a) Assuming variable $g$,

$$d(PE) = d(mgz)$$

$$\Delta PE = m\int_0^h g\,dz = m\int_0^h \left(9.806\ 65 - 3.3\times 10^{-6}z\right)dz$$

$$\Delta PE = m\left[9.806\ 65h - 1.65\times 10^{-6}h^2\right] = 100\left[9.806\ 65(1000) - 1.65\times 10^{-6}(1000)^2\right]$$

$$\Delta PE = 980\ 500\,\text{J}$$

**Answer:** For (a) constant gravitational acceleration $g$, potential energy changes by 980 655 J, whereas for (b) variable $g$, potential energy changes by 980 500 J. ∎

## 2.4.2   Kinetic Energy

A force ($F$) acts on a body of constant mass ($m$), initially at a position ($x_1$) and moving with a velocity ($\mathbf{V}_1$), and accelerates it until it is at a new position ($x_2$) moving with a higher velocity ($\mathbf{V}_2$; see Figure 2.4).

From Newton's law, the force applied is $F = ma$. The work done during this movement is

$$W = \int_{x_1}^{x_2} F\,dx = \int_{x_1}^{x_2} ma\,dx = m\int_{x_1}^{x_2} a\,dx. \tag{2.12}$$

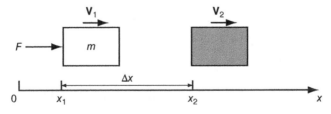

**Figure 2.4**   A force ($F$) accelerates a mass ($m$) from an initial velocity ($\mathbf{V}_1$) to a final velocity ($\mathbf{V}_2$).

Using the definition of acceleration,

$$a = \frac{d\mathbf{V}}{dt} = \frac{d\mathbf{V}}{dx}\frac{dx}{dt} = \frac{d\mathbf{V}}{dx}\mathbf{V} = \mathbf{V}\frac{d\mathbf{V}}{dx},$$

(2.13)

and substituting this expression for acceleration ($a$) into Equation (2.12), we get

$$W = m\int_{\mathbf{V}_1}^{\mathbf{V}_2} \mathbf{V}\,d\mathbf{V} = \frac{1}{2}m\mathbf{V}_2^2 - \frac{1}{2}m\mathbf{V}_1^2.$$

(2.14)

The work done in increasing the velocity of the mass is stored in it in the form of *kinetic energy* (*KE*), defined by

$$KE = \frac{1}{2}m\mathbf{V}^2.$$

(2.15)

The change in kinetic energy when the velocity of a mass ($m$) changes from $\mathbf{V}_1$ to $\mathbf{V}_2$ is

$$\Delta KE = \frac{1}{2}m\left(\mathbf{V}_2^2 - \mathbf{V}_1^2\right).$$

(2.16)

The units of kinetic energy are the same as those of work and potential energy, joules. We can recover the kinetic energy stored in the body if we slow it down and use the force required for deceleration to do work.

**Example 2.6**
**Problem:** A body resting on the ground has a weight of 500 N. If it is accelerated to a velocity of 10 m / s, what is its kinetic energy?
**Find:** Kinetic energy *KE* of the body.
**Known:** Weight of the body $F_w$ = 500 N, velocity of the body $\mathbf{V}$ = 10 m / s.
**Assume:** Acceleration due to gravity at earth's surface $g$ = 9.81 m / s$^2$.
**Governing Equations:**

Weight                                                       $F_w = mg$

Kinetic energy                                          $KE = \frac{1}{2}m\mathbf{V}^2$

**Solution:**
Mass of the body,

$$m = \frac{F_w}{g} = \frac{500\,\text{N}}{9.81\,\text{m/s}^2} = 50.968\,\text{kg}.$$

Kinetic energy of the body,

$$KE = \frac{1}{2}m\mathbf{V}^2 = \frac{1}{2}(50.968\,\text{kg})(10\,\text{m/s})^2 = 2548.4\,\text{J}.$$

**Answer:** The kinetic energy of the body is 2550 J.                                         ∎

## 2.5   Thermodynamic Systems

Before we start any analysis in thermodynamics we must clearly identify what we are discussing. A *thermodynamic system*, often simply called the *system*, can be any piece of matter or region of space that we identify for purposes of analysis (Figure 2.5). The *surroundings* are everything outside the system. The system *boundary* is the surface that separates the system from the surroundings.

The boundary may correspond to a real, physical barrier. When analysing heating of a container filled with water we typically define the water as being our system (Figure 2.6a). The inner surface of the walls of the container is the boundary of the system, and the container and everything outside are the surroundings. The container itself cannot be the boundary, because the boundary is a surface without thickness – it cannot have any volume or mass. We can choose to have the boundary corresponding to the outer surface of the container, but then the container also becomes part of the system and this added mass has to be accounted for in our calculations.

The boundary may be an imaginary surface, which does not conform to a real wall but whose coordinates in space we specify. Sometimes, in analysing fluid flow, we focus our attention on only a small part of the fluid (Figure 2.6b). We define our system by drawing an imaginary surface that separates a small portion of fluid from the remainder, and we follow the trajectory of this system. The remaining fluid, outside our selection, forms the surroundings. The boundary of a thermodynamic system may be stationary or may move.

Thermodynamic systems interact with their surroundings by exchanging either mass, energy or both. Our method of analysis will depend on what is crossing the system boundary. We define three types of systems below.

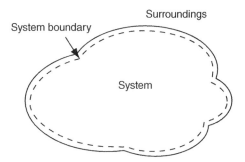

**Figure 2.5**   Definition of system, surroundings and system boundary.

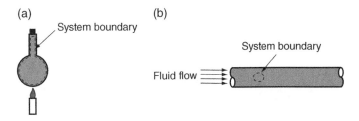

**Figure 2.6**   System boundaries that (a) correspond to a physical boundary and (b) are imaginary surfaces.

### 2.5.1   Closed System

The amount of mass in a *closed system* is fixed and no mass crosses the system boundary. However, energy in the form of either heat or work can enter or leave the system. A closed system is also called a *control mass*. Though the mass of a closed system is fixed, its volume is not and it may be compressed or expanded. A rock (Figure 2.7a) or a sealed bottle (Figure 2.7b) are examples of closed systems, and so is a fixed mass of gas contained in a cylinder fitted with a piston (Figure 2.7c). As the piston moves to compress the gas it does work and the control mass boundary moves, but the gas remains inside the system.

### 2.5.2   Open System

Both energy and mass can cross the boundaries of an *open system*, which is also known as a *control volume*. The boundaries of the control volume may change shape and the system may also move in space. Common examples of control volumes are industrial devices such as heat exchangers, pumps, turbines and boilers. Mass is constantly flowing into and out of these machines while energy in the form of both mass and energy cross the system boundaries.

Figure 2.8a shows a section of a water pipe around which a heater coil is wrapped. The dotted line around the heated section is the control volume. Water flows into and out of this section, while heat is added to it. Figure 2.8b is another example of a control volume, a pump. Water flows through the pump and work is transferred from the surroundings to the control volume as the pump shaft rotates. An operating car, shown in Figure 2.8c, is an example of a moving control volume: the car does work on the surroundings and loses heat while its engine takes in air and emits exhaust gases.

### 2.5.3   Isolated System

If no mass or energy crosses the boundaries of a system it does not interact, in a thermodynamic sense, with its surroundings and is known as an *isolated system*. The properties of an isolated system, once it is at equilibrium, will never change. A sealed, perfectly insulated container with rigid and impervious (through which no matter can pass) walls is an example of an isolated system (Figure 2.9).

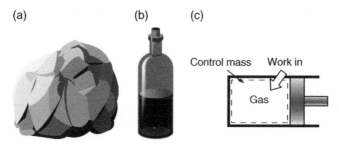

**Figure 2.7**   Closed systems: (a) a rock, (b) a sealed bottle and (c) gas contained in a cylinder fitted with a piston.

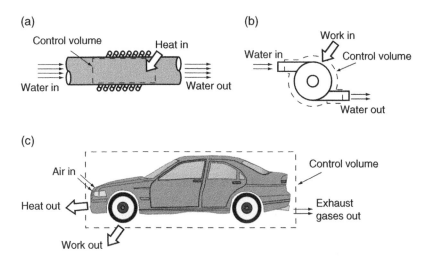

Figure 2.8    Open systems: (a) heated section of a water pipe, (b) a water pump and (c) a car.

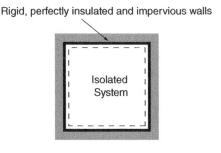

Figure 2.9    Isolated system, surrounded by rigid, perfectly insulated and impervious walls.

## 2.6    Thermodynamic Properties

*A property of a system is any attribute that can be measured without knowing the history of the system.* We will confine our discussion to *thermodynamic properties*, which affect the energy of the system. Although the surface roughness of an object is a property that we can measure, we do not usually consider it a thermodynamic property. There may be circumstances, though, where it becomes important – for example, if we are considering frictional heating. In that case we would need to include it in our analysis. Mass, volume, temperature and energy are all typical examples of thermodynamic properties.

A box resting on the floor at position 1 is pushed until it reaches position 2 (Figure 2.10). Once it has reached the final position, you enter the room. You are asked to describe the position of the box. This is a simple task – first decide on a coordinate system, then measure the distance along both $x$ and $y$ coordinates and pinpoint the location $(x_2, y_2)$. Since you could do this without knowing the original position of the mass or the path it has travelled, position is a property.

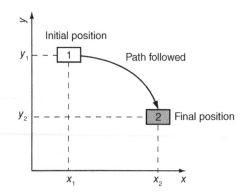

**Figure 2.10**  A box is pushed across the floor from initial position 1 to final position 2.

## 2.6.1  Path Functions

Suppose you are asked how much work was done in moving the box from its original to its final position in Figure 2.10. To determine this you have to know where it started its movement, the trajectory it followed, and the force of friction between the box and the floor. In short, you have to know the history of the system, the box, before you can evaluate the work done on it. Work, therefore, is not a property of the system. Work is something added to the system that alters its properties such as position and energy.

A beaker filled with water at an initial temperature $T_1$ is heated to a final temperature $T_2$. Determining the final temperature is simple – place a thermometer in the water and measure it. No information is required about the initial conditions to find the temperature at any given instant. But to identify how much heat was added we need to know its history – in this case the initial water temperature. If the water were stirred during heating, we would need to know that and distinguish work done on the fluid by the stirring rod from energy added as heat. Temperature is therefore a property of the system, but heat is not – heat added to the system changes properties such as temperature.

The distinction between properties and other quantities that are not properties requires special mathematical notation to make sure that we do not confuse the two. Suppose an object is moved from an initial position $(x_1)$ to a final position $(x_2)$. An infinitesimally small displacement is denoted by $dx$. The total displacement of the object $(\Delta x)$ is calculated by integrating $dx$:

$$\Delta x = \int_{x_1}^{x_2} dx = x_2 - x_1. \tag{2.17}$$

The change in a property depends only on its initial and final values, not on the path followed in getting from one point to another. To move an object through an infinitesimal distance $(dx)$ requires an infinitesimally small amount of work. How do we denote this work mathematically? If we call it $dW$, then following the same reasoning as that behind Equation (2.17) we should be able to integrate the infinitesimal amount of work from the initial to the final position to calculate the work done $(W_{12})$ in moving this distance (from $x_1$ to $x_2$). That is, we should be

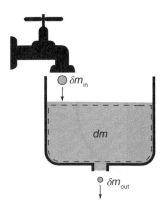

**Figure 2.11**    An infinitesimal mass of water ($\delta m_{in}$) is added to the tank while another infinitesimal mass ($\delta m_{out}$) leaks out. The mass in the tank changes by an amount $dm = \delta m_{in} - \delta m_{out}$.

able to write $W_{12} = W_2 - W_1$. But $W_1$ would imply "the work at position $x_1$", which is a nonsensical statement because work is not a property of the system and cannot be evaluated at a given position. Therefore, infinitesimal amounts of work will be denoted by $\delta W$ rather than $dW$, to remind ourselves that we cannot integrate them in the same way we integrate properties. We can then state that when a force ($F$) moves an object through a distance ($dx$) the work done is $\delta W = F dx$. Similarly, an infinitesimally small amount of heat added to or removed from a system will be denoted by $\delta Q$. Heat and work are known as *path functions*, since they depend on the path taken to get from the initial to the final state.

The mass of a system is a property, but mass added to or taken away from a system is not a property. Water is added to a tank (Figure 2.11) that has a drain at the bottom through which a small amount of water leaks out. The total mass of water ($m$) within the tank is a property: we can measure it at any instant by noting the water level and an infinitesimal change in the mass is represented by $dm$. The mass of water entering or leaving is not a property, since we cannot determine that from instantaneous observations of the tank alone. An infinitesimal mass of water added to the tank is denoted the symbol $\delta m_{in}$ and an infinitesimal amount leaving is $\delta m_{out}$. The relation between these two terms and $dm$ is given by

$$\delta m_{in} - \delta m_{out} = dm. \qquad (2.18)$$

Mass added or removed are not properties of the system, but the difference between the two terms gives us the net change in the mass of the system, which is a property.

## 2.6.2    Intensive and Extensive Properties

An object with fixed mass ($m$), volume ($V$) and temperature ($T$) that is at thermodynamic equilibrium is sliced into two equal portions (Figure 2.12). What happens to its properties? After it is divided mass and volume have both been halved, but temperature remains the same. Obviously, some properties depend on the size of the system, while others are independent of system extent.

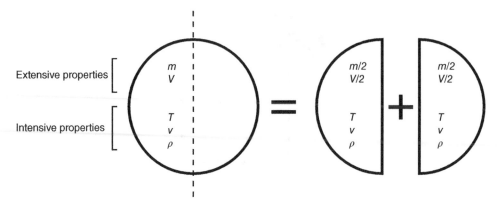

**Figure 2.12**  An object is divided into two equal parts. The *intensive* properties of the two portions [temperature (*T*), specific volume (*v*), density (*ρ*)] are unchanged while the *extensive* properties [mass (*m*), volume (*V*)] have half their original values.

*Intensive* properties are those can be specified at a point within the system and are independent of system mass. Measuring these properties in only a portion of the system does not affect their value. Examples of intensive properties are temperature and pressure.

*Extensive* properties depend on the size of the system. Mass, volume and energy are all extensive properties.

If we take any extensive property and divide it by the mass of the system we obtain a new intensive property. For example, the volume (*V*) of a system is an extensive property. Dividing it by mass (*m*) gives us the volume per unit mass, which is known as the specific volume:

$$v = \frac{V}{m}. \tag{2.19}$$

The units of specific volume (*v*) are m³ / kg. The specific volume is the reciprocal of the density of a system, defined as

$$\rho = \frac{m}{V} = \frac{1}{v}. \tag{2.20}$$

The units of density (*ρ*) are kg / m³ and it is also an intensive property. Similarly, we can define the specific energy of a system:

$$e = \frac{E}{m}. \tag{2.21}$$

The units of specific energy (*e*) are J / kg and it is an intensive property, while the total energy (*E*) is an extensive property. In general, upper case letters will be used to denote extensive properties and lower case letters for the corresponding intensive properties.

## 2.7   Steady State

A system whose properties do not change with time, even though it is exchanging energy or mass with its surroundings, is said to be at *steady state*. When we switch on power to an electric heater its temperature begins to increase. The heater element loses heat to the surrounding air at a rate that increases with its temperature. Eventually, when it is sufficiently hot, the rate of heat loss equals the rate at which electrical energy is supplied. The heater temperature then stays constant and it is in a steady state.

When we place an empty bucket under a running tap it fills with water until it is brimming over. Then, if the tap continues to run, the rate at which water spills out equals the rate at which it flows in and the system has reached steady state. The mass of water in the bucket remains constant even though there is constant flow through it.

## 2.8   Equilibrium

Drop a rubber ball on the ground and it will start bouncing (Figure 2.13a). Each bounce will get progressively smaller and eventually the ball will stop moving and remain at rest until it is disturbed again. Place a hot piece of metal inside a rigid, insulated box (which is an isolated system) and it will start to cool until its temperature is the same as that of the air in the box, after which it does not change further (Figure 2.13b). *All physical systems that are left isolated eventually reach a state of equilibrium where their properties do not change with time.* External intervention is necessary to change the properties of a system once it reaches equilibrium.

The opposite of an equilibrium state is a *non-equilibrium* state, where the properties of an isolated system change spontaneously. A metal piece cooling inside a sealed, insulated box is a isolated system that is in non-equilibrium as long as the temperature of the metal and air continue to change.

If we disturb a system in equilibrium it will pass through a stage of non-equilibrium until it reaches a new equilibrium state. A ball at rest on a table is at equilibrium, but pushed off the edge it will fall and bounce on the floor until it is comes to rest again. The transition from one equilibrium state to another is important because we can use the system to do useful work while it is in non-equilibrium. A ball cannot lift a weight while it is resting on

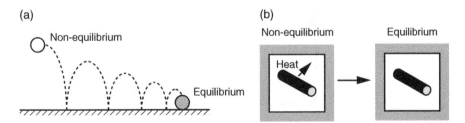

**Figure 2.13**   (a) A ball bounces until it comes to rest and reaches equilibrium. (b) A heated metal rod in an insulated box loses heat until it reaches equilibrium when the temperature everywhere in the box is uniform.

a table or on the floor, but we could design a mechanism that uses the momentum of the ball while it is falling to do work. A tranquil lake cannot drive a hydroelectric turbine but a waterfall can.

Thermodynamics cannot be used to analyse a system whose properties are changing with time. It is possible to study time-varying systems using the principles of dynamics, fluid mechanics and heat transfer, but thermodynamics only describes systems at equilibrium. This information is still valuable: in the case of a ball falling from a table to the floor, we can calculate the potential energy at both the initial and final equilibrium states and their difference will gives us the maximum possible amount of work that the ball could do. This is the approach followed in thermodynamic analysis, where we calculate the difference in properties between two equilibrium states and use that to set bounds on the energy transfers that occur during the intervening process. Therefore, predicting the properties of a system once it has reached equilibrium is a very important task in thermodynamics.

When dealing with control volumes it is often useful to make the assumption of *local thermodynamic equilibrium*, which means that if we take an infinitesimally small volume of fluid from anywhere within the system and isolate it we would not observe any change with time in its properties. With this assumption we can measure the properties of a volume of fluid when it enters the control volume and again when it leaves, and from their difference deduce the processes occurring in the system.

It is important to understand the difference between steady state and equilibrium. A system that is interacting with its surroundings, but whose properties are not changing with time, is at steady state. An *isolated* system with constant properties is at equilibrium. When we open a water tap over a sink the water level in the sink rises until the rate at which water flows out through the drain equals the rate at which it flows in, at which time the system has reached steady state (Figure 2.14a). The mass of water in the sink remains constant even though there is constant flow in and out of it. If we close the drain, turn off the tap and leave the sink isolated from the surroundings, it will be at equilibrium (Figure 2.14b).

There are many different kinds of equilibrium possible. Some that we will be studying are introduced in the following sub-sections.

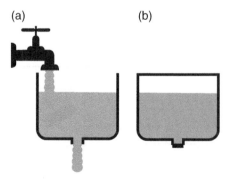

**Figure 2.14** (a) A system at steady state, in which water flows into the sink at the same rate that it drains out. (b) An isolated system at equilibrium.

### 2.8.1 Mechanical Equilibrium

A cylinder is separated into two sections by a sliding piston. The piston is initially held in place while the two compartments are filled with gas at different pressures ($P_1$ and $P_2$), shown in Figure 2.15. If the piston is released it will move to a position where the pressure on both faces is equal. The system is then said to be in *mechanical equilibrium*.

### 2.8.2 Thermal Equilibrium

Two masses, initially at different temperatures ($T_1$ and $T_2$), are brought into contact with each other (Figure 2.16). Heat will flow from the hotter to the cooler mass until eventually they both reach the same equilibrium temperature ($T_{eq}$), which lies somewhere between $T_1$ and $T_2$. The temperature does not change further and the bodies are in *thermal equilibrium*.

### 2.8.3 Phase Equilibrium

A rigid container is evacuated, partially filled with liquid and sealed. Some of the liquid will evaporate and eventually the container will reach equilibrium, with part of it filled with liquid and the remainder filled with vapour. The system is then said to be in a state of *phase*

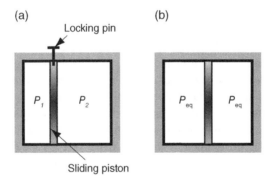

**Figure 2.15** A cylinder filled with gas is divided into two chambers by a sliding piston. (a) Initially the piston is locked and the pressures in the two chambers are different ($P_1 > P_2$). (b) Once the piston is allowed to slide it reaches mechanical equilibrium when the pressure ($P_{eq}$) on both sides of the piston is the same.

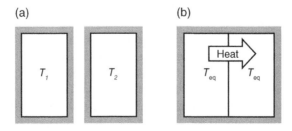

**Figure 2.16** (a) Two systems initially at different temperatures ($T_1$ and $T_2$) are brought into close thermal contact. (b) Once thermal equilibrium is reached both systems will have the same temperature ($T_{eq}$).

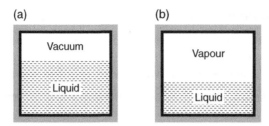

**Figure 2.17** (a) A container partially filled with liquid is evacuated to remove all the air in it and then allowed to come to equilibrium. (b) Some of the liquid will evaporate and fill the empty space, until it reaches phase equilibrium.

*equilibrium*, shown in Figure 2.17. Both liquid and vapour will be at the same temperature, which we call the *saturation temperature* $(T_{sat})$. The pressure in the vapour is known as the *saturation pressure* $(P_{sat})$, and is a function of the saturation temperature. If the temperature of the system is changed the system will shift to a new equilibrium state with different proportions of liquid and vapour. Raising the temperature will make more liquid evaporate so that the pressure of the vapour increases. Cooling the system will condense some of the vapour and decrease its pressure.

## 2.9 State and Process

We describe a system by recording its properties. A complete list of properties describes the *state* of the system. If any property is changed the system shifts to a different state.

The change of a system from one state to another is known as a *process*. We can depict a thermodynamic process on a two-dimensional graph by selecting two properties and showing how they vary during the process. For example, pressure $(P)$ and volume $(V)$ are frequently used as axes to describe the compression or expansion of a gas. The initial state (labelled 1) and final state (labelled 2) are defined by the coordinates $(P_1, V_1)$ and $(P_2, V_2)$ in Figure 2.18. The process path is shown by the line connecting these two points, which defines intermediate values of the two properties in going from the initial to final state.

Often one or more properties are held constant during a thermodynamic process, and specific names are given to such processes as follows.

- *Isothermal* process: The temperature of the system is constant. Slowly compressing gas in a cylinder that is immersed in a very large bath of water closely approximates an isothermal process. The gas will stay at the same temperature as the water, which we assume has such a large mass that its temperature does not change significantly.
- *Isobaric* process: The pressure of the system is constant. A gas being heated in a cylinder with a freely moving piston undergoes isobaric expansion. The pressure inside the cylinder will always be the same as that of the atmosphere outside, which we assume constant.
- *Isochoric* process: The volume of the system is constant. Heating or cooling gases sealed in rigid container are examples of isochoric processes. The terms *isometric* or *isovolumetric* are also used for constant volume processes.

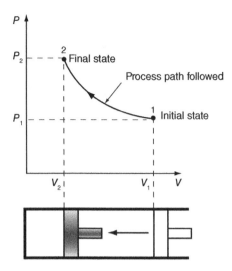

**Figure 2.18**   A fixed mass of gas in a cylinder is compressed by a piston from initial state 1 to final state 2. The process path shows the state of the system during the compression.

- *Adiabatic* process: No heat is added to or removed from the system during the process. Any surface that completely prevents the transmission of heat is known as an *adiabatic wall*. In reality no surface is perfectly adiabatic, but many insulating materials can reduce heat transfer to negligible amounts and are very good approximations of adiabatic walls. A system whose boundaries are perfectly insulated will undergo an adiabatic process. Note that this is not the same as an isothermal process. If a gas is placed in a well-insulated cylinder and compressed, its temperature will rise. The process is not isothermal, but it is adiabatic.

## 2.10   Quasi-Equilibrium Process

A piston rapidly compresses a gas confined in a cylinder. Can we calculate the variation of gas pressure ($P$) as a function of the cylinder volume ($V$)? Surprisingly, we cannot do this very precisely for any real process. If the piston advances quickly, molecules of gas will tend to accumulate on its face since they will not have enough time to move out of the way (Figure 2.19a). The frequency of molecules striking the piston will be greater than those hitting the other faces, and it will experience a higher pressure. If the piston was moving in the other direction, expanding the gas, fewer molecules would strike it and the pressure on the piston would be lower than on the other surfaces of the cylinder (Figure 2.19b). The pressure on the piston therefore depends on the speed of the cylinder and the direction in which it is moving.

Suppose that instead of moving the piston rapidly through the length of the cylinder we slowly advance it by a small distance and then wait until the gas reaches equilibrium (Figure 2.19c). Once the gas is at equilibrium the pressure on all the faces will be equal and can be measured accurately. We then repeat this process until we have completed the compression, measuring the pressure each time the piston has come to rest. The process can now be depicted

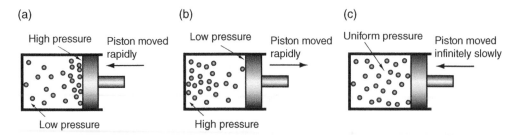

Figure 2.19    (a) During rapid compression molecules cluster near the piston. (b) During rapid expansion there are fewer molecules near the piston. (c) During quasi-equilibrium compression molecules are distributed uniformly.

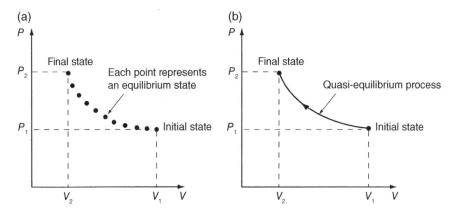

Figure 2.20    Quasi-equilibrium compression of a gas (a) approximated by a finite number of points and (b) found precisely by an infinite number of points.

as a series of points on a *P-V* diagram, as shown in Figure 2.20a. The greater the number of stages we take to complete the process the greater the number of points we can put on the graph and the more accurate our depiction of the process.

If we were to make the compression infinitely slow, advancing the piston by an infinitesimal amount at each step, the gas will be at equilibrium throughout the process and we can draw a continuous line to show the variation of pressure with volume as shown in Figure 2.20b. This is known as a *quasi-equilibrium* process, and is an idealisation that allows us to analyse thermodynamic processes since we can define the properties of the system at every stage in the process. Whenever we represent a process by a curve on thermodynamic axes, we are implicitly making the assumption that the process is quasi-equilibrium.

Heat transfer to a system can also be done is a quasi-equilibrium process. Suppose we want to heat a mass from temperature $T_1$ to $T_2$. We can do this by bringing it into contact with a *thermal reservoir* at temperature $T_2$ and waiting until they both come to equilibrium. *A thermal reservoir is defined as a system whose temperature remains constant in spite of heat transfer to or from it.* This is an idealisation since in reality no system is a perfect thermal reservoir, but very large masses such as the atmosphere, an ocean or a lake are close approximations of

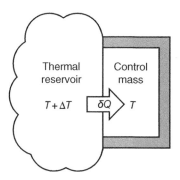

**Figure 2.21** Quasi-equilibrium heat transfer to a control mass at temperature $T$ that is brought into contact with a thermal reservoir at temperature $T + \Delta T$ where $\Delta T \to 0$. An infinitesimal amount of heat transfer ($\delta Q$) occurs from the reservoir to the system until equilibrium is reached.

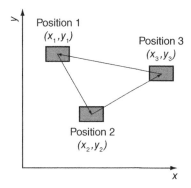

**Figure 2.22** An object is moved from position 1 to position 2, then to position 3 and finally returned to the initial position 1.

thermal reservoirs. Thermal reservoirs from which we withdraw heat are *heat sources* and those to which we transfer heat are *heat sinks*.

While the system is being heated its temperature will vary from point to point, so that we cannot define a system temperature. If however we bring a control mass at temperature $T$ into contact with a thermal reservoir at temperature $T + \Delta T$ where $\Delta T \to 0$, the temperature will rise by an infinitesimal amount until equilibrium is re-established (Figure 2.21). By repeating this process with a series of thermal reservoirs at successively higher temperatures, we can raise the temperature of the control mass from $T_1$ to $T_2$ in a quasi-equilibrium manner.

## 2.11 Cycle

Any process, or series of processes, that results in the system being restored to its initial state is known as a *cycle*. Figure 2.22 shows an object that is moved from position 1 to position 2, then to position 3 and finally returned to its initial position. These series of moves constitute a cycle since system properties are restored to their initial state.

The symbol $\oint$ means "integration over a cycle". If we integrate any property ($X$) over a cycle, $\oint X = 0$ since the value of the property depends only on the state of the system, not on its previous history. For example, if the property is the $x$-coordinate of the object in Figure 2.22, over the cycle,

$$\oint dx = \int_{x_1}^{x_2} dx + \int_{x_2}^{x_3} dx + \int_{x_3}^{x_1} dx = \left(x_2 - x_1\right) + \left(x_3 - x_2\right) + \left(x_1 - x_3\right) = 0. \tag{2.22}$$

If we bring the system back to its original state all its properties will return to their initial values. However, the work done during the cycle will not be zero, since the work is not a property but depends on the path taken during the cycle:

$$W = \oint F dx \neq 0. \tag{2.23}$$

We may choose any two properties as axes to show a process. Figure 2.23 shows an example of a cycle plotted on $P$-$V$ axes. A mass of gas contained in a piston–cylinder system is taken through four processes, which constitute a cycle since they result in the system being restored to its initial state. The four processes are:

| | |
|---|---|
| $1 \rightarrow 2$ | Adiabatic compression |
| $2 \rightarrow 3$ | Isochoric heating |
| $3 \rightarrow 4$ | Adiabatic expansion |
| $4 \rightarrow 1$ | Isochoric cooling. |

Integrating a property such as pressure ($P$) or volume ($V$) over the cycle gives

$$\oint dP = 0 \text{ and } \oint dV = 0, \tag{2.24}$$

but the total heat transfer ($Q$) and work done ($W$) done during the cycle are not zero. Analysis of cycles is important in thermodynamics because any device that has to work continuously must operate in a cycle. All heat engines, by definition, work in cycles.

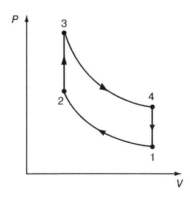

**Figure 2.23**   A fixed mass of gas in a cylinder is taken through processes 1, 2, 3 and 4 that make up a cycle, so that it is brought back to the initial state 1.

## 2.12    Solving Problems in Thermodynamics

You will test your understanding of thermodynamics by solving the problems given in this book. Adopt a systematic approach when tackling problems to avoid getting confused by a lot of information. The following steps are helpful.

1. *Find:* Read the problem and state what you are asked to find in your own words. This step may seem obvious, but it is surprising how often people misunderstand the question or miss essential information.
2. *Known:* List all the information and properties provided, as well as information about properties that remain constant (e.g. the volume in a isochoric process or the temperature in an isothermal process).
3. *Diagrams:* Draw a schematic diagram of the system. Mark the system boundary and decide whether the system is a control mass or volume. Show energy and mass transfers between the system and surroundings by arrows. Draw a process diagram if necessary.
4. *Assumptions:* Decide how you are going to model the system. List all assumptions that you make.
5. *Governing Equations:* Which conservation law are you going to apply? Depending on what you are trying to find, you may use principles of conservation of mass or energy. Write down the governing equations.
6. *Properties:* List all property values that are not given in the problem statement. This includes information extracted from tables or other sources.
7. *Solution:* Substitute known values of variables in the governing equations and solve them to find unknowns.
8. *Answer:* State the answer and confirm that it is what was asked for.
9. *Discussion:* Are your results reasonable? Can you draw any conclusions from them?

## 2.13    Significant Digits and Decimal Places

Adding significant digits and decimal places to numbers cost engineers time and money. Suppose you ask a machine shop to make 10 mm long steel rods. A machinist will interpret this as meaning that pieces with dimensions between 9.5 and 10.5 mm are acceptable, which are easy and cheap to produce since they can be cut off with a saw. Rods specified as 10.00 mm long will require a precision tool to manufacture and be more expensive. Fabricating a 10.000 000 mm long rod is a major project since machining the part and measuring it with sufficient accuracy are both difficult and expensive tasks. An engineer who persistently puts unnecessary decimal places on dimensions in drawings is not going to stay employed for long.

Digital calculators give answers with a large number of decimal places, but that does not necessarily make them meaningful. If we want to cut a 10 mm long rod into three equal pieces, an 8-digit calculator will tell us that each piece is 3.333 333 3 mm long. However, if the uncertainty in the length of the original piece was $\pm 0.5$ mm, we cannot expect greater accuracy in the pieces we cut from it and we should round off our answer to 3.3 mm. Similarly, if a 10.4 N force is applied to a 3.7 kg mass the acceleration, calculated by dividing force by mass, is $2.810\,810\,8$ m$/$s$^2$. We certainly cannot predict acceleration with such precision: 2.8 m$/$s$^2$ would be a more realistic answer.

The science of specifying uncertainties and tolerances in engineering calculations is a subject in its own right that we do not have time to cover in detail. Some simple rules of thumb for calculations are given here.

All non-zero digits in a number are considered significant. For example, 415 and 4.15 both have three significant digits. Leading zeros are not considered significant so 0.415 and 0.0415 both have the same number of significant digits – three. Trailing zeros are significant if they are after a decimal place: 415.0 has four significant figures and 415.00 has five. The significance of trailing zeros can be ambiguous if there is no decimal: 3300.0 has five significant digits, while 3300 may have two or four. The true number of significant digits is clear when we write in scientific notation. In that case $330 \times 10^3$ has three significant digits while $330.00 \times 10^3$ has five.

The number of digits in a number after the decimal are called decimal places. Examples of numbers with two decimal places are 32.88, 24.80 and 65.05. The number 45.3577 rounded off to three decimal places is 45.358, while rounded off to three significant digits it is 45.4.

When adding or subtracting, the number of *decimal places* in the answer should be the same as the least number of decimal places in those being added or subtracted. If a 10.0 mm block is placed on top of a 15.23 mm block, the resulting height is 25.2 mm, not 25.23 mm.

When multiplying or dividing, check the least number of *significant digits* in numbers used in calculations. Your final answer should have this same number of significant digits. So, if we cut a 10.0 mm long rod into three pieces, each will be 3.33 mm long. If a tank with 2.3 $m^3$ volume is filled with a liquid with density of 0.842 kg / $m^3$, the mass of liquid, given by multiplying the two values, is 1.9 kg.

During a long calculation you should not round off numbers in each step. Keep whatever accuracy your calculator provides you in intermediate steps and round off the final answer to the appropriate value.

## Further Reading

1. D. Halliday, R. Resnick, J. Walker (**2015**) Fundamentals of Physics, John Wiley & Sons, Ltd, London.
2. Y. A. Cengel, M. A. Boles (**2015**) Thermodynamics – An Engineering Approach, McGraw Hill, New York.
3. M. J. Moran, H. N. Shapiro, D. D. Boettner, M. B. Bailey (**2014**) Fundamentals of Engineering Thermodynamics, John Wiley & Sons, Ltd, London.

## Summary

*Fundamental concepts* are those that cannot be explained in simpler terms. Examples are length (area, volume), mass and time. *Derived quantities* are those defined in terms of fundamental concepts, such as velocity and acceleration.

*Postulates* are statements that we cannot prove from any more fundamental principle, but believe to be true based on observation. An example of a postulate is Newton's second law:

$$\sum_i F_i = ma.$$

The *weight* ($F_w$) of a body is the force it experiences when placed in a gravitational field. If the acceleration due to gravity is $g$, the weight of an object with a mass ($m$) is

$$F_w = mg.$$

The *work* ($W$) done when a force ($F$) moves an object from an initial position ($x_1$) to a final position ($x_2$) is

$$W = \int_{x_1}^{x_2} F dx.$$

The *gravitational potential energy* ($PE$) of a body with mass ($m$) at a height ($z$) is

$$PE = mgz.$$

The *kinetic energy* ($KE$) of a body with mass ($m$) moving with velocity (**V**) is:

$$KE = \frac{1}{2} m\mathbf{V}^2.$$

A *thermodynamic system* is any piece of matter or region of space that we identify for purposes of analysis. The *surroundings* are everything outside the system. The system *boundary* is the surface that separates the system from the surroundings. In a *closed system* (or *control mass*) the amount of mass is fixed and no mass enters or leaves the system though energy in the form of either heat or work can cross the system boundary. In an *open system* (or *control volume*) both energy and mass can cross the system boundary. No mass or energy crosses the boundary of an *isolated system*, which does not interact with its surroundings.

A *property* of a system is any attribute that can be measured without knowing the history of the system. *Intensive* properties are those that can be specified at a point within the system and are independent of system mass, whereas *extensive* properties depend on the size and mass of the system. If we take any extensive property and divide it by the mass of the system we get an intensive property.

A system whose properties do not change with time, even though it is interacting with its surroundings, is said to be at *steady state*. Isolated systems whose properties do not change with time are in a state of *equilibrium*. *Mechanical equilibrium* is reached when the pressure in a system is the same everywhere. *Thermal equilibrium* is reached when the temperature is uniform everywhere. *Phase equilibrium* is reached when there is no more phase change within a system.

In a *thermodynamic process* a system changes from one state to another. In an *isothermal* process the temperature of the system is constant. In an *isobaric* process the pressure of the system is constant. In an *isochoric* process the volume of the system is constant. No heat is added to or removed from a system during an *adiabatic* process. A *quasi-equilibrium* process is an infinitely slow one, in which the system approaches equilibrium at each stage. A *cycle* is any process, or series of processes, that result in the system being restored to its initial state. The integral of any property over a cycle equals zero.

## Problems

2.1 The velocity of a particle is given by $V = 10t^2 - 40t - 100$, where $V$ is in m / s and $t$ is in s. When is its acceleration zero?

2.2 An object has acceleration $a = 9t^2$. At time $t = 0$, its position $x_0 = 10$ m and velocity $V_0 = -5$ m / s. What are its position and velocity when $t = 1$ s?

2.3 The acceleration of a mass is given by $a = 4t - 10$ where $a$ is in m / s$^2$ and $t$ is in s. If the mass starts at $t = 0$ with displacement $x = 0$ and velocity $V = 2$ m / s, find the displacement as a function of time.

2.4 How much mass can a 1 N force lift vertically?

2.5 What is the weight of an object with a mass of 150 kg on a planet where $g = 4.1$ m / s$^2$?

2.6 Humans subjected to accelerations greater than $5g$ may lose consciousness. What is the force acting on a 60 kg person at this acceleration?

2.7 A cylindrical water tank, 3 m high and 3 m in diameter is filled with water. If the density of water is 1000 kg / m$^3$, what is the mass of the contained water? If the acceleration due to gravity is 9.81 m / s$^2$, what is the weight of the water?

2.8 A 5 kg box sliding across the floor with an initial velocity of 8 m / s is decelerated by friction to 3 m / s over 5 s. What is the force of friction acting on it?

2.9 A 500 N force accelerates a 50 kg mass moving at 10 m / s for 3 s. What is its final velocity?

2.10 A spaceship weighs 10 000 N on Earth. How much will it weigh on the moon where $g = 1.64$ m / s$^2$?

2.11 A linear spring is one whose extension is proportional to the force applied. When a mass is suspended from a linear spring on Earth its extension is 6.3 mm. When the same mass is suspended from the spring on Mars, its extension is 2.5 mm. What is the acceleration due to gravity on Mars?

2.12 Acceleration due to gravity at Earth's surface is $g = 9.80665$ m / s$^2$ and decreases by approximately $3.3 \times 10^{-6}$ m / s$^2$ for each metre of height above the ground. What is the potential energy of a 100 kg mass raised to an altitude of 1000 m: (a) assuming constant $g$, (b) accounting for the decrease in $g$ with height?

2.13 Acceleration due to gravity varies as $g = 9.806\,65 - 3.3 \times 10^{-6} z$, where $z$ is the altitude in metres. What is the reduction in weight of a space shuttle as it rises from the Earth's surface to orbit at an altitude of 400 km?

2.14 When a 6 kg mass is suspended from a spring whose extension ($x$) is proportional to the force applied ($F$), so that $F = Cx$, it stretches the spring by 30 mm. What is the proportionality constant $C$ of the spring, in units of newtons per millimetre?

2.15 A car travelling at 40 km / h collides with a wall and comes to rest in 0.3 s. What is the force exerted by the seatbelt on the car driver, who has a mass of 75 kg?

2.16 An 1100 kg car undergoes constant acceleration from rest to a speed of 100 km / h in 8 s. What is its acceleration and how much force was applied to it?

2.17 What is the kinetic energy of an object with a mass of 50 kg and a velocity of 10 m / s?

2.18 How fast would a mass lifted to a height of 100 m have to move for its kinetic energy to equal its potential energy?

2.19 An aircraft with a mass of 8000 kg has engines that have a total thrust of 100 kN. If it requires a speed of 50 m / s for take-off, what is the length of runway required?

2.20 A force $F = Cx^2$ where $C$ is a constant and $x$ the distance it acts through. The force has a value of 120 N when $x = 2$ m. Determine the work done when it moves an object from $x = 1$ m to $x = 4$ m.

2.21 A 1500 kg steel container with 5 m³ volume is filled with water and lifted vertically by a hoist that exerts a force of 80 kN. What is its acceleration?

2.22 A mass slides without friction down a ramp inclined at 45° to the horizontal. What is its acceleration along the direction of the ramp surface?

2.23 Find the acceleration of the two masses connected by a rope in Figure P2.23. Neglect friction in the pulleys.

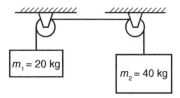

**Figure P2.23** Two masses connected by a rope and pulleys.

2.24 Water enters a tank from two pipes, one with a flow rate of 0.3 kg / s and the other with a flow rate of 0.1 kg / s. The tank has a small hole through which water leaks out at a rate of 0.03 kg / s. If the tank initially contained 40 kg of water, how much will it have after 2 minutes?

2.25 What is the weight of air in a room 5 × 10 × 3 m in size? Assume the density of air is 1.2 kg / m³.

2.26 What is the weight of a gas that has 60 m³ volume and 3 m³ / kg specific volume?

2.27 Tank $A$ has a volume of 0.5 m³ and contains air with density 1.2 kg / m³ while tank $B$ has a volume of 0.8 m³ and contains air with density 0.9 kg / m³. The two tanks are connected to each other and their contents mixed. What is the final air density in the tanks?

2.28 A 0.5 m³ container is filled with a mixture of 10% by volume ethanol and 90% by volume water at 25 °C. Find the weight of the liquid.

2.29  A bucket of volume 0.2 m³ is filled with sand in which the pores between particles constitute 40% of the total volume. Water is poured into the sand until all the pores are filled. What is the weight of this mixture?

2.30  A 300 kg mass at the bottom of a mine shaft is suspended from a 200 m long chain that has a mass of 3 kg / m. How much work is done in lifting the mass to the top of the shaft?

2.31  A 2 m long chain with a mass of 10 kg hangs from the ceiling of a room. The bottom end the chain is lifted until it touches the ceiling. How much work is done?

2.32  A cylindrical water tank with a height of 10 m and a diameter of 14 m is half full of water. How much work is required to pump all the water to the top of the tank? Assume the density of water to be 1000 kg / m³.

2.33  Find the specific kinetic energy and specific potential energy of an aircraft flying at an altitude of 10 km with a velocity of 800 km / h.

2.34  A pot of water is heated on an electric resistance heater, as shown in Figure P2.34. Each of the two cases (a, b) show the energy transfer crossing the system boundary.

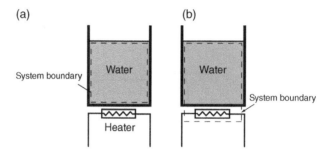

**Figure P2.34**   A pot of water warming on an electric resistance heater. (a) The system excludes the heater. (b) The system includes the heater.

2.35  A jet aircraft takes off from the ground and climbs to cruising altitude. Place a system boundary around it and mark the transfers of work and mass across the system boundary.

2.36  A hair dryer is switched on inside a room. For each of the following systems list the energy and mass transfers across the system boundary: (a) the hair dryer, (b) the room, excluding the hair dryer, and (c) the room and hair dryer combined.

2.37  The inside of a house is at 20 °C while the exterior air is at –5 °C. The temperature of the walls of the house does not change with time. Are the walls at equilibrium?

2.38  A light bulb is switched on and within a few minutes its temperature becomes constant. Is it at equilibrium or steady state?

2.39  Would you treat each of the following as a control mass or a control volume? (a) A car engine, (b) an open bottle, (c) a sealed bottle, (d) a refrigerator, (e) the pump in a refrigerator, (f) a human heart.

2.40  Classify the following quantities as (a) intensive properties, (b) extensive properties or (c) not properties: $T, x, a, W, KE, PE, P, m, e, v, \rho$.

# 3

# Thermodynamic System Properties

---

**In this chapter you will:**

- Become familiar with terms used to describe thermodynamic systems, including pure substance, homogeneous substance and phase.
- Use molar units to measure mass.
- Learn about pressure and temperature, including the absolute temperature scale.
- Use the ideal gas equation to calculate gas properties.
- Develop a simple kinetic model for ideal gases.
- Study internal energy, temperature and heat at a microscopic level.

---

## 3.1 Describing a Thermodynamic System

Before we can analyse a thermodynamic system we must describe it fully. To begin, we have to determine what state of matter it contains: solid, liquid or gas? Does it consist of only a single substance or is it a mixture? If it has several components, how are they distributed within it? In the next section we will discuss the terminology used to describe systems.

Next, we quantify how much matter we have in the system by specifying its mass or volume. Then, to fix the state of the system, we have to list other properties. Which properties besides mass and volume can we measure easily? We have instruments to directly measure pressure and temperature, so these are usually selected as independent variables.

Mass, volume, pressure and temperature are relatively simple to measure, but they are not all independent properties. If we heat a given mass of gas in a constant volume container its pressure rises along with temperature. We will derive relationships, known as equations of state, between the system pressure, temperature, mass and volume.

---

*Energy, Entropy and Engines: An Introduction to Thermodynamics*, First Edition. Sanjeev Chandra.
© 2016 John Wiley & Sons, Ltd. Published 2016 by John Wiley & Sons, Ltd.
Companion website: www.wiley.com/go/chandraSol16

Properties such as energy cannot be measured directly. The energy of a gas has to be calculated indirectly from measurements of temperature. We will discuss how to do this and show that simple models of the molecular structure of gases are useful in giving us insight into the relationship between energy and temperature.

## 3.2   States of Pure Substances

A *pure substance* is one that has the same, distinct chemical composition everywhere: it may be a pure element or chemical compound. Examples of pure substances are oxygen, nitrogen, water and copper. A *homogeneous substance* has the same composition and properties throughout, but may be a mixture of several pure substances. Air, which is mixture of several gases, is treated as a homogeneous substance, as are metal alloys such as steel or brass. A solution of carbon dioxide dissolved in water is a homogeneous substance (Figure 3.1a), but if you shake the flask so that the carbon dioxide comes out of solution and forms bubbles in the liquid (Figure 3.1b), the system is no longer homogeneous since its composition is not uniform throughout. Samples taken from different locations in the flask will give different ratios of carbon dioxide and water and we call it a *heterogeneous substance*.

All matter consists of atoms or molecules that interact with each other by exerting forces known as interatomic or intermolecular bonds. The magnitude of these forces determines the *state* of any material. Pure substances are commonly found in one of three different states: solid, liquid or gas. *Solids* have very strong intermolecular forces so the molecules are close together and their positions relative to each other are fixed (Figure 3.2a). *Liquids* have weaker intermolecular forces and the molecules, though still close, are free to move so that liquids,

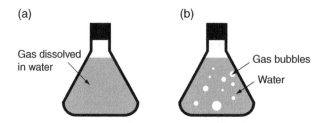

**Figure 3.1**   Examples of (a) a homogeneous solution and (b) a heterogeneous mixture.

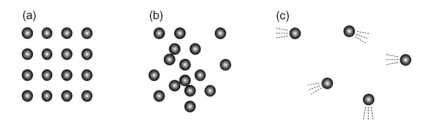

**Figure 3.2**   Molecular arrangements for different states of matter: (a) solid, (b) liquid and (c) gas.

unlike solids, can flow (Figure 3.2b). *Gases* have very large spacing between the molecules, which are free to move randomly. Intermolecular forces in gases are so weak as to be negligible (Figure 3.2c).

A *phase* is a quantity of matter with uniform chemical composition and physical properties. The three states of matter, solids, liquids and gases, therefore constitute different phases because they have different physical properties, even though they may have the same chemical structure. We can have different phases with the same chemical composition and state: snowflakes and icicles are both solid forms of water, but have quite distinct molecular structures and properties. Similarly, diamond and graphite are different phases of carbon.

A *phase transition* occurs when a material is transformed from one phase to another. The most common phase transitions are those between liquid and solid phases (freezing or melting) or liquid and gas phases (evaporation or condensation). Altering the intermolecular forces and spacing, typically by changing the temperature or pressure of a substance, produces phase transitions. As a solid is heated the molecules gain energy and begin to vibrate about their mean position. Eventually they break loose from their fixed positions and begin to move randomly – the material melts and becomes liquid. Further heating increases the velocity of molecules and the spacing between them. The substance expands in volume and the spacing between molecules increases. Intermolecular forces decrease in magnitude until they become negligible and the material becomes a gas in which molecules interact only when they collide with each other.

## 3.3   Mass and Volume

How do we describe matter? *Mass* and *volume* are the most obvious properties to describe the quantity of a substance and the amount of space that it occupies. The units of mass in the SI system are kilograms (kg) and those of volume are cubic metres ($m^3$). Since 1 $m^3$ is a rather large volume, smaller units frequently used to measure volume are the litre (l), where 1 l = $10^{-3}$ $m^3$, and the millilitre (ml), which equals $10^{-3}$ l. Smaller units of mass are the gram (g), equal to $10^{-3}$ kg, and the milligram (mg), which is $10^{-6}$ kg.

Molar units are often used to measure mass. Chemists first used this unit system to describe chemical reactions. The French chemist Joseph Gay-Lussac had observed in the late eighteenth century that hydrogen gas burning in oxygen gives water vapour according to the reaction

$$2H_2 + O_2 \rightarrow 2H_2O,$$

the volumes of gases required were surprisingly regular: 2 l of hydrogen react with 1 l of oxygen to produce 2 l of steam. When this was first observed no one understood why such exact ratios were required, but an Italian scientist, Amadeo Avogadro, speculated in 1811 that equal volumes of gases are made up of an equal number of molecules. Therefore, 2 l of $H_2$ contained twice the number of molecules as 1 l of $O_2$ and they would combine in exactly the required proportion. There was no way of either proving or disproving this idea at the time he proposed it, and Avogadro's hypothesis was almost forgotten for over half a century.

Avogadro's ideas were used in the late nineteenth and early twentieth century by a new generation of scientists who had a much better understanding of the molecular nature of gases and the structure of atoms. Atoms consist of a heavy nucleus surrounded by much lighter electrons. The mass of an atom is almost entirely that of its nucleus, consisting of positively charged protons, and uncharged neutrons, which have near equal masses – approximately $1.67 \times 10^{-27}$ kg. Electrons are much smaller, with a mass of only $9.11 \times 10^{-31}$ kg. The *atomic mass unit* is defined as being one-twelfth the mass of a carbon-12 atom that contains six protons and six neutrons, so that the protons and neutrons both have masses of almost exactly one atomic mass unit. The masses of atoms are specified in terms of atomic mass units.

Many elements do not exist naturally as individual atoms, but combine with others of the same species to form molecules. Atoms of different species may also combine to form molecules of chemical compounds. The mass of a molecule, specified in atomic mass units, is known as its *molar mass* ($M$). Table 3.1 gives values for the molar masses of several elements and compounds. The approximate values listed are sufficiently accurate for most practical purposes.

Very careful experimentation has established that 2 g of diatomic hydrogen contains $6.022 \times 10^{23}$ molecules, which is known as Avogadro's number ($N_A$). A molecule of diatomic oxygen has a mass 16 times that of a molecule of hydrogen, so 32 g of oxygen will also contain $N_A$ molecules. The mass of any substance that contains $N_A$ molecules is known as a *gram-mole* (abbreviated *gmol*) or simply *mole* (abbreviated *mol*). A mole is frequently used as a unit of mass: a mole of any substance is its molar mass in grams. We can also define a *kilogram-mole* (*kmol*), which is the molar mass in kilograms. The units of molar mass then become g / mol or kg / kmol. We can easily convert between mass in kg ($m$) and mass in kmol ($N$) using the relation

$$m(\text{kg}) = N(\text{kmol}) \times M(\text{kg} / \text{kmol}).$$

During the combustion of hydrogen two molecules of hydrogen react with one molecule of oxygen to produce two molecules of water. To carry out this reaction in the laboratory a chemist would burn 2 mol of $H_2$ (4 g) in 1 mol of $O_2$ (32 g) to produce 2 mol of $H_2O$ (36 g). These masses will give exactly the right ratio of oxygen and hydrogen molecules to complete the

**Table 3.1** Molar masses of common substances.

| Name | Formula | Molar mass ($M$, kg/kmol) | Approximate molar mass |
|---|---|---|---|
| Hydrogen | $H_2$ | 2.016 | 2 |
| Oxygen | $O_2$ | 31.999 | 32 |
| Water | $H_2O$ | 18.015 | 18 |
| Ammonia | $NH_3$ | 17.031 | 17 |
| Nitrogen | $N_2$ | 28.013 | 28 |
| Carbon | C | 12.01 | 12 |
| Carbon dioxide | $CO_2$ | 44.01 | 44 |
| Methane | $CH_4$ | 16.043 | 16 |
| Helium | He | 4.0026 | 4 |

reaction without having any gas left over.

Avogadro's hypothesis can be expressed mathematically by saying that if all other conditions (such as temperature and pressure) are held constant, the volume of a gas is

$$V = C_1 N, \quad \text{at constant } T \text{ and } P \tag{3.1}$$

where $C_1$ is a constant of proportionality and $N$ is the number of moles in the gas. One mole of any gas has $N_A = 6.022 \times 10^{23}$ molecules and therefore occupies the same volume as a mole any other gas. One mole of gas at $0\,°C$ and atmospheric pressure occupies a volume of approximately 22.4 l.

## Example 3.1
**Problem:** What is the mass in kg of 6 kmol of: (a) water vapour and (b) methane?

**Find:** Mass $m$ of each gas.

**Known:** Mass $N = 6$ kmol for each gas.

**Governing Equation:**

Molar mass
$$m(\text{kg}) = N(\text{kmol}) \times M(\text{kg} / \text{kmol})$$

**Properties:** Molar mass of water $M_{water} = 18.015$ kg / kmol (Table 3.1), molar mass of methane $M_{methane} = 16.043$ kg / kmol (Table 3.1).

**Solution:**
(a)  For water vapour $m = 6$ kmol $\times$ 18.015 kg / kmol = 108.09 kg.
(b)  For methane $m = 6$ kmol $\times$ 16.043 kg / kmol = 96.258 kg.

**Answer:** The mass of 6 kmol of water vapour is 108 kg, and 6 kmol of methane is 96.3 kg. ■

## Example 3.2
**Problem:** How many molecules are there in a mixture of 1 g of $N_2$ and 1 g of $CO_2$?

**Find:** Number of molecules in mixture of $N_2$ and $CO_2$.

**Known:** Mass of each gas is $m = 1$ g.

**Governing Equation:**

Molecules in 1 gmol
$$N_A = 6.022 \times 10^{23} \text{ molecules}$$

**Properties:** Molar mass of $CO_2$ is $M_{CO2} = 12.01$ g / gmol (Table 3.1), molar mass of $N_2$ is $M_{N2} = 28.013$ g / gmol (Table 3.1).

**Solution:**
1 gmol of $CO_2 = 12.01$ g contains $6.022 \times 10^{23}$ molecules,

$$1\text{g of } CO_2 \text{ contains} \frac{6.022 \times 10^{23} \text{ molecules} / \text{mol}}{12.01 \text{ g} / \text{mol}} = 5.0142 \times 10^{22} \text{ molecules} / \text{g}.$$

1 gmol of $N_2 = 28.013$ g contains $6.022 \times 10^{23}$ molecules,

$$1\,\mathrm{g\ of\ N_2\ contains} \frac{6.022 \times 10^{23}\ \mathrm{molecules\,/\,mol}}{28.013\ \mathrm{g\,/\,mol}} = 2.1497 \times 10^{22}\ \mathrm{molecules\,/\,g.}$$

Total number of molecules in 1 g of each gas mixed together is

$$N = 5.0142 \times 10^{22} + 2.1497 \times 10^{22} = 7.1639 \times 10^{22}.$$

**Answer:** There are $7.16 \times 10^{23}$ molecules in the mixture.                                          ∎

## 3.4  Pressure

A gas confined to a closed vessel exerts a force ($F$) on it due to gas molecules hitting the walls of the vessel and rebounding (Figure 3.3). This force, divided by the area ($A$) on which it is acting, is known as the gas pressure:

$$P = \frac{F}{A}. \tag{3.2}$$

Pressure has units of N / m², which is known as a *pascal* (Pa). Since 1 Pa is a rather low pressure, it is often more convenient to use *kilopascal* (kPa), which equals $10^3$ Pa, or *megapascal* (MPa), $10^6$ Pa. Another unit of pressure is the *bar*, defined as being $10^5$ Pa. We also specify pressure relative to that of the atmosphere, where 1 *atmosphere* (atm) is defined as 1.01325 bar, or approximately 0.1 MPa.

In defining pressure by Equation (3.2), we have assumed that the area ($A$) is large enough that the force of molecules striking it remains constant. This is an excellent assumption if the area is much larger than the dimensions of a single molecule. If a very large number of molecules strike the area during the interval of time we are averaging the force, fluctuations created by individual molecules impacting and rebounding will be too small to measure. We are making the *continuum assumption* that properties are obtained by averaging over lengths and times that are much greater than molecular scales.

Pressures within fluids can be defined by imagining that an infinitesimally small cube is inserted into the fluid at the required location, as shown in Figure 3.4. The force exerted by molecules impinging on a wall of this cube, divided by its area, will give the pressure.

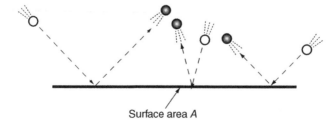

Surface area $A$

**Figure 3.3**   Force exerted by molecules striking a surface and rebounding.

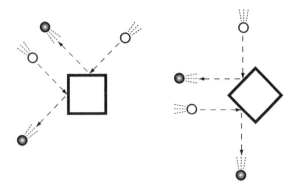

**Figure 3.4**  Pressure within a fluid is independent of direction.

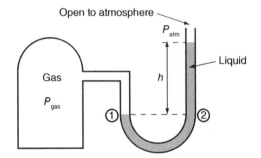

**Figure 3.5**  Pressure measurement with a manometer.

Since molecular motion is random it does not matter which wall is selected for measurement; pressure is independent of direction.

Pressure can be determined using a manometer, which in its simplest form is a flexible tube partially filled with liquid (Figure 3.5). To measure pressure one end of the manometer is connected to the container filled with gas while the other is left exposed to the atmosphere. If the pressure in the container is greater than that of the surrounding air, the column of liquid will rise by a height ($h$) in the open branch. If the cross-sectional area of the tube in Figure 3.5 is $A_c$, the force exerted by the gas on the liquid column at point 1 is $P_{gas}A_c$. The force exerted by the atmosphere on the liquid is $P_{atm}A_c$ while the weight of the liquid column with height $h$ is $F_w = \rho g h A_c$. By symmetry the pressures at points 1 and 2 must be the same at equilibrium so that

$$P_{gas}A_c = \rho g h A_c + P_{atm}A_c. \tag{3.3}$$

Simplifying Equation (3.3) we get

$$P_{gas} = \rho g h + P_{atm}. \tag{3.4}$$

The gas pressure is also known as the *absolute pressure* ($P_{abs}$),

$$P_{abs} - P_{atm} = \rho g h. \tag{3.5}$$

Manometers, and most other pressure gauges, measure the difference between absolute pressure and atmospheric pressure. This is known as gauge pressure, defined by

$$P_{gauge} = P_{abs} - P_{atm}. \tag{3.6}$$

It is important to determine whether any pressure measurement instrument is recording absolute or gauge pressure. If it is the latter, you must measure atmospheric pressure separately and add that to the gauge pressure to get absolute pressure. All calculations of gas properties are done using absolute pressure values.

### Example 3.3
**Problem:** What is the absolute pressure on the hull of a submarine (a) floating on the surface of the sea and (b) at a depth of 100 m underwater?
**Find:** Absolute pressure $P$ on the submarine hull.
**Known:** (a) Depth $h = 0$ m, (b) depth $h = 100$ m.
**Governing Equation:**

Pressure under a liquid of depth $h$ $\qquad\qquad\qquad\qquad P = P_{atm} + \rho g h$

**Properties:** Density of water $\rho_{water} = 1000$ kg / m³ (Appendix 3), atmospheric pressure $P_{atm} =$ 1.013 25 × 10⁵ Pa, gravitational acceleration at Earth's surface is $g = 9.81$ m / s².

**Solution:**
(a) $P = P_{atm} = 1.013\ 25 \times 10^5$ Pa $= 0.101\ 33$ MPa
(b) $P = P_{atm} + \rho g h = 1.01325 \times 10^5$ Pa $+ 1000 \times 9.81 \times 100$ Pa $= 1.0823$ MPa

**Answer:** On the surface the pressure is 0.101 MPa, while at a depth of 100 m pressure is 1.08 MPa. ∎

## 3.5 Temperature

Temperature is a property we use to determine if two systems are in thermal equilibrium. If we take two objects at different temperatures and bring them into close contact with each other, heat will flow from the region of higher temperature to that of lower temperature. The greater the temperature difference the faster the heat transfer. Eventually, when the two objects reach the same temperature, there will be no more heat transfer and they are then said to be in thermal equilibrium.

Temperature can be measured using a thermometer, which in its simplest form is a sealed glass tube with a bulb at one end filled with liquid, usually mercury or alcohol (Figure 3.6). To measure the temperature of an object we place the bulb in contact with it until they are both at the same temperature. Since liquids expand in volume when heated, the length of the liquid column in the tube is an indicator of its temperature. To establish a temperature scale we need two reference temperatures that can be used to mark a high and low temperature on the thermometer. In the Celsius scale, the freezing point of water is taken as 0 °C and the boiling point

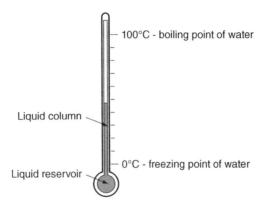

**Figure 3.6**   Liquid in glass thermometer.

as 100 °C. The length between them is divided equally into 100 intervals of 1° each. Though this is a simple definition of a temperature scale, it is not a very satisfactory one. It assumes that the expansion of liquids with temperature is perfectly linear: that is, a liquid expands by the same amount when heated from 0 to 1 °C that it does when heated from 99 to 100 °C. Though this approximation is adequate for many everyday temperature measurements it is not good enough for high precision work. If really accurate temperature measurements are required, down to one-millionth of a degree, we will have to develop a better calibration method, something to which we will return very shortly.

## 3.6   Ideal Gas Equation

When energy is added or removed from a gas its intensive properties such as pressure, temperature and specific volume may change. The changes are not all independent: if a gas in a container of fixed volume is heated so that its temperature increases, its pressure will also increase. Given the change in some of the properties it is possible to develop equations that tell us how other properties will change, so that we can write a function of the form

$$f(P, v, T) = 0. \tag{3.7}$$

Equations that give relations between intensive properties are known as *equations of state*. The earliest equations of state were developed from experiments. Robert Boyle observed as early as 1662 that the pressure of a fixed mass of gas held at constant temperature varied inversely with the volume of the container it is confined in:

$$P = \frac{C_2}{V}, \ \left(\text{at constant } T\right) \tag{3.8}$$

where $C_2$ is a constant that depends on the mass and temperature of the gas. More than a century later, Joseph Gay-Lussac observed that the volume of a fixed mass of gas that is free to expand and contract at constant pressure is proportional to its temperature:

$$V = C_3 T. \ \left(\text{at constant } P\right) \tag{3.9}$$

Gay-Lussac mentioned in his paper that a certain J. A. C. Charles had made similar observations some years previously, though these had never been published. On the strength of this comment English texts began to refer to Charles's law, though in most European countries it is known as Gay-Lussac's law.

Combining the laws of Avogadro [Equation (3.1)], Boyle [Equation (3.8)] and Charles [Equation (3.9)], we obtain the *ideal gas equation*:

$$PV = NR_uT. \tag{3.10}$$

$R_u$ is a constant, known as the *universal gas constant*, which experiments have shown has the same value for all gases at sufficiently low density,

$$R_u = 8.314 \, \text{kJ} / \text{kmol K}.$$

If we prefer to use kilograms instead of moles as our unit of mass, we can use the relation $m = NM$ to rewrite the ideal gas equation as

$$PV = \frac{m}{M} R_u T. \tag{3.11}$$

Defining a gas constant for each particular gas, $R = R_u / M$, we have the ideal gas equation in mass units:

$$PV = mRT. \tag{3.12}$$

The gas constant $R$ is different for every gas and values for some common substances are listed in Appendix 1. All gases behave as ideal gases when their densities are low, that is, when they are at sufficiently low pressure and high temperature. Most common gases, such as air, nitrogen, oxygen and carbon dioxide, behave as ideal gases at atmospheric pressures and temperatures. Steam also behaves as an ideal gas if it is heated above its boiling point.

## 3.7 Absolute Temperature Scale

Applying the ideal gas equation immediately raises questions about our temperature scale. If we use the Celsius scale, then at a temperature $T = 0$ °C Equation (3.10) predicts that the pressure will be zero, which we know from experience is plainly wrong. If true, we would find a vacuum in every freezer. Even worse, if the temperature becomes negative, so does the pressure, which is meaningless. Obviously we need to rethink how we measure temperature.

We can define a new temperature scale to use with the ideal gas equation by making a constant volume gas thermometer, shown in Figure 3.7. It consists of a rigid bulb filled with gas, connected by tubing to a manometer. The bulb is placed in a bath whose temperature is to be measured, and the gas is allowed to expand or contract as it comes to thermal equilibrium with its surroundings. The flexible manometer tube is then raised or lowered to bring the liquid level in the manometer to its original position, so that the volume of the gas remains the

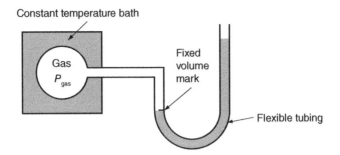

**Figure 3.7**   Constant volume gas thermometer.

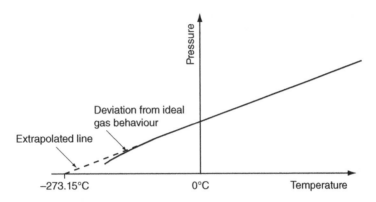

**Figure 3.8**   Variation of pressure with temperature for a gas.

same, and the gas pressure ($P$) is measured from the manometer. By lowering the temperature of the bath, we can plot the decrease in pressure as the gas becomes colder. Eventually, if the gas becomes very cold, it will stop behaving as an ideal gas because as its volume shrinks intermolecular forces become stronger and the pressure will be less than that predicted by the ideal gas equation. If, however, we extrapolate the ideal gas equation line it will cross the temperature axis at $T = -273.15$ °C (see Figure 3.8). If we change the mass of gas in the thermometer bulb and repeat this experiment we will find that the extrapolated pressure line still goes to zero at the same temperature. Trying a different gas will also give the same result.

To define a temperature scale that satisfies the ideal gas equation, we set zero temperature to correspond to zero pressure. This is known as the Kelvin temperature scale, where a temperature difference of 1 kelvin (abbreviated K) equals a temperature difference of 1 °C, and 0 K equals $-273.15$ °C. We can convert temperature from the Celsius to the Kelvin scale by setting

$$T(\mathrm{K}) = T(^{\circ}\mathrm{C}) + 273.15. \tag{3.13}$$

We can use the gas thermometer to define a more accurate temperature scale than we could using the liquid-in-glass thermometer. To set up this temperature scale we need only one reference temperature ($T_{ref}$). The gas bulb is held at $T_{ref}$ while the corresponding gas pressure ($P_{ref}$) is measured. Having recorded these values, the pressure and temperature of the gas bulb will then be related by the equation

$$T = \frac{P}{P_{ref}} T_{ref}.$$ (3.14)

By measuring gas pressure, it is possible to determine its temperature from Equation (3.14). The reference temperature used to define the absolute temperature scale is the *triple point* of water. At a temperature of 0.01 °C and a pressure of 611.73 Pa, liquid water, ice and water vapour can all co-exist in equilibrium (we will study this in detail in Chapter 7). These are the only conditions when all three phases are simultaneously present, so just by observation it is easy to confirm that the temperature is $T_{ref} = 0.01$ °C, which equals 273.16 K. A constant volume gas thermometer, as shown in Figure 3.7, is placed in a constant temperature bath at the triple point of water and the pressure ($P_{ref}$) recorded. Then, if the thermometer is placed in a bath at a different temperature, the pressure ($P$) varies with temperature ($T$) as

$$T = \frac{P}{P_{ref}} \times 273.16 \text{ K}.$$ (3.15)

This temperature scale is an improvement on the one based on a liquid in glass thermometer, because it requires only one reference temperature and is, by definition, linear. However, though we have an intuitive understanding of what temperature is and how to measure it, we have still not given a rigorous definition of it. In thermodynamics temperature is defined in terms of another fundamental property – entropy. We will therefore continue to use the term temperature, in the belief that we all understand what it means. A precise mathematical definition will have to wait until we are familiar with the concept of entropy, at which time we will have to confirm that whatever definition we adopt agrees with our everyday notion of temperature.

**Example 3.4**
**Problem:** What is the pressure of 2 kmol of an ideal gas that has a volume of 5700 l and a temperature of 300 K?
**Find:** Gas pressure $P$.
**Known:** Ideal gas, mass $N = 2$ kmol, volume $V = 5700$ l, temperature $T = 300$ K.
**Governing Equation:**
Ideal gas equation $PV = NR_u T$
**Properties:** Ideal gas constant $R_u = 8.314 \times 10^3$ J / kmol K.

**Solution:**

$$P = \frac{NR_u T}{V} = \frac{2 \text{ kmol} \times 8.314 \times 10^3 \text{ J / kmol K} \times 300 \text{ K}}{57001 \times 10^{-3} \text{ m}^3 / 1} = 0.875 \times 10^6 \text{ Pa}$$

**Answer:** The pressure is 0.875 MPa.                                                    ■

**Example 3.5**
**Problem:** Find the specific volume of nitrogen at a temperature of 110 °C and a pressure of 0.2 MPa.
**Find:** Specific volume of nitrogen $v$.
**Known:** Temperature $T = 110$ °C $= 110 + 273.15$ K $= 383.15$ K, pressure $P = 0.2$ MPa $= 0.2 \times 10^6$ Pa.
**Assumptions:** Nitrogen at 110 °C and 0.2 MPa behaves as an ideal gas.
**Governing Equation:**
Ideal gas equation $PV = mRT$
**Properties:** Gas constant for nitrogen $R = 0.2968$ kJ / kgK (Appendix 1).

**Solution:**
Rearranging the ideal gas equation,

$$v = \frac{V}{m} = \frac{RT}{P} = \frac{0.2968 \times 10^3 \, J/kgK \times 383.15 K}{0.2 \times 10^6 \, Pa} = 0.568 \; 59 \, m^3 / kg.$$

**Answer:** The specific volume of nitrogen is 0.569 m³ / kg.                                  ∎

**Example 3.6**
**Problem:** Two tanks of methane, one with a volume 1 m³, temperature 20 °C and pressure 300 kPa and the other with volume 0.2 m³, temperature 30 °C and pressure 800 kPa are connected by a valve that is opened. The two tanks are allowed to come to equilibrium, when their temperature is 27 °C. What is the equilibrium pressure?
**Find:** Equilibrium pressure in connected tanks.
**Known:** Tank 1 pressure $P_1 = 300$ kPa, temperature $T_1 = 20$ °C $= 273.15 + 20$ K $= 293.15$ K, volume $V_1 = 1$ m³.
Tank 2 pressure $P_2 = 800$ kPa, temperature $T_2 = 30$ °C $= 273.15 + 30$ K $= 303.15$ K, volume $V_2 = 0.2$ m³.
Final temperature $T_f = 27$ °C $= 273.15 + 27$ K $= 300.15$ K.
Final volume $V_f = V_1 + V_2 = 1$ m³ $+ 0.2$ m³ $= 1.2$ m³.
**Assumptions:** Methane behaves as an ideal gas at the temperatures and pressures in this problem.
**Governing Equation:**
Ideal gas equation $PV = mRT$
**Properties:** Gas constant for methane $R = 0.5184$ kJ / kgK (Appendix 1).

**Solution:**
In the final state we know the volume and temperature. If we also know the mass we can calculate the pressure from the ideal gas equation. The final mass must equal the mass of methane in each tank, so $m_f = m_1 + m_2$:

$$m_1 = \frac{P_1 V_1}{RT_1} = \frac{300 \times 10^3 \, Pa \times 1 m^3}{0.5184 \times 10^3 \, J/kgK \times 293.15 K} = 1.9741 kg,$$

$$m_2 = \frac{P_2 V_2}{RT_2} = \frac{800 \times 10^3 \, Pa \times 0.2 m^3}{0.5184 \times 10^3 \, J/kgK \times 303.15 K} = 1.0181 \ kg.$$

Final mass,

$$m_f = m_1 + m_2 = 1.9741\,\text{kg} + 1.0181\,\text{kg} = 2.9922\,\text{kg}.$$

Final volume,

$$V_f = V_1 + V_2 = 1\,\text{m}^3 + 0.2\,\text{m}^3 = 1.2\,\text{m}^3.$$

Then the final pressure can be found with the ideal gas equation:

$$P_f = \frac{m_f R T_f}{V_f} = \frac{2.9922\,\text{kg} \times 0.5184 \times 10^3\,\text{J/kgK} \times 300.15\,\text{K}}{1.2\,\text{m}^3} = 387.983\,\text{kPa}$$

**Answer:** The final pressure in the tanks is 388.0 kPa.                                              ∎

## 3.8   Modelling Ideal Gases

To understand why low-density gases follow the ideal gas equation, it helps to create a simple model of gas behaviour. The *kinetic theory* of gases is based on several simplifying assumptions:

1. A gas is made of a very large number of elementary particles called molecules that are in constant, random motion. All molecules have the same mass ($m_e$).
2. The total number of molecules ($n$) is very large.
3. Molecules collide perfectly elastically with each other and with the walls of the container. They obey Newton's laws during collisions.
4. There is no force acting on molecules except during collisions.
5. The volume of molecules is negligible – they are considered to be point masses.

A gas is confined in a cubical container with sides of length $L$. A molecule with velocity ($c$) hits the wall normal to the $x$-axis and rebounds (Figure 3.9). The molecular velocity can be resolved into components ($c_x$, $c_y$, $c_z$). After colliding with the surface the $x$-component of the velocity becomes $-c_x$. Since the only force acting on the molecule is in the $x$-direction the

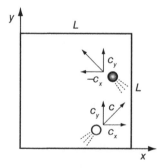

**Figure 3.9**   A molecule contained in a box of side $L$ hits a wall with $x$-velocity component $c_x$ and rebounds. The $x$-velocity component becomes $-c_x$.

other velocity components are unchanged. The change in momentum of the molecule is

$$\Delta p = \left(-m_e c_x\right) - \left(m_e c_x\right) = -2m_e c_x. \tag{3.16}$$

The change in momentum of the container wall is $-\Delta p = 2m_e c_x$, equal in magnitude but opposite in direction to the change in momentum of the gas molecule, so that total momentum is conserved.

The molecule travels to the opposing face, rebounds off it and then crosses the length of the box again. The time taken to cross the length $(L)$ of the box is $\Delta t = L / c_x$, so the interval between a molecule travelling twice across the box and hitting the same face again is $2\Delta t = 2L / c_x$. The force exerted by the molecule on the wall equals the rate of momentum transfer to the wall,

$$-\frac{\Delta p}{2\Delta t} = \left(2m_e c_x\right)\left(\frac{c_x}{2L}\right) = \frac{m_e c_x^2}{L}. \tag{3.17}$$

Summing the force exerted by all the $n$ molecules in the box and dividing by the area of the surface $(L^2)$ gives the total pressure $(P)$ on the wall:

$$P = \left(\frac{m_e c_{x,1}^2 + m_e c_{x,2}^2 + m_e c_{x,3}^2 + \ldots + m_e c_{x,n}^2}{L}\right)\left(\frac{1}{L^2}\right). \tag{3.18}$$

Since the mass of all molecules $(m_e)$ is the same,

$$P = \frac{m_e}{L^3}\left(c_{x,1}^2 + c_{x,2}^2 + c_{x,3}^2 + \ldots + c_{x,n}^2\right) = \frac{nm_e}{L^3}\frac{\left(c_{x,1}^2 + c_{x,2}^2 + c_{x,3}^2 + \ldots + c_{x,n}^2\right)}{n}. \tag{3.19}$$

The total mass of gas in the container is $m = nm_e$ and the volume of the gas is $V = L^3$. We can define an average velocity for all the molecules, known as the root mean square (rms) velocity:

$$c_{x,\mathrm{rms}} = \left[\frac{\left(c_{x,1}^2 + c_{x,2}^2 + c_{x,3}^2 + \ldots + c_{x,n}^2\right)}{n}\right]^{1/2}.$$

Then,

$$P = \frac{m}{V}c_{x,\mathrm{rms}}^2. \tag{3.20}$$

The magnitude of the total velocity of a molecule $(c)$ is obtained by summing the components in all three directions so that $c^2 = c_x^2 + c_y^2 + c_z^2$. Since velocities are random the average value of their components in all directions should be the same so that $c_{x,\mathrm{rms}}^2 = c_{y,\mathrm{rms}}^2 = c_{z,\mathrm{rms}}^2$. In that case, $c_{\mathrm{rms}}^2 = 3c_{x,\mathrm{rms}}^2$ and

$$P = \frac{1}{3}\frac{m}{V}c_{\mathrm{rms}}^2. \tag{3.21}$$

Since $m = NM$ where $N$ is the number of moles of gas and $M$ is its molecular mass, we can rewrite Equation (3.21) as

$$PV = \frac{1}{3} NM c_{rms}^2 . \tag{3.22}$$

This equation helps us understand Boyle's law. If gas temperature is fixed the average molecular speed does not change and the right hand side of Equation (3.22) is a constant. Gas pressure will vary inversely with volume in this case.

## 3.9 Internal Energy

If we combine Equation (3.22) with the ideal gas equation, $PV = NR_uT$, we get

$$\frac{1}{2} M c_{rms}^2 = \frac{3}{2} R_u T. \tag{3.23}$$

Dividing Equation (3.23) by Avogadro's number ($N_A$),

$$\frac{1}{2} \frac{M}{N_A} c_{rms}^2 = \frac{3}{2} \frac{R_u}{N_A} T, \tag{3.24}$$

and remembering that the mass of a single molecule is $m_e = M / N_A$, we can rewrite this as

$$\frac{1}{2} m_e c_{rms}^2 = \frac{3}{2} kT, \tag{3.25}$$

where $k = R_u / N_A$ is known as the *Boltzmann constant* and has a value of $1.38 \times 10^{-23}$ J / K.

The left hand side of Equation (3.25) is the kinetic energy of a single molecule and is proportional to the temperature of the gas. This gives us an interpretation for what temperature means at the molecular level: it is a measure of the energy stored in the motion of molecules. Multiplying both sides of Equation (3.25) by the number of molecules ($n$) gives the total molecular kinetic energy of a gas:

$$\frac{1}{2} m_e n c_{rms}^2 = \frac{3}{2} nkT = \frac{3}{2} \left( \frac{n}{N_A} \right) (N_A k) T = \frac{3}{2} NR_u T. \tag{3.26}$$

Even though the bulk gas itself is not moving, and has no macroscopic kinetic energy, it can store energy at a microscopic level in the motion of its molecules. *The total microscopic energy of all the molecules is known as internal energy.* The internal energy ($U$), is an extensive property with units of joules, given by

$$U = \frac{3}{2} NR_u T. \tag{3.27}$$

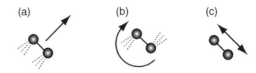

**Figure 3.10** Energy storage modes in molecules with more than one atom: (a) translation, (b) rotation and (c) vibration.

If we change the temperature of an ideal gas from $T_1$ to $T_2$ its energy changes by

$$U_2 - U_1 = \frac{3}{2}NR_u\left(T_2 - T_1\right). \tag{3.28}$$

Dividing by the mass of the gas ($m$) gives the specific internal energy change

$$u_2 - u_1 = \frac{3}{2}R\left(T_2 - T_1\right). \tag{3.29}$$

The internal energy of an ideal gas is a function only of its temperature: as the temperature increases so does the energy. Equation (3.29) was derived assuming that molecules are point masses, so that they can only store translational kinetic energy. This is a good assumption for molecules that have only a single atom, such as helium or argon. Equation (3.29) predicts the internal energy of monoatomic gases very accurately, but does not work well for larger molecules. More complex gases with more than one atom, such as oxygen, nitrogen, or water vapour can store energy in other modes such as rotation or vibration. In general, the internal energy of any system includes all the energy stored at a molecular level, including molecular kinetic energy (due to translation, vibration or rotation), molecular potential energy (due to electric and magnetic fields) and the energy stored in chemical bonds (Figure 3.10).

Calculating internal energy changes for molecules with multiple atoms is more difficult since we have to account for all possible energy storage modes. We can still make use of one very important conclusion from our simple model: *the internal energy of a unit mass of an ideal gas is a function only of its temperature*. Molecules in an ideal gas are so far apart that long-distance forces between them are negligible. Increasing the pressure and bringing the molecules closer together has no effect on internal energy because, as long we have ideal gas behaviour, intermolecular forces can still be disregarded. For gases that do not behave ideally internal energy may, in general, depend on both pressure and temperature. We can state that for any ideal gas

$$u_2 - u_1 = c\left(T_2 - T_1\right), \tag{3.30}$$

where $c$ is a gas property known as the specific heat that can be measured experimentally. In the next chapter we will learn how to determine and use specific heat values.

**Example 3.7**
**Problem:** What is the internal energy of a kilomole of a monoatomic ideal gas at 0 °C?
**Find:** Internal energy $U$.
**Known:** Ideal gas, mass $N = 1$ kmol, temperature $T = 0\,°C = 273.15$ K.

**Governing Equation:**

Internal energy of monoatomic ideal gas $\qquad U = \dfrac{3}{2} NR_u T$

**Properties:** Ideal gas constant $R_u = 8.314 \ kJ / kmolK$.

**Solution:**

$$U = \frac{3}{2} \times 1\,kmol \times 8.314\,kJ/kmolK \times 273.15\,K = 3406.45\,kJ$$

**Answer:** The internal energy of 1 kmol of a monoatomic ideal gas at 0 °C is 3406 kJ. ∎

**Example 3.8**

**Problem:** How much energy does it take to raise the temperature of 1 kmol of argon by 1 K?

**Find:** Energy $\Delta U$ to raise temperature of a monoatomic ideal gas.

**Known:** Argon is a monoatomic ideal gas, mass $N = 1$ kmol, temperature change $\Delta T = 1\,K$, ideal gas constant $R_u = 8.314 \ kJ / kmolK$.

**Governing Equation:**

Internal energy change $\qquad U_2 - U_1 = \dfrac{3}{2} NR_u (T_2 - T_1)$

**Solution:**

$$\Delta U = \frac{3}{2} NR_u \Delta T$$

$$\Delta U = \frac{3}{2} \times 1\,kmol \times 8.314\,kJ/kmolK \times 1\,K = 12.471\,kJ$$

**Answer:** It takes 12.5 kJ of energy to raise the temperature of 1 kmol of argon by 1 K. ∎

## 3.10   Properties of Liquids and Solids

The pressure and temperature of gases can generally be assumed to behave as predicted by the ideal gas equation. It is only under extreme conditions of high pressure or low temperature that the ideal gas approximation fails, which will be discussed in Chapter 7. Until then we will assume that all gases behave ideally.

What about the properties of liquids and solids? How can we model those? In general, the volume of liquids and solids changes very little with pressure or temperature. If we heat a block of steel by 100 °C, its volume increases by approximately 0.4%. Water heated by the same amount expands by about 2%. Pressure has even less effect: if we increase the pressure on a mass of water by 1 atm (~$10^5$ Pa), its volume decreases by 0.005%. We can reasonably assume that all liquids and solids are incompressible and their equation of state reduces to:

$$v = \text{constant.} \tag{3.31}$$

## Further Reading

1.  D. Halliday, R. Resnick, J. Walker (**2015**) *Fundamentals of Physics*, John Wiley & Sons, Ltd, London.
2.  Y. A. Cengel, M. A. Boles (**2015**) *Thermodynamics – An Engineering Approach*, McGraw Hill, New York.

3.  M. J. Moran, H. N. Shapiro, D. D. Boettner, M. B. Bailey (**2014**) *Fundamentals of Engineering Thermodynamics*, John Wiley & Sons, Ltd, London.

4.  C. Borgnakke, R. E. Sonntag (**2012**) *Fundamentals of Thermodynamics*, John Wiley & Sons, Ltd, London.

## Summary

A *pure substance* is one that has the same, distinct chemical composition everywhere. A *homogeneous substance* has the same composition and properties throughout while a *heterogeneous substance* does not have uniform composition. A *phase* is a quantity of matter with uniform chemical composition and physical properties. A *phase transition* occurs when a material is transformed from one phase to another.

In the SI system the units of mass are kg and those of volume are m$^3$. The *atomic mass unit* is defined as being one-twelfth the mass of a carbon-12 atom. The mass of a molecule, specified in atomic mass units, is known as its *molar mass* ($M$). The mass of any substance that contains Avogadro's number ($N_A$ = 6.022 × 10$^{23}$) of molecules is known as a *gram-mole* (gmol) or simply *mole* (mol). A *kilogram-mole* (kmol) is the molar mass in kg. The units of molar mass are g / mol or kg / kmol. The relation between the mass in kg ($m$) and mass in kmol ($N$) is

$$m \ (\text{kg}) = N \ (\text{kmol}) \times M \ (\text{kg} / \text{kmol}).$$

A fluid confined to a vessel exerts a force ($F$) on it which, divided by the total area ($A$) on which it is acting, is known as the pressure:

$$P = \frac{F}{A}.$$

Pressure has units of N / m$^2$, which is known as a *pascal* (Pa). We are making the *continuum assumption* that properties are obtained by averaging over lengths and times that are much greater than molecular scales. Manometers measure the difference between absolute pressure and atmospheric pressure:

$$P_{abs} - P_{atm} = \rho g h.$$

The gauge pressure is defined as

$$P_{gauge} = P_{abs} - P_{atm}.$$

An *ideal gas* is one that obeys the *equation of state*

$$PV = NR_u T,$$

where the *universal gas constant* $R_u$ = 8.314 kJ / kmol K. Defining a gas constant for each particular gas, $R = R_u / M$, where $M$ is the molar mass, we have the ideal gas equation in mass units:

$$PV = mRT.$$

A temperature scale that gives zero temperature at zero pressure, is the Kelvin scale, where

$$T(\text{K}) = T(°\text{C}) + 273.15.$$

The *internal energy* ($U$) is the total energy of all the molecules in a substance. The internal energy is an extensive property with units of joules, given for a monoatomic, ideal gas by

$$U = \frac{3}{2} NR_u T.$$

The internal energy of an ideal gases depends only on temperature. The change in specific internal energy of an ideal gas is

$$u_2 - u_1 = c(T_2 - T_1),$$

where $c$ is the specific heat. For gases that do not behave ideally, or for liquids and solids, internal energy can, in general, depend on both pressure and temperature. Liquids and solids can be assumed to be incompressible with constant specific volume so that

$$v = \text{constant}.$$

## Problems

3.1    What is the specific volume of a gas that has a weight of 500 N and a volume of 300 m$^3$?

3.2    A gas mixture contains 10 kmol of $H_2$ and 5 kmol of $O_2$. What is the mass of each gas in the mixture?

3.3    If 5 g of methane is burned in the chemical reaction $CH_4 + 2O_2 \rightarrow CO_2 + 2H_2O$, what mass of oxygen is consumed and what are the masses of the combustion products?

3.4    An ultra high vacuum is defined as a pressure lower than $10^{-7}$ Pa. How many molecules will there be in 1 m$^3$ volume of gas at this pressure at 25 °C?

3.5    Nitrogen in a partially evacuated vessel contains $10^{10}$ molecules per mm$^3$ of gas. What is the pressure of the gas if the temperature is 25 °C?

3.6    What is the absolute pressure of a gas whose gauge pressure is 200 cm of water?

3.7    A cubical container, 10 cm long along each edge, is filled with a gas at a pressure of 350 kPa. Determine the force that the gas exerts on each wall of the container.

3.8    Gas is kept in a 0.1 m diameter cylinder under the weight of a 100 kg piston that is held down by a spring with a stiffness $k = 5$ kN / m. If the gauge pressure of the gas is 300 kPa, how much is the spring compressed?

3.9    A cylindrical drum, 1 m high, is filled with water ($\rho = 1000$ kg / m$^3$) to a depth of 0.2 m. The rest of the drum is then filled with oil ($\rho = 850$ kg / m$^3$). What is the gauge pressure on the bottom of the drum?

3.10   A manometer is filled with equal volumes of water ($\rho = 1000$ kg / m$^3$) and oil ($\rho = 850$ kg / m$^3$). The end of the manometer nearest to the side filled with oil is connected to a pressurised container filled with gas and it is observed that the level of oil and water is exactly the same in both sides of the manometer, equal to 10 cm. What is the gauge pressure of gas in the container?

3.11 A flat lid with a mass of 4 kg is placed on a 0.1 m diameter cylindrical vessel. The vessel is evacuated until the absolute pressure inside it is 10 kPa. Find (a) the gauge pressure in the vessel and (b) the force required to lift the lid.

3.12 A gas pressure of 300 MPa inside a rifle barrel acts upon a bullet with a diameter of 9 mm and a mass of 15 g. What is the acceleration of the bullet?

3.13 The valve blocking the opening to a gas cylinder has a diameter of 50 mm. The pressure in the cylinder is 2.5 MPa while the atmospheric pressure outside is 100 kPa. What is the force $F$ required to open the valve?

**Figure P3.13** The valve blocking the opening to a gas cylinder.

3.14 Calculate the gas constant $R$ for (a) diatomic hydrogen and (b) carbon dioxide.

3.15 What is the volume in litres of 1 gmol of an ideal gas at standard temperature and pressure, defined as 0 °C and 101.325 kPa?

3.16 At what temperature would 1.1 gmol of helium gas contained in a 24 l tank have a pressure of 130 kPa?

3.17 A gas sample has a volume of 8.3 l and a mass of 5.94 g at a temperature of 0 °C and pressure of 1 atm. What gas is it?

3.18 What is the volume of 5 kmol of a gas at 25 °C and 110 kPa?

3.19 A 10 m³ tank containing carbon dioxide at 500 kPa and 20 °C begins to leak. What mass of gas has escaped when the tank pressure drops to 200 kPa?

3.20 An air bubble, 1 mm in diameter, is released at the bottom of a lake 20 m deep where the temperature is 10 °C. It rises to the surface of the lake where the temperature is 25 °C. What will the radius of the bubble be at the surface?

3.21 What is the specific volume of superheated steam at 300 °C and a pressure of 1.2 MPa?

3.22 What is the mass of air in a room 5 m long, 4 m wide and 3.5 m high if the atmospheric pressure is 100 kPa and the temperature is 23 °C? Assume that the average molar mass of air is 28.97 kg / kmol.

3.23  A helium filled weather balloon has a diameter of 3 m at a temperature of 18 °C and pressure of 100 kPa. What is the mass of helium in it? If the balloon can lift a mass equal to the mass of air it displaces, what payload can it carry?

3.24  What is the density of 0.1 kg of nitrogen contained in a cylinder at 5 °C and 200 kPa? What would the density of hydrogen be under the same conditions?

3.25  Assume that air consists of 75% nitrogen and 25% oxygen by volume. What is the density of air at 1 atm pressure and 20 °C?

3.26  Estimate the density of air in (a) a house at 20 °C and (b) an oven at 220 °C.

3.27  An evacuated 20 l container is filled with gas until the pressure inside reaches 800 kPa at 25 °C. By weighing the container before and after filling it is determined that its mass increased by 2.6 g. What gas was it filled with?

3.28  An air-filled balloon with a volume of 220 ml is placed inside a glass jar at atmospheric pressure (101.325 kPa). The jar is connected to a vacuum pump and evacuated so that the balloon expands until its volume reaches 350 ml. What is the pressure in the jar?

3.29  A car tyre is filled with air when the temperature is 0 °C. After driving for some time, the temperature in the tyre rises to 20 °C. What is the percentage increase in tyre pressure? Assume the tyre volume remains constant.

3.30  A spherical balloon is filled with 0.2 kg of helium at 20 °C until its pressure reaches 200 kPa. What is the diameter of the balloon? The balloon is released and rises high in the atmosphere where the pressure is 50 kPa and the temperature is –30 °C. What is the diameter now?

3.31  A compressed air tank with a volume of 3 m$^3$ has a pressure of 300 kPa and a temperature of 20 °C. More air is added until its pressure and temperature increase to 400 kPa and 50 °C. What is the mass of air added?

3.32  Given a list of six numbers: –8, –4, –3, 2, 4, 6, find the mean and the root mean square average.

3.33  A gas has three molecules with velocities of 1, 3 and 5 m / s. What is (a) the mean velocity and (b) the root mean square velocity?

3.34  What is the root mean square velocity of (a) hydrogen and (b) oxygen molecules at 0 °C?

3.35  What is the root mean square velocity of a nitrogen molecule at 25 °C? To what temperature would we have to heat the gas to double this velocity?

3.36  What is the kinetic energy of a helium atom at 400 K?

3.37  The root mean square velocity of $CH_4$ molecules at a given temperature is 442 m / s. What is the root mean square velocity of He molecules at the same temperature?

3.38   Three kilograms of helium are heated from 20 to 100 °C. Find the increase in internal energy.

3.39   Five kilograms of air, initially at 0 °C, are heated until the temperature reaches 50 °C. The internal energy increases by 180 kJ during this process. Find the specific heat of air.

3.40   Six kilograms of nitrogen at 30 °C are cooled so that the internal energy decreases by 60 kJ. Find the final temperature of the gas. Assume that the specific heat of nitrogen is 0.745 kJ / kg °C.

# 4

# Energy and the First Law of Thermodynamics

---

**In this chapter you will:**

- Study energy transfer as work and heat.
- Identify different work modes, such as boundary work, flow work, shaft work, spring work and electrical work.
- Postulate the first law of thermodynamics for both open and closed systems.
- Perform energy balance calculations.
- Define new properties: enthalpy and specific heat.
- Calculate changes in internal energy and enthalpy for liquids, solids and gases.
- Analyse steady flow devices such as turbines, pumps, compressors, nozzles and diffusers.

---

## 4.1 Energy

News headlines regularly announce that the world faces an energy crisis. Modern societies use vast amounts of energy to heat and light homes, power factories and fuel vehicles. As oil and gas resources dwindle we explore ever more inhospitable terrain and drill further under the sea to find new ways to satisfy global needs, often at the risk of serious environmental damage. The search for energy is one of the most important endeavours in the world today and it drives technology, business and politics. At the same time we are told that power can be extracted from fossils, plants, sunlight, tides and wind. It seems that we are surrounded by energy, so why is it so hard to find?

What is energy? For a term that we use so often, it is surprisingly hard to pin down. This is because energy is a fundamental concept in science that we cannot describe using any simpler ideas. We grasp fundamental concepts, such as mass, length and time, by observing the natural

---

*Energy, Entropy and Engines: An Introduction to Thermodynamics*, First Edition. Sanjeev Chandra.
© 2016 John Wiley & Sons, Ltd. Published 2016 by John Wiley & Sons, Ltd.
Companion website: www.wiley.com/go/chandraSol16

world and we use them to formulate laws about how the universe works. The meaning of mass is understood by noting that when gravity acts on an object it exerts a downward force, proportional to its mass, that we call weight. We measure the weight of an object and deduce its mass without ever explicitly stating the meaning of mass.

We will tackle the concept of energy in the same way and postulate that energy is an extensive property of all thermodynamic systems. *A system possesses energy if it is capable of lifting a weight.* This is not really a definition of energy, but gives us a test to determine how much energy a system possesses: the higher it can lift a given weight, the greater its energy.

Our criterion encompasses all types of energy. If a mass has kinetic or potential energy, we can design a mechanism that will use its motion to raise a weight (Figure 4.1a, b). If a system possesses internal energy, stored in molecular motion or chemical bonds, we can extract the energy as heat and use it to expand a gas and raise a piston on which a weight is placed (Figure 4.1c).

## 4.2   Forms of Energy

There are three different ways in which energy can be stored in a system. *Potential energy* and *kinetic energy* are macroscopic forms, and altering them requires a change in the position or velocity of the system. *Internal energy* includes all microscopic forms of energy storage. Variations in internal energy are not the result of a displacement of the system, but correspond to changes in its temperature, pressure, chemical composition, electrical or magnetic state.

The gravitational potential energy (*PE*) of a mass $m$ placed in a gravitational field that produces an acceleration $g$ is given by

$$PE = mgz. \tag{4.1}$$

The height $z$ of the system is measured above an arbitrary datum plane. Changing the height by an amount $\Delta z$ produces a change in the potential energy:

$$\Delta PE = mg\Delta z. \tag{4.2}$$

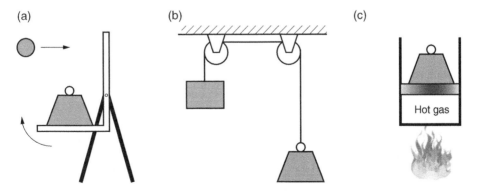

**Figure 4.1**   Lifting a weight using: (a) the kinetic energy of a moving object, (b) the potential energy of a raised mass, (c) the internal energy in the chemical bonds of a fuel.

The kinetic energy of a mass $m$ moving with velocity $\mathbf{V}$ is

$$KE = \frac{1}{2}m\mathbf{V}^2. \tag{4.3}$$

The total energy ($E$) of a system is the sum of its potential ($PE$), kinetic ($KE$) and internal ($U$) energies:

$$E = PE + KE + U. \tag{4.4}$$

Energy is an extensive property and has units of joules (J). We can also define an intensive property, the specific energy

$$e = \frac{E}{m}, \tag{4.5}$$

which has units of J / kg. Other intensive properties are the potential energy per unit mass of the system

$$pe = \frac{PE}{m} = gz, \tag{4.6}$$

the kinetic energy per unit mass

$$ke = \frac{KE}{m} = \frac{1}{2}\mathbf{V}^2 \tag{4.7}$$

and the specific internal energy

$$u = \frac{U}{m}. \tag{4.8}$$

All these intensive properties have units of J / kg. The specific energy of a system is the sum of its specific kinetic, potential and internal energies:

$$e = pe + ke + u \tag{4.9}$$

or

$$e = gz + \frac{1}{2}\mathbf{V}^2 + u. \tag{4.10}$$

For a stationary system, which does not move, the potential and kinetic energies are constant and only the internal energy changes.

## Example 4.1
**Problem:** A 5 kg mass is dragged up a slope through a vertical height of 8 m while its velocity decreases from 12 m / s to 2 m / s. Find the changes in kinetic and potential energy.
**Find:** Change in kinetic energy $\Delta KE$ and change in potential energy $\Delta PE$.
**Known:** Mass $m = 5$ kg, initial height $z_1 = 0$ m, final height $z_2 = 8$ m, initial velocity $\mathbf{V}_1 = 12$ m / s, final velocity $\mathbf{V}_2 = 2$ m / s.

**Assumptions:** Acceleration due to gravity is at earth's surface $g = 9.81$ m / s$^2$.
**Governing Equations:**

Change in potential energy $\qquad \Delta PE = mg\left(z_2 - z_1\right)$

Change in kinetic energy $\qquad \Delta KE = \dfrac{1}{2}mV_2^2 - \dfrac{1}{2}mV_1^2$

**Solution:**

$$\Delta PE = mg\left(z_2 - z_1\right) = 5\,\text{kg} \times 9.81\,\text{m / s}^2 \times \left(8\,\text{m} - 0\,\text{m}\right) = 392.4\,\text{J}$$

$$\Delta KE = \frac{1}{2}m\left(V_2^2 - V_1^2\right) = \frac{1}{2} \times 5\,\text{kg}\left(\left(2\,\text{m / s}\right)^2 - \left(12\,\text{m / s}\right)^2\right) = -350.0\,\text{J}$$

**Answer:** The potential energy increases by 392 J while the kinetic energy decreases by 350 J. ∎

## 4.3 Energy Transfer

Energy can be transferred to or from a system in two ways: as *work (W)* or *heat (Q)*. *Heat transfer is an exchange of energy that occurs due to a temperature difference between the system and its surroundings.* All other forms of energy transfer are classified as work. Work does not have to be mechanical: an electrical current entering a system is considered a form of work.

To make an energy balance, start by defining the system and marking its boundary with a dashed line as shown in Figure 4.2. Show all energy transfers across the system boundary, either as heat or work, with arrows. *We will adopt the sign convention that energy transfers to a system from the surroundings, either as heat or work, are positive. Energy transfers from a system to the surroundings are negative.* Therefore in Figure 4.2 $Q_{in}$ and $W_{in}$ are positive while $Q_{out}$ and $W_{out}$ are negative.

If you are not sure whether energy is added to or removed from the system in a particular problem, make a guess. If you guess incorrectly, it will not make the analysis wrong; you will

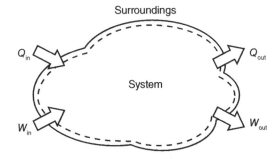

**Figure 4.2** Heat and work transfers between the system and its surroundings.

simply get a negative answer, showing that the actual direction is the opposite of what you assumed.

The *net heat* added to a system in which heat transfer occurs through several streams labelled $Q_1$, $Q_2$, $Q_3$, ..., is given by summing all the heat transfer terms:

$$Q_{net} = \sum_i Q_i. \tag{4.11}$$

A positive value of $Q_{net}$ means that more heat is added to the system than it loses to the surroundings. Similarly the *net work* done on the system is

$$W_{net} = \sum_i W_i. \tag{4.12}$$

A positive value of $W_{net}$ implies that more work is supplied to the system than it does on the surroundings.

The *rate of doing work* is also known as *power* ($\dot{W}$):

$$\dot{W} = \frac{\delta W}{dt}. \tag{4.13}$$

Power has units of J / s, which is known as a watt (W). One watt is a rather low rate of energy transfer so larger units are the kilowatt (kW), equal to $10^3$ W, and the megawatt (MW), which is $10^6$ W.

Mechanical work done by a constant force $F$ acting through an infinitesimal distance $dx$ is

$$\delta W = Fdx. \tag{4.14}$$

The power expended in applying the force is

$$\dot{W} = \frac{\delta W}{dt} = F\frac{dx}{dt} = F\mathbf{V}, \tag{4.15}$$

where $\mathbf{V}$ is the velocity of the point of application of the force.

The *rate of heat transfer* is denoted by $\dot{Q}$ and also has units of watts:

$$\dot{Q} = \frac{\delta Q}{dt}. \tag{4.16}$$

It is sometimes useful to talk about the *heat transfer per unit mass of the system*

$$q = \frac{Q}{m} \tag{4.17}$$

or the *work done per unit mass of the system*

$$w = \frac{W}{m}. \tag{4.18}$$

Both $q$ and $w$ have units of J / kg. Remember, though, that heat and work are not properties of the system, so that $q$ and $w$ are not intensive properties.

**Example 4.2**
**Problem:** A 40 kW engine powers a car driving at a constant speed of 100 km / h. What is the force resisting the motion of the car?
**Find:** Force $F$ resisting motion of car.
**Known:** Power $\dot{W} = 40\,\text{kW}$, velocity $\mathbf{V} = 100\,\text{km / h}$.
**Governing Equation:**

Power
$$\dot{W} = F\mathbf{V}$$

**Solution:**

$$F = \frac{\dot{W}}{\mathbf{V}}$$

$$\mathbf{V} = 100\,\text{km / h} \times 10^3\,\text{m / km} \times \frac{1}{3600\,\text{s / h}} = 27.778\,\text{m / s}$$

$$F = \frac{40 \times 10^3\,\text{W}}{27.778\,\text{m / s}} = 1440.0\,\text{N}$$

**Answer:** The force resisting the motion of the car is 1.44 kN. ∎

## 4.4 Heat

A gas is confined between two surfaces, one insulated and the other at a constant temperature $(T_1)$ so that at equilibrium the gas is uniformly at $T_1$. The temperature of the lower plate is suddenly increased to $T_2$ (Figure 4.3a). Molecules of the gas in contact with this plate will be raised to the higher temperature and will begin to move with a greater velocity. These faster moving molecules will collide with those next to them and, after an exchange of momentum, the previously slower molecules will begin to move faster. The increase in internal energy will gradually propagate through the gas, until the average velocity everywhere is higher and the temperature of the gas reaches a new equilibrium value $(T_2)$. If we are making macroscopic

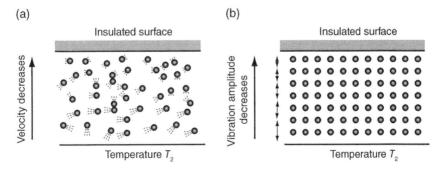

**Figure 4.3**   Heat conduction in (a) a gas and (b) a solid.

measurements of gas properties, we will observe a progressive increase in gas temperature, starting at the lower surface and gradually propagating upwards. This is known as heat conduction. *Heat transfer is a mode of energy transport that occurs when a temperature difference exists.* Heat transfer occurs at the microscopic level, with no macroscopic displacement. Heat conduction occurs in all materials, not just gases. In solids, where molecules are not free to move, raising the temperature increases the amplitude of vibration of molecules about their equilibrium position. This vibration is passed on to neighbouring molecules and therefore energy is propagated through the entire solid (Figure 4.3b).

## 4.5  Work

*Work is energy transfer across the boundary of a closed system in the absence of any temperature difference.* When an adiabatic system is taken from one state to another, the work ($W$) done on the system equals its change in energy ($\Delta E$). A piston compressing a gas, a rotating turbine shaft, the compression of a spring and an electric current passing through a heater are all examples of work.

### 4.5.1  Boundary Work

When a force acts on the boundaries of a system and deforms them, so that the system is either compressed or expanded, work is done since the point of application of the force moves through a finite distance. This is known as *boundary work*.

A piston with cross-sectional area $A$ compresses a gas in a cylinder as shown in Figure 4.4. The pressure ($P$) of the gas resists the forward movement of the piston, so a force ($F$) has to be exerted on it. The force must be just sufficient to overcome the resistance offered by the gas, so $F = PA$. The work ($\delta W$) done in moving the piston through distance $dx$ is

$$\delta W = Fdx = PAdx. \tag{4.19}$$

We have neglected any friction that may exist between the piston and cylinder and assumed that the process is quasi-equilibrium. The force $F$ is only infinitesimally greater than the force exerted by the gas on the piston, so the piston moves infinitely slowly. If $F > PA$ the piston would accelerate as it moved, and the gas pressure on it would increase as gas molecules accumulated on the piston face. The work calculated is therefore the theoretical minimum required

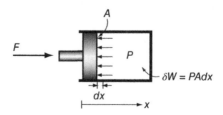

**Figure 4.4**   Boundary work done during compression of a gas.

to compress the gas – in reality it would always be greater. *How* much greater will depend on the speed with which the piston is moved and the magnitude of friction forces.

When the piston advances through a distance $dx$ the volume of the gas decreases by $dV$, so that $dV = -Adx$. The boundary work done is

$$\delta W = -PdV. \tag{4.20}$$

The work done in compressing or expanding a gas from state 1 to state 2, in which the volume changes from $V_1$ to $V_2$, is

$$W_{12} = -\int_{V_1}^{V_2} PdV. \tag{4.21}$$

Note the negative sign before the integral: we are assuming that work is done *by* the surroundings *on* the system. If a gas is compressed its volume is reduced so that $dV < 0$ and therefore $W > 0$. When we compress a system we do work on it. Conversely, when a gas expands (so that $dV > 0$) it does work on the surroundings and $W < 0$.

If a gas is compressed in an adiabatic cylinder, it cannot lose energy in the form of heat. The work done during the compression process is transferred to the molecules of gas, where it is stored as internal energy, and the gas temperature and pressure rise accordingly. If the piston is released the gas will expand, pushing out the piston. The movement of the piston can be used to do work on the surroundings and the energy stored in the gas recovered. *A simple, compressible system is one on which we can do work only by expanding or compressing it.*

Representing a compression or expansion process on a graph with $P$ and $V$ as axes is a convenient way of visualising the work done. The term $PdV$ represents an infinitesimal amount of work done in changing the system volume by $dV$ (see Figure 4.5). Integrating $PdV$ from initial state 1 to final state 2 gives the area under the $P$-$V$ curve, which is the total amount of work done during the process.

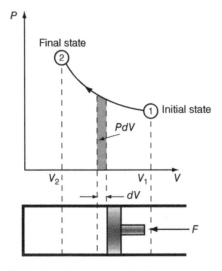

**Figure 4.5**   Work done is the area under the curve on a $P$-$V$ diagram.

It is possible to have two processes that begin and finish with the same pressure and volume, but require very different amounts of work. Figure 4.6 shows two processes on *P-V* axes, both starting at 1 and ending at 2. In the first process (Figure 4.6a) a gas is compressed from 1 to *a*. The pressure is kept constant during the process 1–*a* by cooling the gas as it is compressed. The gas is then heated while being held at constant volume, in the process *a*–2. Alternately, the gas can be heated first in a constant volume process, as shown in Figure 4.6b, to state *b*. The volume of the hot gas is then reduced in a constant pressure process to state 2. The work done during process 1–*b*–2 is greater than that during 1–*a*–2, as seen by comparing the areas under the two process curves. It takes more work to compress a hot gas than the same amount of gas at a lower temperature: that is the reason industrial gas compressors have large cooling fins to release heat to the atmosphere.

Since the work done during an expansion or compression process depends on pressure variation during the process we need to know how system pressure varies with volume by specifying the function $P(V)$. Some common processes are listed below.

*Constant volume:* An example of a constant volume process is one in which heat is added to a system confined within rigid walls (see Figure 4.7). In this case $dV = 0$, so

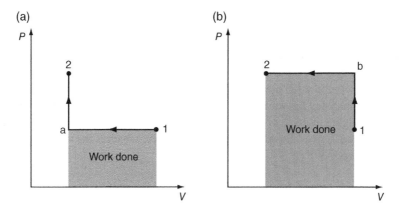

**Figure 4.6**  Two processes with the same starting and ending points, but a greater amount of work is done in the second process.

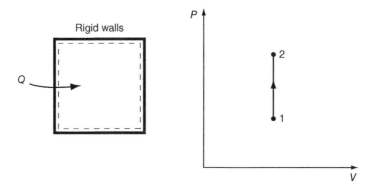

**Figure 4.7**  Constant volume process.

$$W_{12} = -\int_{V_1}^{V_2} PdV = 0. \tag{4.22}$$

No boundary work is done on a system whose volume remains constant.

**Example 4.3**
**Problem:** A rigid container contains 0.5 kg of air at 20 °C and a volume of 0.3 m³. The air is heated until its temperature reaches 150 °C. What is the work done by the air on the surroundings?
**Find:** Work $W$ done by system.
**Known:** Mass of air $m = 0.5$ kg, initial volume $V_1 = 0.3$ m³, initial temperature $T_1 = 20$ °C, final temperature $T_2 = 150$ °C.
**Assumptions:** The container is rigid, so its volume remains constant $V_2 = V_1$.
**Governing Equation:**

Work for volume change $\qquad W_{12} = -\int_{V_1}^{V_2} PdV$

**Solution:**

$$W_{12} = -\int_{V_1}^{V_2} PdV = -P_1 \int_{V_1}^{V_2} dV = P_1(V_1 - V_2) = 0$$

**Answer:** There is no work done by the system.  ∎

*Constant pressure:* A system in a cylinder with a piston that is free to move (see Figure 4.8) always stays at constant pressure. The piston will rise or fall during any process, maintaining the pressure $P_1 = P_2 =$ constant throughout. We can select the pressure that we want by placing weights on the piston – the greater the weight the higher the system pressure. The work done during the process is

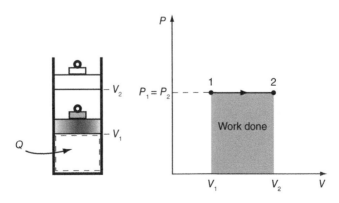

**Figure 4.8**  Constant pressure process.

$$W_{12} = -\int_{V_1}^{V_2} PdV = -P_1\int_{V_1}^{V_2} dV = P_1(V_1 - V_2) = P_1V_1 - P_2V_2. \tag{4.23}$$

## Example 4.4

**Problem:** A piston driven by compressed air at 200 kPa in a cylinder is required to do 500 J of work while moving through a distance of 0.25 m. What should the diameter of the cylinder be?

**Find:** Diameter $D$ of the cylinder.

**Known:** Gas pressure is constant at $P = 200$ kPa, work done by system $W_{12} = -500$ kPa, distance moved by piston $\Delta x = -0.25$ m. Since the system expands and does work on the surroundings, work and displacement are negative.

**Governing Equation:**

Work                                    $\delta W = Fdx = PAdx$

**Solution:**

$$W_{12} = \int_{x_1}^{x_2} PAdx = PA\int_{x_1}^{x_2} dx = PA(x_2 - x_1) = PA\Delta x,$$

where $A$ is the cross-sectional area of the piston and $\Delta x$ is the distance moved by the piston.

$$A = \frac{W_{12}}{P\Delta x} = \frac{-500\,\text{J}}{200 \times 10^3\,\text{Pa} \times -0.25\,\text{m}} = 0.01\,\text{m}^2$$

The diameter of a circular piston is

$$D = \sqrt{\frac{4A}{\pi}} = \sqrt{\frac{4 \times 0.01\,\text{m}^2}{\pi}} = 0.112\ 84\,\text{m}$$

**Answer:** The piston diameter should be 0.113 m.                                    ∎

## Example 4.5

**Problem:** A cylinder with a frictionless piston contains 0.5 kg of air with a volume of 0.3 m³ at 20 °C. The air is allowed to expand freely while being heated until its temperature reaches 150 °C. What is the work done by the piston on the surroundings?

**Find:** Work $W$ done by piston.

**Known:** Mass of air $m = 0.5$ kg, initial volume $V_1 = 0.3$ m³, initial temperature $T_1 = 20$ °C, final temperature $T_2 = 150$ °C.

**Diagrams:**

**Assumptions:** Air behaves as an ideal gas, air pressure remains constant since the piston is free to move $P_2 = P_1$.

**Governing Equations:**

Ideal gas equation                          $PV = mRT$

Work under constant pressure          $W_{12} = P_1(V_1 - V_2)$

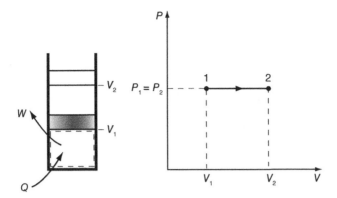

**Figure E4.5**   Constant pressure expansion in a cylinder with a frictionless piston.

**Properties:** The gas constant for air $R = 0.2870$ kJ / kgK.

**Solution:**
To find initial pressure:

$$P_1 = \frac{mRT_1}{V_1} = \frac{0.5\,\text{kg} \times 0.2870 \times 10^3\,\text{J/kgK} \times (273.15 + 20)\,\text{K}}{0.3\,\text{m}^3} = 140.22\,\text{kPa}.$$

To find final volume:

$$V_2 = \frac{mRT_2}{P_2} = \frac{0.5\,\text{kg} \times 0.2870 \times 10^3\,\text{J/kgK} \times (273.15 + 150)\,\text{K}}{140.22 \times 10^3\,\text{Pa}} = 0.433\,05\,\text{m}^3.$$

Work done by system:

$$W_{12} = P_1(V_1 - V_2) = 140.22 \times 10^3\,\text{Pa} \times (0.3\,\text{m}^3 - 0.433\,05\,\text{m}^3) = -18.656\,\text{kJ}.$$

**Answer:** The system does 18.7 kJ of work on the surroundings while expanding.   ■

*Constant temperature, ideal gas:* If a cylinder filled with gas is immersed in a bath through which water at a regulated temperature circulates (Figure 4.9), the system stays at constant temperature. From the equation of state for an ideal gas, $P = mRT / V$, so the work done is

$$W_{12} = -\int_{V_1}^{V_2} PdV = -\int_{V_1}^{V_2} \frac{mRT}{V}\,dV. \tag{4.24}$$

Since $m$, $R$ and $T$ are all constants for a closed, isothermal system, we can take them outside the integral:

$$W_{12} = -mRT \int_{V_1}^{V_2} \frac{dV}{V} = mRT \ln \frac{V_1}{V_2}. \tag{4.25}$$

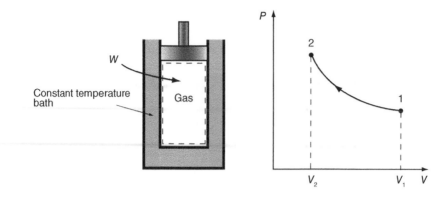

**Figure 4.9**   Constant temperature process for an ideal gas.

## Example 4.6

**Problem:** Air is contained in a 1 m³ volume cylinder at a pressure of 100 kPa. It is compressed until its volume is halved while keeping the temperature constant. How much work is required for compression?

**Find:** Work $W$ required for compression.

**Known:** Initial volume $V_1$ = 1 m³, final volume $V_2 = 0.5V_1$, initial pressure $P_1$ = 100 kPa, temperature is constant during compression.

**Assumptions:** Air behaves as an ideal gas.

**Governing Equations:**

Ideal gas equation

$$PV = mRT$$

Work under constant temperature (ideal gas)

$$W_{12} = mRT \ln \frac{V_1}{V_2}$$

**Solution:**

Using the ideal gas equation,

$$P_1 V_1 = mRT_1,$$

therefore,

$$W_{12} = P_1 V_1 \ln \frac{V_1}{V_2}.$$

$$W_{12} = 100 \times 10^3 \, \text{Pa} \times 1 \, \text{m}^3 \times \ln\left(\frac{1 \, \text{m}^3}{0.5 \, \text{m}^3}\right) = 69.315 \, \text{kJ}$$

**Answer:** The work required to compress the gas is 69.3 kJ.                              ∎

*Polytropic Process:* A variety of common thermodynamic processes can be modeled by a curve of the form

$$PV^n = C, \tag{4.26}$$

where $C$ and $n$ are both constants. Any process that follows Equation (4.26) is known as a *polytropic process*, applicable for ideal gases. Specific values of $n$ correspond to different processes. For example, setting $n = 0$ give a constant pressure process, $n = 1$ for an ideal gas undergoing an isothermal process and $n = \infty$ a constant volume process. In general,

$$C = P_1 V_1^n = P_2 V_2^n. \tag{4.27}$$

The work done during a polytropic process is

$$W_{12} = -\int_{V_1}^{V_2} P dV = -C \int_{V_1}^{V_2} \frac{dV}{V^n}. \tag{4.28}$$

If $n \neq 1$,

$$W_{12} = -\int_{V_1}^{V_2} P dV = -C \int_{V_1}^{V_2} \frac{dV}{V^n}. \tag{4.29}$$

Substituting for the constant $C$ from Equation (4.27),

$$W_{12} = \frac{P_2 V_2 - P_1 V_1}{n-1}, \qquad \text{for } n \neq 1 \tag{4.30}$$

Alternately, if $n = 1$, as is the case in an isothermal process,

$$W_{12} = -\int_{V_1}^{V_2} P dV = -C \int_{V_1}^{V_2} \frac{dV}{V^n} = C \ln \frac{V_1}{V_2}, \qquad \text{for } n = 1 \tag{4.31}$$

Then substituting values for $C$ from Equation (4.27) yields

$$W_{12} = P_1 V_1 \ln \frac{V_1}{V_2} = P_2 V_2 \ln \frac{V_1}{V_2}, \qquad \text{for } n = 1 \tag{4.32}$$

## Example 4.7
**Problem:** Gas in a cylinder is expanded by a piston in a process for which $PV^n = C$, where $C$ and $n$ are constants. The initial pressure and volume are 3 bar and 0.2 m³ respectively and the final volume is 0.6 m³. Determine the work done by the gas if (a) $n = 1.4$ and (b) $n = 1.0$.
**Find:** Work $W$ done by the gas.
**Known:** Initial pressure $P_1 = 3$ bar, initial volume $V_1 = 0.2$ m³, final volume $V_2 = 0.6$ m³, gas expands in a polytropic process where $PV^n = C$.
**Diagrams:**
**Governing Equations:**
For a polytropic process

$$\Delta PE = mg\Delta z$$

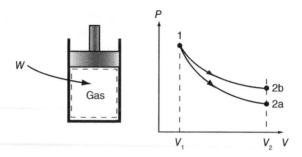

**Figure E4.7**   Polytropic expansion of gas in a cylinder.

$$W_{12} = P_1 V_1 \ln \frac{V_1}{V_2} \text{ for } n = 1$$

**Solution:**

(a) For $n = 1.4$

To find the final pressure,

$$P_1 V_1^{1.4} = P_2 V_2^{1.4}$$

$$P_2 = P_1 \left( \frac{V_1}{V_2} \right)^{1.4} = 3 \text{ bar} \times \left( \frac{0.2 \text{ m}^3}{0.6 \text{ m}^3} \right)^{1.4} = 0.644 \ 39 \text{ bar}$$

$$W_{12} = \frac{P_2 V_2 - P_1 V_1}{n - 1} = \frac{0.644 \ 39 \times 10^5 \text{ Pa} \times 0.6 \text{ m}^3 - 3 \times 10^5 \text{ Pa} \times 0.2 \text{ m}^3}{1.4 - 1} = -53.342 \text{ kJ}$$

(b) For $n = 1$

$$W_{12} = P_1 V_1 \ln \frac{V_1}{V_2} = 3 \times 10^5 \text{ Pa} \times 0.2 \text{ m}^3 \times \ln \left( \frac{0.2 \text{ m}^3}{0.6 \text{ m}^3} \right) = -65.917 \text{ kJ}$$

**Answer:** The work done by the system on the surroundings during a polytropic process with (a) $n = 1.4$ is 53.3 kJ and with (b) $n = 1$ is 65.9 kJ.                                                      ∎

## 4.5.2   Flow Work

Before fluid can enter a control volume it has to push back the matter already filling that space. Forcing additional fluid into a control volume requires work: think of the effort required to inflate a bicycle tyre. In Figure 4.10a, a fluid element with mass $(m)$ and volume $(V)$ is about to be pushed into a control volume through an inlet with cross-sectional area $A$. The force required is

$$F = PA, \tag{4.33}$$

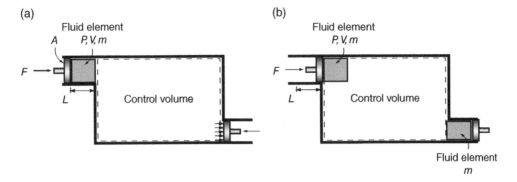

**Figure 4.10**   Fluid element (a) before and (b) after entering the control volume.

where $P$ is the pressure of fluid at the entrance of the control volume resisting the entry of additional material. In Figure 4.10b, the mass ($m$) has entered the control volume, and, since the process is steady state, an equal amount of mass has left the system. To completely drive the fluid element into the control volume a force ($F$) must act through a distance ($L$), which requires *flow work*:

$$W_{\text{flow}} = FL = PAL = PV. \qquad (4.34)$$

The flow work per unit mass of fluid is

$$w_{\text{flow}} = \frac{W_{\text{flow}}}{m} = Pv. \qquad (4.35)$$

When fluid enters a control volume the surroundings do work on it and this flow work increases the total energy of the system. Conversely, when fluid leaves a control volume the system has to do work on the fluid to force it out. The fluid transports the flow work with it and the system energy decreases.

### 4.5.3   Shaft Work

Industrial turbines use a flowing fluid such as water or steam to turn a shaft (Figure 4.11). Energy from the fluid flowing through the control volume is transferred to the surroundings by the rotating shaft. These transfers of energy are known as *shaft work* ($W_{\text{shaft}}$). There are many ways to make use of a rotating shaft – it can drive wheels, propellers or electric generators. Shaft work is negative in turbines, where work is done by the system on the surroundings. Positive shaft work is seen in devices such as pumps or compressors where the shaft transfers energy in the form of work from the surroundings to the fluid in the control volume.

When a constant force ($F$) acting at a radius ($r$) about a shaft makes it rotate (Figure 4.12), the *torque* produced is

$$\tau = Fr. \qquad (4.36)$$

Steam in

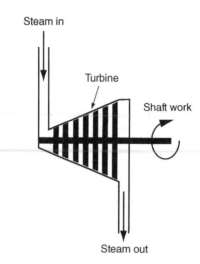

Turbine

Shaft work

Steam out

**Figure 4.11**   Shaft work in a steam turbine.

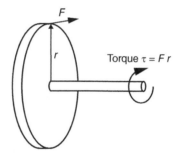

$F$

$r$

Torque $\tau = F\,r$

**Figure 4.12**   Torque produced in a shaft by a force ($F$) acting at radius ($r$).

The units of torque are Nm. If the shaft revolves $N_r$ times, the force acts through a distance $2\pi r N_r$. The work done by the shaft is

$$W_{shaft} = 2\pi N_r \tau. \tag{4.37}$$

The rotational speed in number of revolutions per second is $\omega = dN_r / dt$ (the units of $\omega$ are simply s$^{-1}$ since the number of revolutions is dimensionless), the power transmitted by the shaft, in watts, is

$$\dot{W}_{shaft} = 2\pi\omega\tau. \tag{4.38}$$

**Example 4.8**
**Problem:** A car engine generates 60 kW of power when rotating at 3000 revolutions per minute (RPM). Determine the torque on the engine shaft.
**Find:** Engine shaft torque $\tau$.
**Known:** Shaft power $\dot{W}_{shaft} = 60\,\text{kW}$, engine rotational speed $\omega = 3000$ RPM.

## Governing Equation:

Power from rotating shaft $\qquad \dot{W}_{shaft} = 2\pi\omega\tau$

## Solution:

$$\tau = \frac{\dot{W}_{shaft}}{2\pi\omega} = \frac{60\times10^3 \text{ J/s}}{2\pi \times 3000\left(1/\min\right) \times \left(1\min/60\text{s}\right)} = 190.986 \text{ Nm}$$

**Answer:** The torque on the engine shaft is 191.0 Nm.                                       ■

### 4.5.4   Spring Work

Compressing or expanding a spring requires work. The force required to change the length of a linear spring by a distance $x$, measured from its uncompressed position, is

$$F = Kx,\tag{4.39}$$

where $K$ is the *spring constant* or *spring stiffness* with units of N / m or kN / m. The work required to displace the end of the spring from $x_1$ to $x_2$ (Figure 4.13) is

$$W_{spring} = \int_{x_1}^{x_2} F\,dx = \int_{x_1}^{x_2} Kx\,dx = \frac{1}{2}K\left(x_2^2 - x_1^2\right).\tag{4.40}$$

### Example 4.9
**Problem:** A cylinder fitted with a frictionless piston with 0.5 m² cross-sectional area contains 0.10 m³ of air. The piston is in contact with an uncompressed spring with a spring constant of 800 kN / m. The air is heated so that the piston rises and pushes against the spring until the air volume increases to 0.15 m³. Find the work done in compressing the spring and the final air pressure.
**Find:** Spring work $W$ and final air pressure $P_2$.
**Known:** Initial volume of air $V_1 = 0.10$ m³, final volume of air $V_2 = 0.15$ m³, spring constant $K = 800$ kN / m, cross-sectional area of piston $A = 0.5$ m², initial compression of spring $x_1 = 0$.

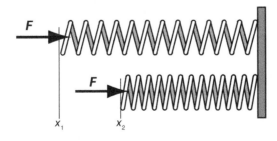

**Figure 4.13**   A spring is compressed from position $x_1$ to $x_2$ by a force ($F$).

**Diagrams:**

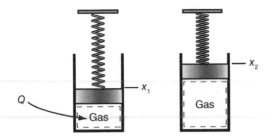

**Figure E4.9**   Gas in a cylinder fitted with a frictionless piston expands while pushing against a spring.

**Governing Equations:**

Work done by spring          $W_{spring} = \frac{1}{2}K\left(x_2^2 - x_1^2\right)$

Volume                       $V = Ax$

**Solution:**
To find work we first need the distance moved by the piston, since $V = Ax$:

$$\Delta x = x_2 - x_1 = \frac{V_2 - V_1}{A} = \frac{0.15\,\mathrm{m}^3 - 0.10\,\mathrm{m}^3}{0.5\,\mathrm{m}^2} = 0.1\,\mathrm{m}.$$

Then the work done on the spring is

$$W_{spring} = \frac{1}{2}K\left(x_2^2 - x_1^2\right) = \frac{1}{2} \times 800\,\mathrm{kN/m} \times \left[\left(0.1\,\mathrm{m}\right)^2 - 0\right] = 4\,\mathrm{kJ}.$$

Next, to find the final pressure we need the force exerted by the compressed spring,

$$F = Kx_2 = 800\,\mathrm{kN/m} \times 0.1\,\mathrm{m} = 80\,\mathrm{kN}.$$

Then the pressure in the cylinder must balance the force of the spring at equilibrium:

$$P = \frac{F}{A} = \frac{80\,\mathrm{kN}}{0.5\,\mathrm{m}^2} = 160\,\mathrm{kN}.$$

**Answer:** The work done on the spring is 4 kJ and the final pressure in the cylinder is 160 kN.   ∎

## 4.5.5   Electrical Work

A cable carrying an electric current ($I$) enters a control mass in which there is an electrical resistance ($R_e$) across which a voltage ($V_e$) is applied (Figure 4.14). Energy enters the control mass, and since it is not the result of a temperature difference we consider it to be work. The rate at which energy enters is

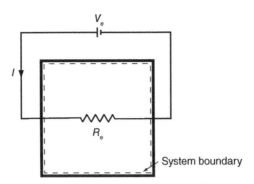

**Figure 4.14**   An electric current ($I$) enters a system enclosing a resistance ($R_e$).

$$\dot{W}_{\text{elec}} = V_e I = I^2 R_e = \frac{V_e^2}{R_e}. \qquad (4.41)$$

## 4.6   The First Law for a Control Mass

We have postulated the existence of an extensive property in all thermodynamic systems, the energy ($E$). The energy of a system can be measured, relative to a reference state whose energy we arbitrarily assume to be zero. Heat and work are the only two ways of transferring energy across the boundaries of a closed system (Figure 4.15). By definition the work ($W$) done on a system in an adiabatic process equals its change in energy ($\Delta E$). Similarly, the heat ($Q$) transferred to a closed system is defined as being equal to its change in energy, less any work done during the process.

Combining both of these definitions into a single statement gives the *first law of thermodynamics:The change in energy of a closed system equals the net energy transferred to it in the form of work and heat.* Stated mathematically,

$$\underbrace{Q + W}_{\text{Energy addition}} = \underbrace{\Delta E,}_{\text{Change in energy}} \qquad (4.42)$$

In writing Equation (4.42) we have postulated that the energy of an isolated system is conserved and cannot be created or destroyed. Following our sign convention, heat transfer ($Q$) from the surroundings to the system and work ($W$) done by the surroundings on the system are both positive. The total energy of a system is the sum of its internal, kinetic and potential energies, so that

$$\Delta E = \Delta U + \Delta PE + \Delta KE, \qquad (4.43)$$

with the change in energy of the system as

$$\Delta E = E_{\text{final}} - E_{\text{initial}}. \qquad (4.44)$$

We can also write the first law for infinitesimal amounts of heat transfer ($\delta Q$) and work ($\delta W$) which produce an infinitesimal change in system energy ($dE$):

$$\delta Q + \delta W = dE. \qquad (4.45)$$

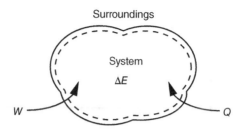

**Figure 4.15**   Energy transfer to a system by heat ($Q$) and work ($W$).

If the infinitesimal energy change ($dE$) takes place over an infinitesimal interval of time ($dt$), we can divide Equation (4.45) by $dt$ and write it as a rate equation:

$$\dot{Q} + \dot{W} = \frac{dE}{dt}. \tag{4.46}$$

$\dot{Q}$ and $\dot{W}$ are the rates of energy transfer to the system in the form of heat and work respectively. The amount of energy stored in the system is a property, and it changes at a rate $dE / dt$.

**Example 4.10**
**Problem:** Five kilograms of gas are contained in a rigid cylinder. An impeller in the gas, driven by an electrical motor, does 20 kJ / kg of work on the gas. At the same time the gas loses 80 kJ of heat to the surrounding. What is the change in specific internal energy of the gas?
**Find:** Change in specific internal energy of gas.
**Known:** Mass of gas $m = 5$ kg, heat loss $Q = -80$ kJ, work added per unit mass of gas $w = 20$ kJ / kg.
**Diagram:**

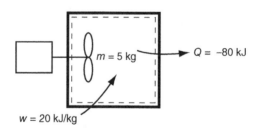

$w = 20$ kJ/kg

**Figure E4.10**   An impeller does work on gas contained in a rigid cylinder.

**Assumptions:** Changes in kinetic and potential energy are negligible, $\Delta KE = 0$ and $\Delta PE = 0$.
**Governing Equation:**
    First Law            $Q + W = \Delta U + \Delta PE + \Delta KE$

**Solution:**
Since $\Delta KE = 0$ and $\Delta PE = 0$, the energy balance reduces to

$$\Delta U = Q + W.$$

The work done on the system is

$$W = mw = 5\,\text{kg} \times 20\,\text{kJ} / \text{kg} = 100\,\text{kJ}.$$

Substituting work into the energy balance,

$$\Delta U = -80\,\text{kJ} + 100\,\text{kJ} = 20\,\text{kJ},$$

$$\Delta u = \frac{\Delta U}{m} = \frac{20\,\text{kJ}}{5\,\text{kg}} = 4\,\text{kJ} / \text{kg}.$$

**Answer:** The specific internal energy of the gas increases by 4 kJ / kg. ■

### Example 4.11

**Problem:** A gas initially at a pressure of 40 kPa and a volume of 0.1 m³ is compressed until the pressure doubles and its volume is 0.04 m³. The internal energy of the gas increases by 2.1 kJ. During compression gas pressure varies linearly with volume so that $P = a + bV$. What is the heat transfer during this process?

**Find:** Amount of heat transferred $Q$.

**Known:** Initial pressure $P_1$ = 40 kPa, initial volume $V_1$ = 0.1 m³, final pressure $P_2$ = 80 kPa, final volume $V_2$ = 0.04 m³, change in internal energy $\Delta U$ = 2.1 kJ, during compression $P = a + bV$.

**Diagrams:**

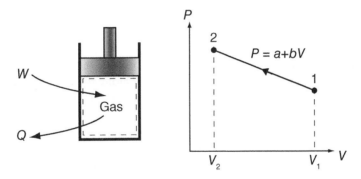

**Figure E4.11** A gas is compressed until the pressure doubles.

**Assumptions:** Changes in kinetic and potential energy are negligible, $\Delta KE = 0$ and $\Delta PE = 0$.

**Governing Equations:**

First Law $\qquad Q + W = \Delta U + \Delta PE + \Delta KE$

Boundary Work $\qquad W_{12} = -\int_{V_1}^{V_2} P dV$

**Solution:**

Since $\Delta KE = 0$ and $\Delta PE = 0$, the energy balance reduces to

$$Q = \Delta U - W.$$

To find the work done during the process,

$$W = -\int_{V_1}^{V_2} PdV = -\int_{V_1}^{V_2} (a + bV) dV = -\left[ aV + \frac{bV^2}{2} \right]_{V_1}^{V_2} = -a(V_2 - V_1) - \frac{b}{2}(V_2^2 - V_1^2).$$

To find the constants $a$ and $b$, since $P_1 = a + bV_1$ and $P_2 = a + bV_2$:

$$b = \frac{P_2 - P_1}{V_2 - V_1} = \frac{80 \text{ kPa} - 40 \text{ kPa}}{0.04 \text{ m}^3 - 0.1 \text{ m}^3} = -666.67 \text{ kPa} / \text{m}^3$$

$$a = P - bV = 40 \text{ kPa} - \left(-666.67 \text{ kPa} / \text{m}^3\right)\left(0.1 \text{ m}^3\right) = 106.67 \text{ kPa}$$

Therefore,

$$W = -(106.67 \text{ kPa})(0.04 \text{ m}^3 - 0.1 \text{ m}^3) - \frac{\left(-666.67 \text{ kPa} / \text{m}^3\right)}{2}\left[\left(0.04 \text{ m}^3\right)^2 - \left(0.1 \text{ m}^3\right)^2\right]$$

$$W = 6.40 \text{ kJ} - 2.80 \text{ kJ} = 3.60 \text{ kJ}$$

Work is positive since done on the system by the surroundings in compressing it, and

$$Q = \Delta U - W = 2.10 \text{ kJ} - 3.60 \text{ kJ} = -1.50 \text{ kJ}.$$

**Answer:** The system loses 1.50 kJ of heat to the surroundings. ∎

### Example 4.12

**Problem:** A gas is compressed by a piston in a cylinder from an initial pressure of 1.0 bar and initial volume of 0.4 m³, to a final pressure of 1.4 bar. During the compression process, the product of the gas pressure $P$ and volume $V$ remains constant. Determine the work done on the gas. If during the compression the gas loses 8.1 kJ of heat to the surroundings, determine the change in internal energy of the gas.

**Find:** Work done on the gas $W$ and change in internal energy of the gas $\Delta U$.

**Known:** Initial pressure $P_1 = 1$ bar, initial volume $V_1 = 0.4$ m³, final pressure $P_2 = 1.4$ bar, heat loss $Q = -8.1$ kJ, during compression $PV$ = constant.

**Diagrams:**

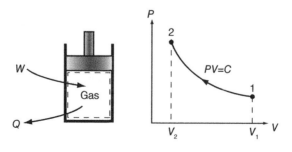

**Figure E4.12**   A gas is compressed by a piston in a cylinder.

**Assumptions:** Changes in kinetic and potential energy are negligible, $\Delta KE = 0$ and $\Delta PE = 0$.
**Governing Equations:**

First Law $\qquad\qquad Q + W = \Delta U + \Delta PE + \Delta KE$

Boundary work $\qquad W_{12} = -\int_{V_1}^{V_2} PdV$

**Solution:**
Since $\Delta KE = 0$ and $\Delta PE = 0$, the energy balance reduces to

$$\Delta U = Q + W.$$

During the compression $PV = \text{constant} = C$, therefore $P_1V_1 = P_2V_2 = C$.
The final volume is

$$V_2 = \frac{P_1V_1}{P_2} = \frac{1.0\,\text{bar} \times 0.4\,\text{m}^3}{1.4\,\text{bar}} = 0.285\ 71\,\text{m}^3.$$

Work done during the compression is

$$W = -\int_{V_1}^{V_2} PdV = -C\int_{V_1}^{V_2}\frac{dV}{V} = C\ln\frac{V_1}{V_2},$$

Where the constant $C$ can be evaluated:

$$C = P_1V_1 = 1\times10^5\,\left(\text{N}/\text{m}^2\right) \times 0.4\left(\text{m}^3\right) = 0.4\times10^5\ \text{Nm}.$$

So, the work done is

$$W = C\ln\frac{V_1}{V_2} = 0.4\times10^5\,(\text{Nm}) \times \ln\left(\frac{0.4\,\text{m}^3}{0.285\ 71\,\text{m}^3}\right) = 13\ 459.5\,\text{J} = 13.460\,\text{kJ}.$$

Then, the change in internal energy is

$$\Delta U = Q + W = -8.10\,\text{kJ} + 13.460\,\text{kJ} = 5.360\,\text{kJ}.$$

**Answer:** The work done on the system during compression is 13.5 kJ. The internal energy of the system increases by 5.36 kJ.  ∎

## 4.7  Enthalpy

Water jets out of an outlet near the bottom of a full tank, driven by the pressure of the liquid, and turns a turbine wheel (Figure 4.16a). As the tank empties and the level of liquid becomes lower the pressure of the water diminishes until it is trickling out and can no longer do any work (Figure 4.16b).

What property of a liquid describes its capacity to do work? The velocity of water inside the tank is very low so the difference in kinetic energy between liquid in a full or almost empty

(a)                                                    (b)

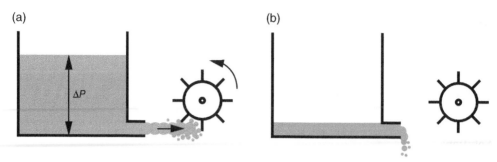

**Figure 4.16** (a) Water comes out of a full tank at high pressure and can be used to do work. (b) Water drips out of a tank that is almost empty.

tank is negligible. Water is at the same temperature throughout the process and therefore has constant internal energy. The only property that changes as the water drains is its pressure ($P$): a fluid can do more work as its pressure increases. But $P$ alone does not establish the amount of work that the fluid does: if two liquids are at the same pressure but have different temperatures, we can extract more work from the higher temperature fluid. A fluid's ability to do work depends on some combination of its internal energy and pressure.

When fluid with specific internal energy ($u$) leaves a control volume the system has to do flow work on it ($Pv$ per unit mass) and this energy is transported along with the fluid so that its energy per unit mass becomes $u + Pv$. This combination of properties is so useful in thermodynamics that we give it a name, *enthalpy*, with the symbol $H$. We define enthalpy as

$$H \equiv U + PV. \tag{4.47}$$

Since $U$ and $PV$ are both extensive properties $H$, their sum, is also an extensive property. The corresponding intensive property is the *specific enthalpy*,

$$h = \frac{H}{m} = \frac{U}{m} + P\frac{V}{m} = u + Pv. \tag{4.48}$$

Enthalpy ($H$) has the same units as energy, joules. Specific enthalpy ($h$) has units of J / kg. The enthalpy is not a fundamental property in the way that mass, volume and internal energy are. We could carry out our analysis without defining it, but it is a convenient way of referring to a combination of other properties that we encounter frequently. *Enthalpy measures the capacity of a fluid to do work.*

The definition of enthalpy is also useful in other applications, such as when we are studying heating of gases. When we heat a constant-volume system its internal energy increases, which we observe as a rise in temperature and pressure. If the system is free to expand part of the energy supplied is lost in doing work to push back the surroundings. It therefore takes more energy to heat a system at constant pressure than at constant volume. The difference can be important when gases are heated because the work done by them while expanding is a significant portion of the total energy required for heating. We can use the first law to calculate the expansion work. Suppose that we are heating gas contained in a cylinder (Figure 4.17), which expands from $V_1$ to $V_2$ while its pressure remains constant, equal to that of the surrounding atmosphere.

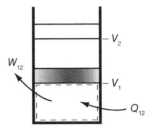

**Figure 4.17**   Heat ($Q_{12}$) is transferred to a constant pressure system that does work ($W_{12}$) on the surroundings as it expands from volume $V_1$ to $V_2$.

Assuming the piston to have negligible mass, there is no change in potential or kinetic energy during the expansion process—only the internal energy changes. Applying an energy balance,

$$Q_{12} + W_{12} = \Delta U = U_2 - U_1.  \tag{4.49}$$

The work done during an isobaric expansion is given by

$$W_{12} = P_1 V_1 - P_2 V_2.$$

Then substituting into Equation (4.49),

$$Q_{12} = \left(U_2 + P_2 V_2\right) - \left(U_1 + P_1 V_1\right).  \tag{4.50}$$

*The heat supplied to a constant pressure system equals its change in enthalpy:*

$$Q_{12} = H_2 - H_1.  \tag{4.51}$$

A mass in contact with the atmosphere is an example of a system that is free to expand under constant pressure. The heat required per unit mass of the system is

$$q_{12} = \frac{Q_{12}}{m} = \frac{H_2}{m} - \frac{H_1}{m} = h_2 - h_1.  \tag{4.52}$$

## 4.8   Specific Heats

We have defined two extensive properties, the internal energy ($U$) and the enthalpy ($H$), but we still do not know how to measure them. We can measure the temperature, pressure and volume of a system but there are no energy or enthalpy meters. We have to find a way of relating $U$ and $H$ to $P$, $T$ and $V$, so by measuring the latter we can calculate the former.

We know from experience that if we add energy to a material its temperature increases. How much the temperature rises for a given amount of heat is a property of the material: it takes much more energy to heat 1 kg of water than 1 kg of air. We define the *specific heat* ($c$) *as the amount of energy required to raise the temperature of a unit mass of a substance by one degree.* If the temperature of a system with mass $m$ (kg) increases by $\Delta T$ (°C) when $Q$ (J) of heat are added to it, the average specific heat of the system is

$$c_{avg} = \frac{Q}{m\Delta T} = \frac{q}{\Delta T}.  \tag{4.53}$$

The units of specific heat are J / kg °C or J / kgK. Both these units are identical. When we are talking about a temperature *difference* it does not matter whether we use units of °C or K because $\Delta T = 1$ °C is the same as $\Delta T = 1$ K.

The specific heat may vary with the temperature of the substance: the heat required to raise the temperature of water from 0 to 1 °C is not exactly the same as that required to raise it from 99 to 100 °C. It is more accurate to define $c$ as a function of temperature:

$$c(T) = \lim_{\Delta T \to 0} \frac{\delta q}{\Delta T} = \frac{\delta q}{dT}. \tag{4.54}$$

This definition is still not complete because we have not specified how the material is heated, which may make a significant difference. If the system is allowed to expand while being heated it does work in pushing back the surroundings. Some of the energy added to the system is lost as work. For a material being heated in a rigid container, whose volume is constant (Figure 4.18a), $W = 0$. The heat added is

$$\delta q = du. \tag{4.55}$$

The *specific heat at constant volume* is defined as

$$c_v(T) \equiv \left( \frac{\partial u}{\partial T} \right)_v. \tag{4.56}$$

If the material is heated at constant pressure (Figure 4.18b), the heat added equals the change in enthalpy

$$\delta q = dh. \tag{4.57}$$

The *specific heat at constant pressure* is defined as

$$c_p(T) \equiv \left( \frac{\partial h}{\partial T} \right)_p. \tag{4.58}$$

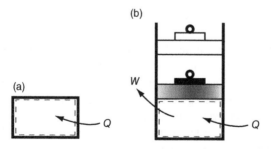

**Figure 4.18**   Control mass being heated at (a) constant volume and (b) constant pressure.

We can also define molar specific heats, using kmol rather than kg as a unit of mass. The *molar specific heat at constant volume* is

$$\bar{c}_v = c_v M = \left(\frac{\partial \bar{u}}{\partial T}\right)_v \tag{4.59}$$

and the *molar specific heat at constant pressure* is

$$\bar{c}_p = c_p M = \left(\frac{\partial \bar{h}}{\partial T}\right)_P . \tag{4.60}$$

The units of both $\bar{c}_v$ and $\bar{c}_p$ are J / kmol °C or J / kmolK.

## 4.9   Specific Heats of Ideal Gases

The internal energy ($U$) of an ideal gas is a function only of its temperature ($T$). The simple model of a monoatomic ideal gas developed in Chapter 3 showed this, and many researchers have confirmed it experimentally for a wide variety of gases. If $u = u(T)$ and not $u = u(T,P)$ we do not require a partial differential in Equation (4.56) and can write

$$c_v(T) = \frac{du}{dT}. \tag{4.61}$$

For an ideal gas, $Pv = RT$. The enthalpy of an ideal gas is therefore

$$h = u + Pv = u + RT. \tag{4.62}$$

Since specific internal energy ($u$) is a function only of temperature and gas constant ($R$) is a constant, specific enthalpy ($h$) for an ideal gas is also a function of temperature alone; therefore,

$$c_P(T) = \frac{dh}{dT}. \tag{4.63}$$

Equations (4.61) and (4.63) are derived from the definitions of specific heats $c_v$ and $c_p$ and are true for all processes. They are not restricted to any particular type of process. They give us a means of calculating changes in internal energy and enthalpy, because by integrating them we obtain

$$\Delta u = u_2 - u_1 = \int_{T_1}^{T_2} c_v(T) dT \tag{4.64}$$

and

$$\Delta h = h_2 - h_1 = \int_{T_1}^{T_2} c_P(T) dT . \tag{4.65}$$

Equations (4.64) and (4.65) relate the internal energy and enthalpy, that we cannot measure directly, to the temperature which we can. To evaluate the integrals we have to know the dependence of $c_p$ and $c_v$ on gas temperature. Such data are available, either from theoretical models for simple gases or from experiments. Values of specific heats of gases, evaluated at 25 °C and 100 kPa, are given in Appendix 1. Values of $c_p$ and $c_v$ as a function of temperature are listed in Appendix 4 for some common gases. If the temperature range $T_1$ to $T_2$ over which the integration is done is not very large we can use average values of the specific heats over this interval. Typically, this is done by using the value of $c_p$ and $c_v$ evaluated at the average temperature $T_{avg} = (T_2 + T_1)/2$. Then we can define

$$c_{v,avg} = c_v, \qquad \text{at } T_{avg}$$

and

$$c_{p,avg} = c_p, \qquad \text{at } T_{avg}$$

Since $c_{v,avg}$ and $c_{p,avg}$ are both constants we can integrate Equations (4.64) and (4.65) to give

$$\Delta u = u_2 - u_1 = c_{v,avg}\left(T_2 - T_1\right) \tag{4.66}$$

and

$$\Delta h = h_2 - h_1 = c_{p,avg}\left(T_2 - T_1\right). \tag{4.67}$$

The two specific heats of an ideal gas are related, as we can show by differentiating Equation (4.62):

$$\frac{dh}{dT} = \frac{du}{dT} + R \tag{4.68}$$

or

$$c_p = c_v + R. \tag{4.69}$$

For molar properties,

$$\bar{c}_p = \bar{c}_v + R_u. \tag{4.70}$$

It was shown in Chapter 3 that the internal energy of a monoatomic ideal gas is

$$U = \frac{3}{2} N R_u T$$

and the specific internal energy is

$$u = \frac{U}{m} = \frac{3}{2} RT. \tag{4.71}$$

For an ideal gas, with $Pv = RT$, the specific enthalpy is

$$h = u + Pv = u + RT = \frac{5}{2} RT. \tag{4.72}$$

Therefore, differentiating Equation (4.71) yields

$$c_v = \frac{du}{dT} = \frac{3}{2}R,$$

(4.73)

and differentiating Equation (4.72) gives

$$c_p = \frac{dh}{dT} = \frac{5}{2}R.$$

(4.74)

For an ideal gas both $c_p$ and $c_v$ are constants that differ by $R$ as predicted by Equation (4.69). We define a new property, the *specific heat ratio*:

$$\gamma = \frac{c_p}{c_v} = \frac{\bar{c}_p}{\bar{c}_v}.$$

(4.75)

Combining Equations (4.73) and (4.74) with Equation (4.75),

$$\gamma = \frac{5}{3} = 1.667.$$

(4.76)

The specific heat ratio of all monoatomic ideal gases, derived from our very simple model of gas kinetics, is equal to 1.667. This is something that can be checked very easily from experimental data. Table 4.1 gives average values for the specific heats of some common gases. For monoatomic gases such as helium, argon and neon, the measured specific heat ratio is exactly that predicted by theory. This was one of the earliest demonstrations of the validity of kinetic gas theory.

For other gases, with molecules consisting more than one atom, the value of $\gamma$ is less than 1.667. More sophisticated models of molecular motion, which account for molecular vibration and spin, can be developed to predict these values as well.

**Table 4.1**   Specific heats of common gases evaluated at 25 °C and 100 kPa.

| Gas | Number of atoms | $R$ (kJ/kgK) | $c_p$ (kJ/kg K) | $c_v$ (kJ/kg K) | $\gamma = \dfrac{c_p}{c_v}$ |
|---|---|---|---|---|---|
| Helium | 1 | 2.0770 | 5.193 | 3.116 | 1.667 |
| Argon | 1 | 0.2081 | 0.520 | 0.312 | 1.667 |
| Neon | 1 | 0.4120 | 1.030 | 0.618 | 1.667 |
| Hydrogen | 2 | 4.1242 | 14.209 | 10.085 | 1.409 |
| Carbon monoxide | 2 | 0.2968 | 1.041 | 0.744 | 1.400 |
| Nitrogen | 2 | 0.2968 | 1.042 | 0.745 | 1.400 |
| Oxygen | 2 | 0.2598 | 0.922 | 0.662 | 1.393 |
| Water (steam) | 3 | 0.4615 | 1.872 | 1.410 | 1.327 |
| Carbon dioxide | 3 | 0.1889 | 0.842 | 0.653 | 1.289 |
| Ammonia | 4 | 0.4882 | 2.130 | 1.642 | 1.297 |
| Air | Mixture | 0.2870 | 1.004 | 0.717 | 1.400 |

## 4.10   Which should you use, $c_p$ or $c_v$?

What if you are dealing with a process that is neither constant volume nor constant pressure? Which specific heat should you use to determine changes in internal energy and enthalpy?

Energy and enthalpy are both properties and to calculate changes in their values during a process we need to know only the initial and final states of the system – the path taken in going from one state to another does not matter. As a simple analogy, think of a box with mass $m$ that is initially resting on the floor then lifted vertically and placed on a table at height ($\Delta z$; Figure 4.19). The change in potential energy is $\Delta PE = mg\Delta z$. Suppose the box is instead raised high above the table and then dropped onto it. The change in potential energy is still the same since it does not depend on the path taken.

Changes in internal energy also depend only on the initial and final state. In Figure 4.20 a gas is compressed at constant temperature from state 1 to 2 and then expanded at constant

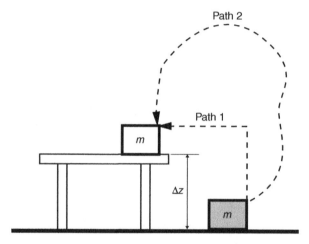

**Figure 4.19**   A box with mass $m$ is lifted onto a table of height $\Delta z$ following either path 1 or path 2. The change in potential energy is $\Delta PE = mg\Delta z$, irrespective of the path taken.

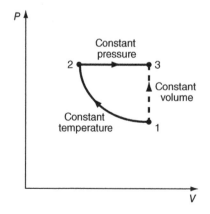

**Figure 4.20**   System undergoing either a constant temperature (1–2) followed by a constant pressure (2–3) process or a constant volume (1–3) process.

pressure from state 2 to 3. What is the change in internal energy for the complete process, going from 1 to 3? We can imagine a process in which we go directly from 1 to 3 by heating the gas while holding it at constant volume. In this case the change in internal energy, if we assume constant specific heat, is given by $\Delta u_{13} = c_{v,avg}(T_3 - T_1)$. But the specific internal energy is a property and its value depends only on the state of the system, not the path followed. Therefore, if we took the system first from state 1 to 2 and then to state 3, by a series of processes that are *not* constant volume, the change in internal energy ($\Delta u_{13}$) would still be the same.

We do not require that the volume or pressure remain constant during a process to use $c_v$ and $c_p$ values to calculate changes in $u$ and $h$. It is *always* true for an ideal gas undergoing *any* thermodynamic process, even if it is not under constant volume or constant pressure conditions that the change in specific internal energy is

$$\Delta u = u_2 - u_1 = \int_{T_1}^{T_2} c_v(T)\,dT$$

and the change in specific enthalpy is

$$\Delta h = h_2 - h_1 = \int_{T_1}^{T_2} c_p(T)\,dT.$$

**Example 4.13**
**Problem:** Oxygen contained in a cylinder is compressed adiabatically by a piston from an initial state with $V_1 = 0.2$ m³, $P_1 = 200$ kPa and $T_1 = 25$ °C to a final temperature $T_2 = 200$ °C. Find the work done during this process.
**Find:** Work $W$ done during compression.
**Known:** Initial volume $V_1 = 0.2$ m³, initial pressure $P_1 = 200$ kPa, initial temperature $T_1 = 25$ °C, final temperature $T_2 = 200$ °C, adiabatic process so $Q_{12} = 0$.
**Diagram:**

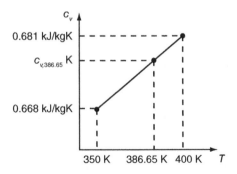

**Figure E4.13**  Linear interpolation to calculate $c_v$ at a temperature of 386.65 K.

**Assumptions:** Oxygen is an ideal gas, no change in KE and PE of system, constant specific heat.

**Governing Equations:**

Ideal gas equation $\quad PV = mRT$

First Law $\qquad\qquad Q_{12} + W_{12} = \Delta U_{12}$

**Properties:** Gas constant of oxygen $R = 0.2598$ kJ / kgK (Appendix 1).

As a first approximation, from Appendix 1 for oxygen at 25 °C, $c_v = 0.662$ kJ / kgK. But we have more accurate information in Appendix 4, where $c_v$ is given as a function of temperature. The average temperature for the process is $T_{avg} = (T_1 + T_2)/2 = (25 °C + 200 °C)/2 = 112.5 °C$ = 385.65 K. From Appendix 4, at 350 K, $c_v = 0.668$ kJ / kgK and at 400 K, $c_v = 0.681$ kJ / kgK. We can do a linear interpolation between these values to get

$$\frac{c_{v,385.65K} - 0.668 \text{ kJ / kgK}}{385.65 \text{ K} - 350 \text{ K}} = \frac{0.681 \text{ kJ / kgK} - 0.668 \text{ kJ / kgK}}{400 \text{ K} - 350 \text{ K}}$$

Solving,

$$c_{v,385.65K} = 0.677 \ 27 \text{ kJ / kgK}.$$

We will use this number, but note that the difference from the value at 25 °C is only about 2%. For most engineering applications such a small difference does not matter. You will need to exercise judgment as to whether it is necessary to interpolate values.

**Solution:**

The mass of oxygen is

$$m = \frac{P_1 V_1}{RT_1} = \frac{200 \times 10^3 \left(N/m^2\right) \times 0.2 \, m^3}{0.2598 \text{ kJ / kgK} \times (273.15 + 25)K} = 0.516 \ 40 \, kg$$

Using an energy balance,

$$\underbrace{Q_{12}}_{=0} + W_{12} = \Delta U_{12} = mc_v(T_2 - T_1),$$

$$W_{12} = 0.516 \ 40 \, kg \times 0.677 \ 27 \text{ kJ / kgK} \times (200 °C - 25 °C) = 61.205 \text{ kJ}$$

**Answer:** The work done during compression is 61.2 kJ.                                    ■

**Example 4.14**

**Problem:** An ideal gas with $c_p = 1.044$ kJ / kgK and $c_v = 0.745$ kJ / kgK contained in a cylinder–piston assembly initially has a pressure of 150 kPa, a temperature of 30 °C, and a volume of 0.22 m³. It is heated slowly at constant volume (process 1–2) until the pressure is doubled. It is then expanded slowly at constant pressure (process 2–3) until the volume is doubled. Determine the work done and heat added in the combined process.

**Find:** Work done $W$ and heat added $Q$.

**Known:** Ideal gas, specific heats $c_p = 1.044$ kJ / kgK and $c_v = 0.745$ kJ / kgK, initial volume

$V_1 = 0.22$ m³, initial pressure $P_1 = 150$ kPa, initial temperature $T_1 = 30$ °C, final pressure $P_2 = 2P_1$, final volume $V_3 = 2V_1$.

**Diagram:**

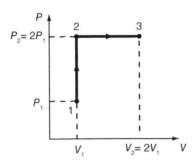

**Figure E4.14**   An ideal gas being heated in a cylinder–piston assembly, first at constant volume (1–2) and then at constant pressure (2–3).

**Assumptions:** No change in KE and PE of system, constant specific heats.

**Governing Equations:**

Ideal gas equation         $PV = mRT$

First Law                      $Q_{12} + W_{12} = \Delta U_{12}$

**Solution:**

The gas constant for this ideal gas is $R = c_p - c_v = 1.044$ kJ / kgK $- 0.745$ kJ / kgK $= 0.299$ kJ / kgK.

To find the mass of gas in the cylinder,

$$m = \frac{P_1 V_1}{RT_1} = \frac{150 \times 10^3 \,\text{N/m}^2 \times\ 0.22\,\text{m}^3}{0.299\,\text{kJ/kgK}\ \times\ (273.15+30)\text{K}} = 0.364\,070\,\text{kg}.$$

The temperature at state 2 is

$$T_2 = T_1 \frac{P_2}{P_1} = 303.15\,\text{K}\ \times\ 2 = 606.300\,\text{K}.$$

The temperature at state 3 is

$$T_3 = T_2 \frac{V_3}{V_2} = 606.3\,\text{K}\ \times\ 2 = 1212.60\,\text{K}.$$

Work done during process 1–2 is

$$W_{12} = -\int_{V_1}^{V_2} P dV = 0,$$

with heat transfer during process 1–2 of

$$Q_{12} = \Delta U_{12} - W_{12} = \Delta U_{12} = mc_v (T_2 - T_1),$$

$$Q_{12} = 0.364\ 07\ \text{kg} \times 0.745\ \text{kJ}/\text{kgK}(606.3\ \text{K} - 303.15\ \text{K}) = 82.2240\ \text{kJ}.$$

Work done during process 2–3 is

$$W_{23} = -\int_{V_2}^{V_3} PdV = -P_3(V_3 - V_2) = -(P_3V_3 - P_2V_2) = -mR(T_3 - T_2),$$

$$W_{23} = -0.364\ 07\ \text{kg} \times 0.299\ \text{kJ}/\text{kgK}(1212.6\ \text{K} - 606.3\ \text{K}) = -66.0000\ \text{kJ}.$$

Heat transfer during process 2–3 is

$$Q_{23} = \Delta U_{23} - W_{23} = mc_v (T_3 - T_2) - W_{23},$$

$$Q_{23} = 0.364\ 07\ \text{kg} \times 0.745\ \text{kJ}/\text{kgK}(1212.6\ \text{K} - 606.3\ \text{K}) + 66\ \text{kJ} = 230.448\ \text{kJ}.$$

Total work done is

$$W_{total} = W_{12} + W_{23} = -66\ \text{kJ}.$$

Total heat transfer is

$$Q_{total} = Q_{12} + Q_{23} = 82.224\ \text{kJ} + 230.448\ \text{kJ} = 312.672\ \text{kJ}.$$

**Answer:** The system does 66.00 kJ of work on the surroundings while 312.7 kJ of heat is added to it.                                                                                 ∎

## 4.11   Ideal Gas Tables

To evaluate changes in internal energy and enthalpy for ideal gases, we need to know how the specific heats change with temperature. For monoatomic gases such as helium, argon and neon $c_p = 5R/2$ is accurate over a very wide range of temperatures. For most other gases specific heats increase with temperature and it is possible to fit a polynomial curve of the form

$$\bar{c}_p = Mc_p = a + bT + cT^2 + dT^3. \tag{4.77}$$

Appendix 5 lists values of the coefficients $a$, $b$, $c$ and $d$ for several gases, valid for temperatures ranging from 300 to 1500 K, with a typical accuracy of ±1%. We can substitute this polynomial in Equation (4.65) and integrate to evaluate changes in enthalpy. We do not need separate tables for $c_v$ since we know that $c_v = c_p - R$.

If we assume that at an arbitrary reference temperature ($T_{ref}$) the enthalpy of a gas is $h(T_{ref})$, we can integrate Equation (4.65)

$$h(T)-h(T_{ref})= \int_{T_{ref}}^{T} c_p(T)dT. \qquad (4.78)$$

If we are interested in changes in enthalpy, rather than absolute values, it does not matter what reference temperature we choose since

$$\Delta h = h(T_2)-h(T_1)= \int_{T_{ref}}^{T_2} c_p(T)dT - \int_{T_{ref}}^{T_1} c_p(T)dT = \int_{T_1}^{T_2} c_p(T)dT. \qquad (4.79)$$

It is conventional to pick $T_{ref} = 0$ K and assume $h = 0$ at $T = 0$ K. Using this reference the values of $h$ and $u$ have been calculated by integrating Equation (4.79) and are listed in the *ideal gas tables*, given in Appendix 7 for air. Using these tables it is possible to directly read values of $u$ and $h$ as functions of temperature.

**Example 4.15**
**Problem:** Air is heated from 300 to 500 K. Find the change in specific enthalpy using (a) $c_p$ evaluated at 25 °C, (b) $c_p$ evaluated at $T_{avg}$, (c) a function $c_p(T)$, (d) ideal gas tables.
**Find:** Change in specific enthalpy $\Delta h$ in four ways.
**Known:** Initial temperature $T_1 = 300$ K, final temperature $T_2 = 500$ K.

**Solution:**
(a) For air at 25 °C, $c_p = 1.004$ kJ / kgK (Appendix 1):

$$\Delta h = c_p(T_2 - T_1) = 1.004 \text{ kJ / kgK} \times (500 \text{ K} - 300 \text{ K}) = 200.800 \text{ kJ / kg}$$

(b) The average temperature during the heating is $T_{avg} = \dfrac{(T_2 + T_1)}{2} = \dfrac{500 \text{ K} + 300 \text{ K}}{2} = 400$ K.
For air at 400 K, $c_p = 1.013$ kJ / kgK (Appendix 4):

$$\Delta h = c_p(T_2 - T_1) = 1.013 \text{ kJ / kgK}(500 \text{ K} - 300 \text{ K}) = 202.600 \text{ kJ / kg}$$

(c) From Appendix 5, for air

$$\overline{c}_p = Mc_p = 28.11 + 0.1967 \times 10^{-2} T + 0.4802 \times 10^{-5} T^2 - 1.966 \times 10^{-9} T^3:$$

$$\Delta \overline{h} = \overline{h}_2 - \overline{h}_1 = \int_{T_1}^{T_2} \overline{c}_p(T)dT = \int_{300 \text{ K}}^{500 \text{ K}} (28.11 + 0.1967 \times 10^{-2} T + 0.4802 \times 10^{-5} T^2 - 1.966 \times 10^{-9} T^3)dT$$

$$\Delta \overline{h} = \left[ 28.11T + 0.9835 \times 10^{-3} T^2 + 0.1601 \times 10^{-5} T^3 - 0.4915 \times 10^{-9} T^4 \right]_{300 \text{ K}}^{500 \text{ K}}$$

$$= 5933.58 \text{ kJ / kmolK}$$

$$\Delta h = \frac{\Delta \overline{h}}{M} = \frac{5933.58 \text{ kJ / kmolK}}{28.97 \text{ kg / kmol}} = 204.818 \text{ kJ / kgK}$$

(d) From Appendix 7, $h(300 \text{ K}) = 300.19$ kJ / kgK and $h(500 \text{ K}) = 503.02$ kJ / kgK:

$$\Delta h = h(500 \text{ K}) - h(300 \text{ K}) = 503.02 \text{ kJ / kgK} - 300.19 \text{ kJ / kgK} = 202.830 \text{ kJ / kgK}$$

**Answer:** The changes in specific enthalpy are: (a) 200.8 kJ / kg, (b) 202.6 kJ / kg, (c) 204.8 kJ / kg, (d) 202.8 kJ / kg. The difference in the specific enthalpy change calculated using these different methods is small. Often, assuming constant specific heat is a reasonable assumption. If ideal gas tables are available they are the easiest way of calculating enthalpy changes.     ∎

## 4.12   Specific Heats of Liquids and Solids

The volume of liquids and solids changes very little as pressure and temperature vary. To a very good approximation, we can assume that they are *incompressible substances*, whose specific volume remains constant. We cannot do compression or expansion work on an incompressible substance, since its volume does not change. The internal energy of an incompressible substance is therefore a function of temperature alone:

$$u = u(T).$$
(4.80)

The specific heat is defined as

$$c_v(T) = \frac{du}{dT}.$$
(4.81)

Changes in internal energy are calculated from

$$du = c_v(T)dT.$$
(4.82)

Enthalpy is defined as $h = u + Pv$. Differentiating this expression,

$$dh = du + Pdv + vdP.$$
(4.83)

For incompressible substances $dv = 0$, so

$$dh = c_v(T)dT + vdP.$$
(4.84)

If we assume pressure is constant $dP = 0$ and

$$c_v(T) = \left(\frac{dh}{dT}\right)_P.$$
(4.85)

The right hand side of Equation (4.85) is the definition of $c_p(T)$. *For an incompressible substance $c_p$ and $c_v$ are always equal.* We no longer need a subscript to differentiate between them, so that

$$c_p(T) = c_v(T) = c(T).$$
(4.86)

Typical values of $c$ for common solids and liquids are given in Appendices 2 and 3. If the specific heat $c$ is constant, independent of temperature, we can integrate Equation (4.82) as

$$\Delta u = u_2 - u_1 = c\int_{T_1}^{T_2} dT = c(T_2 - T_1),$$
(4.87)

and integrate Equation (4.84) to get

$$\Delta h = h_2 - h_1 = c\int_{T_1}^{T_2}dT + v\int_{P_1}^{P_2}dP = c(T_2 - T_1) + v(P_2 - P_1).$$ (4.88)

**Example 4.16**

**Problem:** A 1 cm diameter copper sphere heated to 300 °C is dropped into a well-insulated beaker containing 10 ml of water at 25 °C. What is the final temperature once the water and metal have come to equilibrium?

**Find:** Final temperature $T_2$.

**Known:** Initial temperature of copper $T_{1,Cu} = 300$ °C, initial temperature of water $T_{1,W} = 25$ °C, diameter of copper sphere $D = 0.01$ m, volume of water $V_W = 10$ ml $= 10^{-5}$ m$^3$.

**Diagram:**

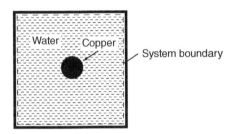

**Figure E4.16** A heated copper sphere is dropped into a well-insulated beaker filled with water.

**Assumptions:** Water and copper are both incompressible, there is no heat loss from the beaker so $Q_{12} = 0$.

**Governing Equation:**

First Law $$Q_{12} + W_{12} = \Delta U_{12}$$

**Properties:** Copper density and specific heat $\rho_{Cu} = 8900$ kg / m$^3$ and $c_{Cu} = 0.386$ kJ / kg°C (Appendix 2), water density and specific heat $\rho_W = 997$ kg / m$^3$ and $c_W = 4.18$ kJ / kg°C (Appendix 3).

**Solution:**

Mass of copper sphere is

$$m_{Cu} = \rho_{Cu}\frac{\pi D^3}{6} = 8900 \text{ kg / m}^3 \times \frac{\pi(0.01\text{ m})^3}{6} = 4.6600\times10^{-3} \text{ kg.}$$

Mass of water is

$$m_W = \rho_W V_W = 997 \text{ kg / m}^3 \times 10^{-5} \text{ m}^3 = 9.97\times10^{-3} \text{ kg.}$$

Since the system is incompressible, $W_{12} = 0$ so,

$$Q_{12} + W_{12} = \Delta U_{12} = 0$$

$$\underset{=0}{\underbrace{Q_{12}}} + \underset{=0}{\underbrace{W_{12}}}$$

$$\Delta U_{12} = m_{Cu}\Delta u_{12,Cu} + m_W \Delta u_{12,W} = m_{Cu}c_{Cu}\left(T_2 - T_{1,Cu}\right) + m_W c_W \left(T_2 - T_{1,W}\right) = 0.$$

Substituting into the energy balance,

$$\Delta U_{12} = 4.66 \times 10^{-3} \text{ kg} \times 0.386 \text{ kJ} / \text{kg}^\circ\text{C} \times \left(T_2 - 300 \text{ }^\circ\text{C}\right)$$
$$+ 9.97 \times 10^{-3} \text{ kg} \times 4.18 \text{ kJ} / \text{kg}^\circ\text{C} \times \left(T_2 - 25 \text{ }^\circ\text{C}\right) = 0.$$

Solving,

$$T_2 = 36.379 \text{ }^\circ\text{C}.$$

**Answer:** The final temperature of the system is 36.4 °C.                                              ∎

## 4.13   Steady Mass Flow Through a Control Volume

Turbines, pumps, compressors and heat exchangers are devices that exchange energy with the surroundings in the form of work or heat while fluids flow through them. They all operate under steady flow conditions in which fluid enters at the same rate as it leaves, so that no mass accumulates inside them. To calculate the mass flow rate through a control volume we need to know the velocity and fluid properties at both the inlet and outlet.

Figure 4.21 shows fluid with average velocity ($\mathbf{V}$) entering a control volume through a duct of cross-section $A$. At time $dt$ a small fluid element with mass $\delta m$ has entered the control volume. The length of the fluid inside the control volume is $dx = \mathbf{V}dt$, and its volume is $V = A dx = A\mathbf{V}dt$. The mass of fluid inside the control volume is

$$\delta m = \rho A \mathbf{V} dt \tag{4.89}$$

and the rate at which fluid enters the control volume is

$$\dot{m} = \frac{\delta m}{dt} = \rho A \mathbf{V}, \tag{4.90}$$

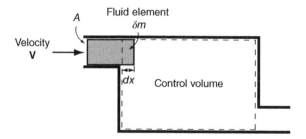

**Figure 4.21**   Fluid entering a control volume.

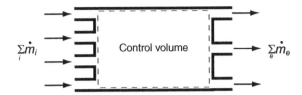

**Figure 4.22**   Multiple streams of fluid entering and leaving a control volume.

where $\rho$ is the density of the fluid. Since the specific volume is $v = 1 / \rho$,

$$\dot{m} = \frac{A\mathbf{V}}{v}. \tag{4.91}$$

At steady state the rates of mass flowing into and out of the control volume must be equal:

$$\underbrace{\dot{m}_1}_{\text{rate of mass inflow}} = \underbrace{\dot{m}_2}_{\text{rate of mass outflow}}. \tag{4.92}$$

The units of $\dot{m}$ are kg / s. If the control volume has multiple inlets and exits (Figure 4.22), we need to sum over all of those to obtain a mass balance:

$$\sum_i \dot{m}_i = \sum_e \dot{m}_e. \tag{4.93}$$

**Example 4.17**
**Problem:** Air at 27 °C flows through a room heating duct with a volume flow rate of 0.5 m³/s. If the duct has a square cross-section with each side 30 cm long, what are the velocity and mass flow rate of air in the duct?
**Find:** Velocity $\mathbf{V}$ and mass flow rate $\dot{m}$ of air.
**Known:** Air temperature $T = 27\,°C = 300.15$ K, air volume flow rate $\dot{V} = 0.5\,\text{m}^3/\text{s}$, duct cross-section 0.3 × 0.3 m.
**Assumptions:** Air is an ideal gas with gas constant $R = 0.2870$ kJ / kgK, atmospheric pressure is 101.325 kPa.
**Governing Equation:**
Mass flow rate $\qquad\qquad \dot{m} = \rho A \mathbf{V}$

**Solution:**

$$\rho = \frac{m}{V} = \frac{P}{RT} = \frac{101.325 \times 10^3 \text{ Pa}}{0.2870 \times 10^3 \text{ J / kgK} \times (273.15 + 27)\text{K}} = 1.1762\,\text{kg / m}^3$$

$$\dot{m} = \rho\dot{V} = 1.1762\,\text{kg / m}^3 \times 0.5\,\text{m}^3/\text{s} = 0.5881\,\text{kg / s}$$

$$\mathbf{V} = \frac{\dot{m}}{\rho A} = \frac{0.5881\,\text{kg / s}}{1.1762\,\text{kg / m}^3 \times (0.3\,\text{m})^2} = 5.5556\,\text{m / s}$$

**Answer:** The mass flow rate is 1.18 kg / m³ and the air velocity is 5.56 m / s.    ∎

## 4.14   The First Law for Steady Mass Flow Through a Control Volume

To apply the first law to control volumes we have to account for the transport of energy with the mass that enters and leaves the system. Energy can be transported with a fluid in several different ways: as internal energy, kinetic energy and potential energy. Kinetic energy and potential energy are both macroscopic forms of energy that depend on the position and velocity of the bulk fluid while internal energy is a microscopic form of energy, determined by measuring the temperature of an ideal gas or incompressible liquid. In addition, when fluid enters a control volume flow work is done on it and it carries this energy into the system with it. When it leaves flow work is done by the system on the fluid and it carries this energy out of the system. The flow work per unit mass ($Pv$) is added to the specific internal energy ($u$) to give the specific enthalpy ($h$).

The energy entering the control volume per unit mass of fluid is $e = h + ke + pe = h + V^2 / 2 + gz$. The heights of the inlet ($z_1$) and outlet ($z_2$) are measured above an arbitrary reference plane (see Figure 4.23).

At steady state, the energy balance for the control volume is

$$\underbrace{\dot{E}_1}_{\text{rate of energy input}} = \underbrace{\dot{E}_2}_{\text{rate of energy output}}. \tag{4.94}$$

$\dot{E}$ includes energy due to the fluid entering and leaving the control volume, and also any heat or work transfers. If the mass flow rate through the control volume is $\dot{m}$, the energy balance for the control volume is

$$\dot{Q} + \dot{W} + \dot{m}\left( h_1 + \frac{V_1^2}{2} + gz_1 \right) = \dot{m}\left( h_2 + \frac{V_2^2}{2} + gz_2 \right). \tag{4.95}$$

$\dot{Q}$ is the rate of heat transfer to the control volume from the surroundings and $\dot{W}$ is the rate at which the surroundings do work on the control volume. $\dot{W}$ is also known as *power*, and has units of J / s or watts (W). Rearranging Equation (4.95) gives

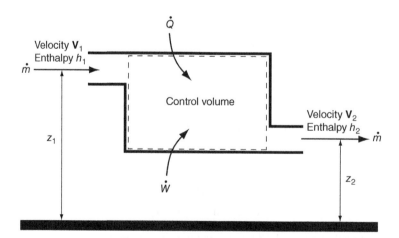

**Figure 4.23**   Energy transfers with fluid flowing through a control volume.

$$\dot{Q} + \dot{W} = \dot{m}\left[\left(h_2 - h_1\right) + \frac{\mathbf{V}_2^2 - \mathbf{V}_1^2}{2} + g\left(z_2 - z_1\right)\right].$$  (4.96)

Dividing Equation (4.96) by the mass flow rate $\dot{m}$ gives an energy balance per unit mass of fluid:

$$q + w = \left(h_2 - h_1\right) + \frac{\mathbf{V}_2^2 - \mathbf{V}_1^2}{2} + g\left(z_2 - z_1\right),$$  (4.97)

where the heat transfer per unit mass of fluid flowing through the control volume is $q = \dot{Q} / \dot{m}$ and the work done per unit mass of fluid is $w = \dot{W} / \dot{m}$.

## 4.15   Steady Flow Devices

### 4.15.1   Turbines and Compressors

A turbine is a device that uses the flow of a fluid to turn a shaft. High-pressure liquid or gas enters a turbine and impinges on blades radiating from a central shaft (Figure 4.24). The change in momentum of the fluid as the blades deflect it produces a reaction force, creating a torque that turns the shaft. The fluid pressure decreases as it flows through the turbine and therefore the cross-sectional area of the turbine has to increase to accommodate gas expansion. Low pressure fluid leaves the turbine through the exhaust duct.

Turbines are well insulated, so heat losses to the surroundings are minimal. Changes in kinetic and potential energy between the inlet and outlet are negligible. Making these simplifying assumptions, we set $\dot{Q} = 0$, $z_2 = z_1$ and $\mathbf{V}_2 = \mathbf{V}_1$ in Equation (4.96). For a turbine the energy balance reduces to

$$\dot{W}_{shaft} = \dot{m}\left(h_2 - h_1\right).$$  (4.98)

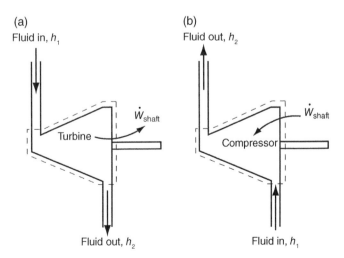

**Figure 4.24**   Fluid flowing through (a) a turbine and (b) a compressor.

A turbine does work on the surroundings so $\dot{W}_{shaft} < 0$, as shown in Figure 4.24a, and $h_2 < h_1$.

A compressor can be treated as a turbine operating in reverse, where the surroundings supply the shaft work required to compress the gas flowing through the device. Gas enters the compressor at low pressure and is forced into narrower passages by the blades turning around the shaft driven by an engine or electric motor. High-pressure gas leaves the compressor. Making the same assumptions that we did for the turbine, that $\dot{Q} = 0$, $z_2 = z_1$ and $\mathbf{V}_2 = \mathbf{V}_1$, the energy balance for a compressor gives

$$\dot{W}_{shaft} = \dot{m}\left(h_2 - h_1\right). \tag{4.99}$$

The enthalpy of the fluid increases as it goes through the compressor and its pressure rises so that $h_2 > h_1$ and $\dot{W}_{shaft} > 0$.

**Example 4.18**
**Problem:** A gas turbine receives compressed air at 800 kPa and 300 °C, which leaves at 120 kPa and 60 °C. What should the mass flow rate of air be if a power output of 5 kW is required?
**Find:** Mass flow rate of air $\dot{m}$.
**Known:** Inlet pressure $P_1$ = 800 kPa, inlet temperature $T_1$ = 300 °C, outlet pressure $P_2$ = 120 kPa, outlet temperature $T_2$ = 60 °C, shaft work $\dot{W}_{shaft}$ = −50 kW.
**Assumptions:** Negligible heat losses from the turbine, negligible changes in kinetic and potential energy between the turbine inlet and outlet, air is an ideal gas with constant specific heat.
**Governing Equation:**
  Rate of shaft work (turbine)        $\dot{W}_{shaft} = \dot{m}\left(h_2 - h_1\right)$
**Properties:** Average temperature $Tavg$ = $(T2 + T1) / 2$ = (300 °C + 60 °C) / 2 = 180 °C = 453.15 K ≈ 450 K, specific heat of air at 450 K $cp$ = 1.020 kJ / kgK (Appendix 4).

**Solution:**

$$\dot{W}_{shaft} = \dot{m}\left(h_2 - h_1\right) = \dot{m}c_p\left(T_2 - T_1\right)$$

$$-50\,\text{kJ}/\text{s} = \dot{m} \times 1.020\,\text{kJ}/\text{kgK} \times \left(60\,°\text{C} - 300\,°\text{C}\right)$$

$$\dot{m} = 0.20425\,\text{kg}/\text{s}$$

**Answer:** The mass flow rate of air should be 0.20 kg / s.                    ∎

**Example 4.19**
**Problem:** Oxygen enters a compressor at a pressure of 100 kPa, temperature of 300 K and a velocity of 5 m / s with a mass flow rate of 0.5 kg / s. The compressor is cooled at a rate of 1 kW. The gas exits at 500 kPa and 400 K with a velocity of 3 m / s. Find the work input to the compressor.
**Find:** Shaft work input to the compressor $\dot{W}$.

**Known:** Inlet pressure $P_1 = 100$ kPa, inlet temperature $T_1 = 300$ K, inlet velocity $\mathbf{V}_1 = 5$ m / s, outlet pressure $P_2 = 500$ kPa, outlet temperature $T_2 = 400$ K, outlet velocity $\mathbf{V}_2 = 3$ m / s, heat loss $\dot{Q} = -1$ kW.

**Assumptions:** Oxygen is an ideal gas with constant specific heat, change in potential energy is negligible, kinetic energy changes and heat losses are not negligible.

**Governing Equation:**

Energy rate balance $\qquad \dot{Q} + \dot{W} = \dot{m}\left[(h_2 - h_1) + \dfrac{\mathbf{V}_2^2 - \mathbf{V}_1^2}{2} + g(z_2 - z_1)\right]$

**Properties:** Average temperature of oxygen is $T_{avg} = (T_2 + T_1)/2 = (300\,\text{K} + 400\,\text{K})/2 = 350\,\text{K}$; at 350 K specific heat of oxygen is $c_p = 0.928$ kJ / kgK (Appendix 4).

**Solution:**

$$\dot{Q} + \dot{W} = \dot{m}\left[c_p(T_2 - T_1) + \frac{\mathbf{V}_2^2 - \mathbf{V}_1^2}{2} + \underbrace{g(z_2 - z_1)}_{=0}\right]$$

$$-1\,\text{kW} + \dot{W} = 0.5\,\text{kg/s}\left[0.928\,\text{kJ/kgK}(400\,\text{K} - 300\,\text{K}) + \frac{(5\,\text{m/s})^2 - (3\,\text{m/s})^2}{2} \times \frac{1}{1000\,\text{J/kJ}}\right]$$

$$\dot{W} = 1\,\text{kW} + 0.5\,\text{kg/s} \times [92.8\,\text{kJ/kg} + 0.008\,\text{kJ/kg}] = 47.404\,\text{kW}$$

**Answer:** The compressor requires 47.4 kW to drive it. Note that the contribution due to the change in kinetic energy was very small. Neglecting it would have made little difference to the final answer. ∎

## 4.15.2   Pumps

Pumps are used to raise liquids from a lower to a higher elevation, or to raise their pressure (Figure 4.25). Heat transfer from a pump to the surroundings is usually negligible, as is the change in kinetic energy, so

$$\dot{W}_{shaft} = \dot{m}\left[(h_2 - h_1) + g(z_2 - z_1)\right]. \qquad (4.100)$$

Assuming the liquid being pumped is incompressible, we can use Equation (4.88) to calculate its change in enthalpy:

$$h_2 - h_1 = c(T_2 - T_1) + v(P_2 - P_1). \qquad (4.101)$$

Temperature changes in the liquid are negligible in a pump, so we can assume that $T_2 \approx T_1$. Substituting into Equation (4.96) gives an expression for the power required to drive a pump:

$$\dot{W}_{shaft} = \dot{m}\left[v(P_2 - P_1) + g(z_2 - z_1)\right]. \qquad (4.102)$$

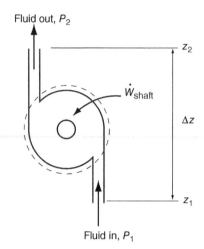

Fluid out, $P_2$

$z_2$

$\dot{W}_{shaft}$

$\Delta z$

$z_1$

Fluid in, $P_1$

**Figure 4.25**   Fluid flowing through a pump.

**Example 4.20**

**Problem:** A water pump raises 5 kg / s of water from a reservoir sunk 10 m underground to a tank located 30 m above ground. The pressure at the reservoir level is 20 kPa and in the tank is 100 kPa. What is the power required to drive the pump?

**Find:** Power to drive the pump $\dot{W}$.

**Known:** Mass flow rate of water $\dot{m} = 5$ kg / s, exit height $z_2 = 30$ m, inlet height $z_1 = -10$ m, exit pressure $P_2 = 100$ kPa, inlet pressure $P_1 = 20$ kPa.

**Assumptions:** Negligible heat losses from the pump, negligible changes in kinetic energy between inlet and outlet, water is incompressible with specific volume $v = 10^{-3}$ m³ / kg, acceleration due to gravity is at earth's surface $g = 9.81$ m / s².

**Governing Equation:**

Rate of shaft work (pump)      $\dot{W}_{shaft} = \dot{m}\left[v\left(P_2 - P_1\right) + g\left(z_2 - z_1\right)\right]$

**Solution:**

$$\dot{W}_{shaft} = 5\,\text{kg}/\text{s}\left[10^{-3}\,\text{m}^3/\text{kg} \times \left(100 \times 10^3\,\text{N}/\text{m}^2 - 20 \times 10^3\,\text{N}/\text{m}^2\right) + 9.81\,\text{m}/\text{s}^2\left(30\,\text{m} - \left(-10\,\text{m}\right)\right)\right]$$

$$\dot{W}_{shaft} = 2362.0\,\text{W} = 2.36\,\text{kW}$$

**Answer:** The pump requires a power input of 2.36 kW.                                                                 ∎

### 4.15.3   Nozzles and Diffusers

A nozzle is a duct whose cross-sectional area is much smaller at the exit than at the entrance (Figure 4.26a). The velocity of a fluid increases as it passes through a nozzle. In steam turbines high pressure steam issues out of nozzles at very high speeds and impinges

(a)                                                                          (b)

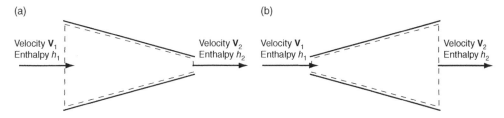

Velocity $V_1$                    Velocity $V_2$     Velocity $V_1$                              Velocity $V_2$
Enthalpy $h_1$                    Enthalpy $h_2$     Enthalpy $h_1$                              Enthalpy $h_2$

**Figure 4.26**   Flow through (a) a nozzle and (b) a diffuser.

on the turbine blades, making the turbine rotor turn. The rear of a jet engine is shaped like a nozzle to accelerate the gases before they are expelled, maximising the thrust produced.

Changes in potential energy between the entrance and exit of a nozzle are negligible ($z_2 \approx z_1$). There is no work done on the fluid ($\dot{W} = 0$), and heat transfer is negligible ($\dot{Q} = 0$). The exit velocity in a nozzle is much greater than the inlet velocity ($V_2 \gg V_1$). With these approximations Equation (4.96) becomes

$$V_2 = \sqrt{2(h_1 - h_2)}.$$                                                           (4.103)

A diffuser is simply a nozzle in reverse (Figure 4.26b) and is used to raise the pressure of high velocity fluids. The front of a jet engine is shaped like a diffuser to increase the pressure of the air entering it as much as possible before it goes into a compressor. A fluid enters at high velocity and leaves with very low velocity and high pressure, so that $V_1 \gg V_2$. Applying the same simplifications that we did for nozzles,

$$V_1 = \sqrt{2(h_2 - h_1)}.$$                                                           (4.104)

**Example 4.21**
**Problem:** Air flowing through an adiabatic diffuser enters at a pressure of 100 kPa and 10 °C with a velocity of 200 m / s and leaves with low velocity at a pressure of 150 kPa. The inlet diameter is 2 cm and the exit diameter is 5 cm. Find the exit temperature and velocity.
**Find:** Exit temperature $T_2$ and exit velocity $V_2$.
**Known:** Inlet pressure $P_1$ = 100 kPa, inlet temperature $T_1$ = 10 °C, inlet velocity $V_{11}$ = 200 m / s, exit Pressure $P_{22}$ = 150 kPa, inlet diameter $D_1$ = 2 cm, exit diameter $D_2$ = 5 cm.
**Assumptions:** Negligible kinetic energy at exit, negligible heat and work transfer in diffuser, negligible change in potential energy, air is an ideal gas with constant specific heat.
**Governing Equations:**

Inlet velocity for diffusers          $$V_1 = \sqrt{2(h_2 - h_1)}$$

Mass flow rate                        $$\dot{m} = \frac{AV}{v}$$

**Properties:** For air at 300 K specific heat capacity $c_p$ = 1.005 kJ / kg°C (Appendix 4).

**Solution:**

$$V_1 = \sqrt{2(h_2 - h_1)} = \sqrt{2c_p(T_2 - T_1)}$$

$$T_2 - T_1 = \frac{V_1^2}{2c_p} = \frac{(200 \text{ m / s})^2}{2 \times 1.005 \times 10^3 \text{ J / kg} ^\circ\text{C}} = 19.900 \text{ }^\circ\text{C}$$

$$T_2 = 19.9 \text{ }^\circ\text{C} + 10 \text{ }^\circ\text{C} = 29.9 \text{ }^\circ\text{C}$$

To find the exit velocity, since mass flow rate into an out of the diffuser are equal:

$$\frac{V_2}{V_1} = \frac{A_1}{A_2} \frac{v_2}{v_1}.$$

For an ideal gas, $v = RT / P$, so

$$\frac{V_2}{V_1} = \frac{A_1}{A_2} \frac{T_2}{T_1} \frac{P_1}{P_2} = \left(\frac{D_1}{D_2}\right)^2 \frac{T_2}{T_1} \frac{P_1}{P_2} = \left(\frac{2 \text{ cm}}{5 \text{ cm}}\right)^2 \left(\frac{303.15 \text{ K}}{283.15 \text{ K}}\right) \left(\frac{100 \text{ kPa}}{150 \text{ kPa}}\right) = 0.11420,$$

$$V_2 = 0.1142 \times 200 \text{ m / s} = 22.84 \text{ m / s}.$$

**Answer:** The exit temperature is 30 °C and the exit velocity is 22.8 m / s. The ratio of the exit kinetic energy to the inlet kinetic energy is $(V_2 / V_1)^2 \approx 0.01$. Our assumption of negligible exit kinetic energy is therefore valid. ∎

## 4.16  Transient Analysis for Control Volumes

If the mass flow rate at the inlet and outlet of a control volume are not the same, the mass contained within it will change. If there is mass both entering ($m_1$) and leaving ($m_2$) the control volume the mass inside it changes by

$$\Delta m = m_1 - m_2, \tag{4.105}$$

where $\Delta m = m_{final} - m_{initial}$.

If infinitesimal amounts of mass enter ($\delta m_1$) and leave ($\delta m_2$) the control volume, the mass inside it changes by

$$dm = \delta m_1 - \delta m_2. \tag{4.106}$$

Dividing by a time interval ($dt$),

$$\frac{dm}{dt} = \dot{m}_1 - \dot{m}_2. \tag{4.107}$$

Similarly, if the energy added to a system and the energy leaving it are not the same, the change in energy stored in the system is

$$\Delta E = \Delta KE + \Delta PE + \Delta U = Q + W + m_1 \left( h_1 + \frac{V_1^2}{2} + gz_1 \right) - m_2 \left( h_2 + \frac{V_2^2}{2} + gz_2 \right). \quad (4.108)$$

If, as is typical, there are no changes in kinetic or potential energy of the system, this reduces to

$$\Delta U = Q + W + m_1 h_1 - m_2 h_2. \quad (4.109)$$

If there is no mass entering or leaving the control volume this reduces to the energy balance for a control mass.

We can write the energy balance as a rate equation:

$$\frac{dE}{dt} = \dot{Q} + \dot{W} + \dot{m}_1 \left( h_1 + \frac{V_1^2}{2} + gz_1 \right) - \dot{m}_2 \left( h_2 + \frac{V_2^2}{2} + gz_2 \right). \quad (4.110)$$

If there are multiple streams leaving and entering the control volume we need to sum over all of these:

$$\frac{dE}{dt} = \dot{Q} + \dot{W} + \sum_i \dot{m}_i \left( h_i + \frac{V_i^2}{2} + gz_i \right) - \sum_e \dot{m}_e \left( h_e + \frac{V_e^2}{2} + gz_e \right). \quad (4.111)$$

### Example 4.22

**Problem:** An evacuated cylinder is connected to a pipeline carrying compressed air at 300 kPa and 52 °C. The cylinder fills until its pressure is the same as that in the pipeline. Determine the final temperature in the tank.

**Find:** Final temperature $T_2$ in the tank.

**Known:** Inlet pressure $P_1 = 200$ kPa, inlet temperature $T_1 = 52$ °C = 325.15 K, final pressure $P_2 = 300$ kPa.

**Assumptions:** Air is an ideal gas, there are no heat losses from the cylinder, no changes in kinetic or potential energy of the cylinder.

**Governing Equation:**

Energy balance $\qquad \Delta U = Q + W + m_1 h_1 - m_2 h_2$

**Solution:**

$$\Delta U = m(u_2 - u_1) = \underset{=0}{Q} + \underset{=0}{W} + m_1 h_1 - m_2 h_2$$

Where $u_1$ and $u_2$ are the initial and final specific internal energies of the gas in the cylinder and $m$ is the final mass of air in the cylinder. There is no exit from the system, so $m_2 = 0$. The cylinder was initially evacuated so $u_1 = 0$. The final mass in the cylinder equals that which entered, so $m = m_1$. Therefore,

$$u_2 = h_1.$$

From the definition of enthalpy, $h_1 = u_1 + P_1 v_1$ where $P_1 v_1$ is the flow work. The final internal energy of the air in the cylinder equals the internal energy of the air in the pipeline plus the work done in filling the cylinder.

Using the ideal gas table (Appendix 7), for a temperature of 325 K (since inlet temperature is 325.15 K), $h_1 = 325.31$ kJ / kgK. Therefore, $u_2 = 325.31$ kJ / kgK. Using this value and interpolating in the ideal gas table for air,

$$T_2 = 453.8 \text{ K}.$$

**Answer:** The final temperature of the cylindrical tank is 454 K.                                         ∎

## Further Reading

1.   Y. A. Cengel, M. A. Boles (**2015**) *Thermodynamics – An Engineering Approach*, McGraw Hill, New York.
2.   M. J. Moran, H. N. Shapiro, D. D. Boettner, M. B. Bailey (**2014**) *Fundamentals of Engineering Thermodynamics*, John Wiley & Sons, Ltd, London.
3.   C. Borgnakke, R. E. Sonntag (**2012**) *Fundamentals of Thermodynamics*, John Wiley & Sons, Ltd, London.
4.   K. A. Kroos, M. C. Potter (**2015**) *Thermodynamics for Engineers*, Cengage Learning, New York.

## Summary

*Energy* is an extensive property of all thermodynamic systems. A system possesses energy if it is capable of lifting a weight. *Potential energy* and *kinetic energy* are macroscopic forms of energy, and altering them requires a change in the position or velocity of the system. *Internal energy* includes all microscopic forms of energy storage. The total energy ($E$) of a system is the sum of its potential ($PE$), kinetic ($KE$) and internal ($U$) energies:

$$E = PE + KE + U.$$

The specific energy is an intensive property and is the sum of its specific kinetic, potential and internal energies:

$$e = \frac{E}{m} = gz + \frac{1}{2}\mathbf{V}^2 + u.$$

The *power* $\left(\dot{W}\right)$ is the rate of doing work:

$$\dot{W} = \frac{\delta W}{dt}.$$

The *rate of heat transfer* $\left(\dot{Q}\right)$ is

$$\dot{Q} = \frac{\delta Q}{dt}.$$

*Boundary work:*

$$W_{12} = -\int_{V_1}^{V_2} PdV.$$

Boundary work for a constant pressure process is

$$W_{12} = -P_1 \int_{V_1}^{V_2} dV = P_1\left(V_1 - V_2\right) = P_1V_1 - P_2V_2.$$

Boundary work for an ideal gas undergoing a constant temperature process is

$$W_{12} = -mRT \int_{V_1}^{V_2} \frac{dV}{V} = mRT \ln \frac{V_1}{V_2}.$$

Boundary work for a polytropic process in which $PV^n = $ constant is

$$W_{12} = \frac{P_2 V_2 - P_1 V_1}{n-1}, \text{ for } n \neq 1.$$

$$W_{12} = P_1 V_1 \ln \frac{V_1}{V_2} = P_2 V_2 \ln \frac{V_1}{V_2}. \text{ for } n = 1.$$

Flow work per unit mass of fluid is $w_{\text{flow}} = \dfrac{W_{\text{flow}}}{m} = Pv.$

Shaft work: $W_{\text{shaft}} = 2\pi n r F = 2\pi n \tau.$

Spring work: $W_{\text{spring}} = \dfrac{1}{2} K \left( x_2^2 - x_1^2 \right).$

Electric power: $\dot{W}_{\text{elec}} = V_e I = I^2 R_e = \dfrac{V_e^2}{R_e}.$

The *first law of thermodynamics* states that the change in energy of a closed system equals the net energy transferred to it in the form of work and heat:

$$\underbrace{Q + W}_{\text{Energy addition}} = \underbrace{\Delta E.}_{\text{Change in energy}}$$

Written as a rate equation,

$$\dot{Q} + \dot{W} = \frac{dE}{dt}.$$

*Enthalpy* is an extensive property defined as

$$H \equiv U + PV.$$

*Specific enthalpy* is an intensive property:

$$h = \frac{H}{m} = u + Pv.$$

The *specific heat at constant volume* is defined as

$$c_v(T) \equiv \left( \frac{\partial u}{\partial T} \right)_v.$$

The *specific heat at constant pressure* is defined as

$$c_p(T) \equiv \left( \frac{\partial h}{\partial T} \right)_p.$$

For an ideal gas,

$$c_p = c_v + R.$$

The *specific heat ratio* is

$$\gamma = \frac{c_p}{c_v} = \frac{\bar{c}_p}{\bar{c}_v}.$$

For an ideal gas undergoing any thermodynamic process the change in specific internal energy is

$$\Delta u = u_2 - u_1 = \int_{T_1}^{T_2} c_v(T) dT,$$

and the change in specific enthalpy is:

$$\Delta h = h_2 - h_1 = \int_{T_1}^{T_2} c_p(T) dT.$$

For an incompressible substance such as a solid or liquid $c_p$ and $c_v$ are always equal:

$$c_p(T) = c_v(T) = c(T).$$

The mass flow rate through a control volume is

$$\dot{m} = \frac{A\mathbf{V}}{v}.$$

At steady state,

$$\sum_i \dot{m}_i = \sum_e \dot{m}_e.$$

The energy balance for a control volume at steady state is

$$\dot{Q} + \dot{W} = \dot{m}\left[ (h_2 - h_1) + \frac{\mathbf{V}_2^2 - \mathbf{V}_1^2}{2} + g(z_2 - z_1) \right].$$

Dividing by the mass flow rate $\dot{m}$ gives an energy balance per unit mass of fluid:

$$q + w = (h_2 - h_1) + \frac{\mathbf{V}_2^2 - \mathbf{V}_1^2}{2} + g(z_2 - z_1).$$

For a turbine or compressor, $\dot{W}_{shaft} = \dot{m}(h_2 - h_1)$.

For a pump, $\dot{W}_{shaft} = \dot{m}[v(P_2 - P_1) + g(z_2 - z_1)]$.

For a nozzle, $\mathbf{V}_2 = \sqrt{2(h_1 - h_2)}$.

For a diffuser, $\mathbf{V}_1 = \sqrt{2(h_2 - h_1)}$.

For transient flow through a control volume,

$$\frac{dE}{dt} = \dot{Q} + \dot{W} + \sum_i \dot{m}_i \left( h_i + \frac{\mathbf{V}_i^2}{2} + gz_i \right) - \sum_e \dot{m}_e \left( h_e + \frac{\mathbf{V}_e^2}{2} + gz_e \right).$$

## Problems

### Work Modes

4.1 A 100 kg box is pushed across the floor by a constant 300 N force. It is then lifted and placed on a 1 m high table. Find the work done.

4.2 A spring with a spring constant $K = 200$ N / m is stretched through a distance of 50 mm. Find the maximum force applied and the work done.

4.3 An engine rated at 200 kW is rotating at 3000 RPM while its engine shaft delivers a torque of 300 Nm. What fraction of full power is the engine operating at?

4.4 A 20 cm diameter pulley is driven by a belt that exerts a net tangential force of 3 kN. What is the power transmitted by the pulley when it is turning at 500 RPM?

4.5 The piston in a cylinder filled with oil at a pressure of 0.8 MPa is advanced 10 cm. What is the work done if the cross-sectional area of the cylinder is 20 cm²?

4.6 A 1 kN force is required to overcome rolling resistance and air drag opposing the motion of a car. What is the power required to keep the car moving at a constant speed of 100 km / h?

4.7 A cylinder contains water filled on top of a piston to a depth of 1 m. As the piston is raised the water drains out of an outlet at the top of the cylinder. Find the work required to empty out all the water if the cross-sectional area of the piston is 0.2 m². Assume the density of water is 1000 kg / m³.

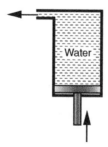

**Figure P4.7**   A cylinder contains water filled on top of a piston.

4.8 A piston–cylinder initially containing 0.05 m³ of air at a pressure of 400 kPa and temperature of 300 K is taken through a constant pressure process in which it does 10 kJ of work on the surroundings. Find the final temperature and volume of the air.

4.9    Two kg of an ideal gas at a pressure of 200 kPa has a specific volume of 0.3 m³ / kg. It goes through a constant pressure process during which 40 kJ of work is done on it. What is the final volume of gas?

4.10   A 2 kg mass of air in a cylinder initially at a pressure of 100 kPa and temperature of 20 °C is compressed by a piston to a pressure of 300 kPa while being kept at constant temperature. Find the work done.

4.11   Air in a cylinder is heated in a process where its pressure varies as shown in Figure P4.11. Calculate the total work done.

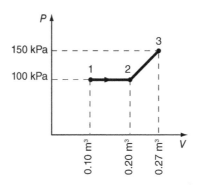

**Figure P4.11**    *P-V* diagram for air being heated in a cylinder.

4.12   A gas is expanded in a polytropic process for which $PV^n$ = constant from 300 kPa and 0.1 m³ to 100 kPa and 0.2 m³. Find the value of $n$.

4.13   Argon at 150 kPa, 320 K and 0.1 m³ expands in a polytropic process for which $n = 1.667$ to a pressure of 100 kPa. What is the work done?

4.14   Air at 800 K and 1 MPa is expanded in a polytropic process for which $PV^{1.6}$ = constant until the pressure reaches 0.1 MPa. Find the final gas temperature and the work done per unit mass of air.

4.15   A cylinder containing 0.05 m³ of carbon dioxide at 200 kPa and 100 °C is expanded in a polytropic process to 100 kPa and 20 °C. Determine the work done.

4.16   The pressure in a balloon increases linearly with its diameter. When it is filled with air at 200 kPa its volume is 1 m³. Find the work done when the balloon is inflated to a volume of 1.5 m³.

4.17   Air in the cylinder in Figure P4.17, initially at 200 kPa and 800 K, is allowed to cool to 300 K. The stops divide the initial volume of the cylinder exactly in half. Calculate the boundary work done per unit mass of air.

**Figure P4.17** Air cools in a cylinder with stops.

4.18 A gas obeys the van der Waals equation of state, given by:

$$P = \frac{RT}{v-b} - \frac{a}{v^2},$$

where $a$ and $b$ are constants. The gas is compressed at constant temperature from $v_1$ to $v_2$. What is the work done per kilogram of gas during this process?

*Control Mass Analysis*

4.19 You throw a ball vertically up in the air so that it leaves your hand from a height 1 m above the ground with a speed of 10 m / s. How high will the ball travel and with what velocity will it hit the ground after it falls?

4.20 Ten kilojoules of energy are added to a 1 kg mass of water initially at rest at 25 °C. (a) If the energy is used to increase internal energy, find the increase in temperature. (b) If the energy is used to increase potential energy, find the increase in height. (c) If the energy is used to increase kinetic energy, find the increase in velocity.

4.21 A 10 kg mass sliding on a horizontal surface with an initial velocity of 10 m / s is decelerated by a constant frictional force of 5 N. Find (a) the initial kinetic energy of the mass, (b) the distance travelled by the mass before it comes to rest and (c) the work done by the frictional force in bringing the mass to rest.

4.22 A 10 kg mass sliding horizontally with a speed of 1 m / s has to be brought to rest over a distance of 5 cm after hitting a spring and compressing it (Figure P4.22). What is the spring constant $K$ required?

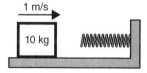

**Figure P4.22** A sliding mass comes to rest after hitting a spring.

4.23 A box slides down a 2 m long ramp inclined at 30° to the horizontal. If the box starts at rest and friction is negligible, what is its velocity when it reaches the bottom of the ramp?

4.24   An electric transformer operating at steady state draws a current of 10 A at 440 V and has an output of 31 A at 120 V. What is the rate of heat loss from the transformer?

4.25   A rechargeable battery draws a current of 1.2 A at 110 V while losing 15 W of heat to the surroundings. If the battery is left to charge for 30 min, how much energy is stored in it?

4.26   An insulated bucket containing 10 kg of water has an immersion heater placed in it. When the heater is plugged into a 110 V electrical outlet it draws a current of 2 A. How much time will it take for the water temperature to rise from 25 to 90 °C?

4.27   A 750 ml bottle of water at 28 °C is placed in a refrigerator. How long will it take for the water to cool to 10 °C if the rate of heat transfer from the bottle surface is 15 W? Neglect the mass of the bottle.

4.28   A 500 W electric heater is embedded in a mild steel block 0.2 × 0.2 × 0.025 m in size that is well insulated on all sides. How long will it take for the block temperature to rise from 25 to 200 °C once the heater is switched on?

4.29   A 20 kg block of aluminium at 150 °C is dropped into an insulated container filled with 45 kg of water at 0 °C. Find the final equilibrium temperature.

4.30   Two kilograms of glycerine at 90 °C are mixed in an insulated tank with 10 kg of water at 25 °C. What is the final equilibrium temperature?

4.31   A 5 kg block of copper heated to 300 °C is allowed to cool in air at 20 °C. Find the time for the temperature to drop to 100 °C. Assume that the rate of heat transfer from the surface of the block is $Q = U(T - T_a)$ where $U = 0.05$ W / °C, $T$ is the temperature of the block and $T_a$ is the temperature of the air, assumed constant.

4.32   A 0.5 kg piece of iron at 300 °C is placed to cool in a sealed, insulated, 1 m³ volume box filled with air initially at 20 °C and 100 kPa. What is the final pressure in the box?

4.33   Two insulated tanks, labelled A and B, are connected by a valve that is initially closed. Tank A contains air at 100 kPa and 350 K and has a volume of 0.75 m³. Tank B contains air at 800 kPa and 1000 K and has a volume of 0.5 m³. Find the final temperature and pressure of the air after the valve is opened and the contents of the two tanks have mixed.

4.34   An insulated cylinder with a volume of 1 m³ is divided into two equal compartments by a piston that is initially locked in place. Compartment A contains oxygen at 100 kPa and 300 K while compartment B contains oxygen at 500 kPa and 800 K. The piston is released and heat transfer takes place through it until equilibrium is reached. Find the final gas temperature and pressure.

4.35   The piston in a 4 cm diameter cylinder filled with compressed air is initially locked in place. When the 5 kg piston is released it accelerates to a velocity of 10 m / s after travelling 5 cm. What is the pressure in the cylinder, assuming it remains constant as the piston moves?

4.36 Nitrogen at 80 kPa and –5 °C is compressed to a pressure of 500 kPa in an isothermal process. Find the work done and heat transfer per unit mass of gas.

4.37 A well-insulated, rigid tank contains 0.5 kg of carbon dioxide. If the gas is stirred with a paddle wheel until its temperature rises by 20 °C, how much work was done on the gas?

4.38 A sealed, perfectly insulated room with a floor measuring 4 × 4 m and a ceiling height of 3 m has a 500 W electric space heater placed inside it. The initial air temperature is 17 °C and the pressure is 100 kPa. What will the room pressure be after 1 h?

4.39 A 200 W computer is left running in a 4 × 5 × 5 m room for 12 h. What will be the increase in temperature of the air in the room that was initially at 100 kPa and 20 °C? Assume that 80% of the energy from the computer is lost through the walls of the room to the surroundings.

4.40 A vertical cylinder has a cross-sectional area of 0.05 m² and a 100 kg piston. The atmospheric pressure acting on the exterior of the piston is 100 kPa. The initial volume of air in the cylinder is 0.05 m³, and this doubles when 10 kJ of heat are added to it. Find the increase in internal energy of the air.

4.41 Four kilograms of gas were heated at a constant pressure of 12 MPa. The gas volumes were 0.005 m³ and 0.006 m³ in the initial and final states, respectively, and 3.9 kJ of heat was transferred to the gas. What is the change in specific internal energy between the initial and final states?

4.42 A cylindrical piston containing 0.1 kg of nitrogen at 200 kPa and 20 °C initially rests on stops. The piston lifts when the pressure in the cylinder reaches 1 MPa. How much heat has to be added to the gas to lift the piston?

4.43 Air in a cylinder is cooled from 800 K to 400 K. The pressure is kept constant at 0.3 MPa during this process. Find the heat transfer during this process given the mass of air is 0.5 kg.

4.44 Five kilograms of air in a cylinder fitted with a piston is heated from 27 to 267 °C by switching on a 200 W electric heater placed under the cylinder. The piston is free to move during this process. How long was the heater kept on?

4.45 One kilogram of oxygen is contained in a cylinder with a piston that moves freely. The initial volume is 0.8 m³ and the pressure is 100 kPa. Heat is added in an isobaric process until the volume of gas increases by 50%. Determine the amount of heat added.

4.46 Two kilograms of methane gas at 120 kPa and 15 °C is compressed in a polytropic process for which $PV^{1.3}$ = constant until its volume is 20% of the initial volume. Find the work done and the heat transfer during this process.

4.47 A rigid tank with a volume of 0.5 m³ contains air at 120 kPa and 300 K. Find the final temperature after 20 kJ of heat is added to the air using (a) constant specific heats and (b) ideal gas tables.

4.48   A rigid, insulated tank contains 0.5 kg of air at 320 K and 100 kPa. An electric heater placed in the tank has a power output that varies with time as $\dot{W} = 2t^2$ where $\dot{W}$ is in watts and $t$ is in seconds. Find the final temperature of the air using the ideal gas tables and the final pressure in the tank, if the heater is switched on for 30 s.

4.49   A tank with a volume of 1 m³ contains 0.2 kg of helium. The gas is initially confined to a section of the tank by a partition, while the remaining portion of the tank is evacuated. The temperature of the helium is 500 K and its pressure is 600 kPa. The partition is removed and the gas allowed to expand and fill the entire tank. The final pressure of the gas is 300 kPa. How much heat transfer occurred during this process?

4.50   The piston in a cylinder containing 0.2 m³ of air at 100 kPa is in contact with a linear spring that is initially not compressed (Figure P4.50). The air is heated until its volume is 0.6 m³ and its pressure is 400 kPa. Show the process on a $P$-$V$ diagram. How much work did the air do during expansion? If the spring was compressed by 0.1 m by the piston, what is the spring constant?

**Figure P4.50**   Air in a cylinder with a piston pushing against a linear spring is heated.

4.51   A cylinder containing 0.8 m³ of air at 300 K and 120 kPa has a piston restrained by a spring so that the gas pressure increases linearly with volume. The air is heated until its temperature reaches 1200 K and the volume is 1.5 m³. How much heat was added?

4.52   Argon gas is contained in a cylinder with an initial volume of 0.2 m³ at 100 kPa and 25 °C. The piston compresses the gas until its pressure is 1 MPa and temperature is 120 °C. During compression 5 kJ of heat were removed from the cylinder. Find the work done during this process.

4.53   An ideal gas is expanded in a polytropic process for which $Pv^\gamma = $ constant, where the specific heat ratio is $\gamma = c_p / c_v$. Prove that this process is adiabatic, assuming $\gamma \neq 1$.

4.54   Air with a mass of 0.2 kg and temperature of 40 °C is contained in a cylinder with a spring-loaded piston, shown in Figure P4.54. The spring constant is $K = 10^6$ N / m and the spring is initially compressed by 10 cm. The gas is cooled by removing 10 kJ of heat, so that its volume is decreased and the spring compression is reduced to 5 cm. What is the final temperature of the air?

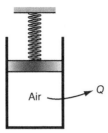

**Figure P4.54**   Air in a cylinder with a spring-loaded piston is cooled.

4.55   Carbon dioxide gas is contained in a cylinder with the piston initially resting on stops, shown in Figure P4.55. The gas pressure is 100 kPa, its temperature is 20 °C and volume is 100 l. The piston lifts off the stops only when the cylinder pressure reaches a pressure of 300 kPa. The gas is heated until its volume becomes 150 l. Find the work done by the gas during expansion, the final gas temperature and the heat added.

**Figure P4.55**   Carbon dioxide gas in a cylinder with the piston initially resting on stops.

*Control Volume Analysis*

4.56   Water flows through a 25 mm diameter pipe with a velocity of 2 m / s, pours out a tap with a 10 mm diameter opening and fills a 50 l tank. What velocity does the water have at the outlet of the tap and how long will it take for the tank to fill?

4.57   Air flows through a pipe with circular cross-section at a mass flow rate of 1.2 kg / s and a velocity of 15 m / s. The gas temperature is 300 °C and its pressure is 0.2 MPa. What is the pipe diameter?

4.58   A room with a floor area of 3 × 4 m and a ceiling height of 3 m is ventilated with an exhaust fan that has a diameter of 20 cm. If the flow velocity at the exit of the fan is 3 m / s how much time will it take for the air in the room to be completely changed?

4.59   During his honeymoon in Switzerland in 1847 James Joule is said to have tried to measure the temperature difference between the water at the top and bottom of the Sallanches waterfall, which has a height of 270 m. What is the maximum temperature rise he would have found?

4.60   Air enters a 34 kW electrical heater at a rate of 0.8 kg / s with negligible velocity and a temperature of 60 °C. The air is discharged at a height 50 m above the inlet at a temperature of 200 °C and a velocity of 50 m / s. What is the work done in the heater?

4.61  An exhaust fan blowing air at 25 °C through a ventilation duct with a square cross-section 0.5 × 0.5 m in size is powered by a 1 kW electric motor. What is the maximum mass flow rate of air in the duct?

4.62  Air enters a hair dryer at a temperature of 20 °C. It is heated by blowing it with a small fan over an electrically heated coil, so that it leaves at 50 °C through an opening with cross-sectional area 50 cm². The heating element of the hair dryer consumes 1 kW and the fan consumes 100 W. What is the maximum exit velocity of the air?

4.63  Air enters a square heating duct, 20 × 20 cm in cross-section, with a pressure of 120 kPa and temperature 35 °C. The volume flow rate of air is 0.2 m³ / s and it loses 1.5 kW of heat as it flows through the duct. Determine the velocity of air and its temperature when it leaves the duct.

4.64  Two streams of air enter a control volume: stream 1 enters at a rate of 0.05 kg / s at 300 kPa and 380 K, while stream 2 enters at 400 kPa and 300 K. Stream 3 leaves the control volume at 150 kPa and 270 K. The control volume does 3 kW of work on the surroundings while losing 5 kW of heat. Find the mass flow rate of stream 2. Neglect changes in kinetic and potential energy.

4.65  An electric heater is wrapped around a 5 cm diameter tube. Air enters the tube at 200 kPa and 290 K with a velocity of 15 m / s. It leaves at 150 kPa and 410 K. Determine the heater power and air exit velocity.

4.66  An air to water heat exchanger has air at 600 °C entering at a rate of 0.5 kg / s and leaving at 300 °C. Water enters at 20 °C and leaves at 80 °C. What is the mass flow rate of water?

4.67  Air entering a nozzle with an inlet area of 45 cm² has a density of 1.9 kg / m³ and a velocity of 20 m / s and it leaves the nozzle with a density of 0.8 kg / m³ and velocity of 95 m / s. Find the mass flow rate of air and the exit area of the nozzle.

4.68  Air enters a nozzle at 300 kPa and 350 K with a velocity of 30 m / s and leaves at 100 kPa with a velocity of 200 m / s. The heat loss from the nozzle is 20 kJ / kg of airflow through it. Find the exit temperature of the air.

4.69  Steam flows through an adiabatic nozzle whose inlet area is 10 times that of its outlet. What are the exit velocity and temperature if at the inlet the velocity is 10 m / s, temperature is 800 °C and pressure is 600 kPa? The outlet pressure is 150 kPa.

4.70  Helium at 100 kPa and 25 °C enters an adiabatic diffuser with a velocity of 200 m / s and mass flow rate of 10 kg / min. It leaves the diffuser with a velocity of 50 m / s. Find the exit temperature and diffuser exit area.

4.71  Air at 90 kPa and 300 K enters a diffuser with a mass flow rate of 20 kg / s and velocity of 260 m / s. The diffuser exit area is 500 cm² and the rate of heat loss through its walls is 40 kW. If the air exit temperature is 320 K, find the exit pressure.

4.72  Air at 80 kPa and 270 K enters an adiabatic diffuser with a velocity of 220 m / s and leaves with a much lower velocity at a pressure of 120 kPa. The inlet diameter of the diffuser is half that of the exit. Find the exit temperature and velocity of the air.

4.73   A pump raises water from the basement of a building, 5 m below street level, to the top floor located 112 m above ground. What is the minimum power required to pump 20 kg of water per second?

4.74   A pump takes oil from a supply tank at a pressure of 200 kPa and delivers it through a 2 cm diameter pipe to a high-pressure hydraulic system at 800 kPa. The mass flow rate of oil is 8 kg / s and its density is 820 kg / m$^3$. What is the power required to drive the pump? Assume the oil is incompressible and the inlet velocity is negligible.

4.75   A 15 kW pump receives water from a 5 cm diameter pipe and sends it with a velocity of 3 m / s into a 10 cm pipe. What is the pressure rise of the water?

4.76   Nitrogen flowing at a rate of 60 kg / min is compressed from 150 kPa and 30 °C to 800 kPa and 150 °C. The compressor is cooled at a rate of 15 kJ / kg of gas during operation. What is the power required to drive the compressor?

4.77   An air compressor takes air at 315 K and compresses it. For every kilogram of air passing through the compressor it does 200 kJ of work and loses 10 kJ of heat. Find the temperature of the air exiting the compressor. Do not assume that the specific heat of air is constant.

4.78   A water-cooled air compressor takes in 0.5 m$^3$ / s of air at 100 kPa and 300 K and delivers it at a pressure of 750 kPa and a temperature of 350 K. The power required to drive the compressor is 120 kW. Due to safety concerns the temperature of the cooling water cannot rise by more than 20 °C. What is the minimum mass flow rate of water required?

4.79   The turbine in a hydroelectric plant takes 30 m$^3$ / s of water from the bottom of a 50 m deep reservoir and discharges it to the atmosphere. What is the maximum power output from the turbine?

4.80   Argon enters an adiabatic turbine at 2 MPa and 600 °C at a rate of 2.5 kg / s and leaves at 150 kPa. Determine the exit temperature of the gas when the turbine is generating 300 kW. Neglect kinetic energy changes.

4.81   Air enters an adiabatic turbine at 800 kPa and 870 K with a velocity of 60 m / s, and leaves at 120 kPa and 520 K with a velocity of 100 m / s. The inlet area of the turbine is 90 cm$^2$. What is the power output?

4.82   A gas turbine operates with 0.1 kg / s of helium that enters at 8 MPa and 600 K and leaves at 200 kPa and 350 K. If the turbine power output is 120 kW find the rate of heat loss from the turbine. Neglect kinetic energy changes.

4.83   A well-insulated tank that was initially evacuated is filled with an ideal gas from a reservoir at temperature $T_1$. Assuming constant specific heats, show that the final temperature in the tank is $\gamma T_1$.

4.84   A 200 l tank is evacuated and then filled through a valve connected to an air reservoir at 1 MPa and 20 °C. The valve is shut off when the pressure in the tank reaches 0.5 MPa. What is the mass of air in the tank?

4.85    A 100 l tank initially contains air at 200 kPa and 300 K. The tank is connected to a
        pipeline in which air at 2 MPa and 300 K is flowing and filled until the pressure in it
        equals that in the pipeline. The tank is cooled during the filling process to keep the gas
        temperature constant at 300 K. How much heat is removed?

4.86    A rigid tank with a volume of 0.8 m³ contains air at 80 kPa and 280 K. It is connected
        through a valve to a supply line in which air at 1 MPa and 280 K is flowing. The valve
        is shut off when the pressure in the tank becomes the same as that in the pipeline, at
        which time the temperature in the tank is 350 K. Find the heat transfer from the tank
        while it is being filled.

4.87    The valve of a nitrogen tank with 0.1 m³ volume was accidentally opened for a few
        minutes, letting 20% of the volume of gas inside leak out before the valve was shut
        again. The initial pressure and temperature of the nitrogen were 1 MPa and 30 °C. Find
        the final temperature and pressure.

# 5

# Entropy

---

**In this chapter you will:**

- Define a new extensive property: entropy.
- State the second law of thermodynamics: the entropy of an isolated system increases until it reaches equilibrium.
- Distinguish between reversible and irreversible processes.
- Relate entropy to the distribution of energy among molecules.
- Understand how the laws of probability determine equilibrium.
- Discuss how entropy changes because of heat and mass transfer across system boundaries.

---

## 5.1 Converting Heat to Work

The air all around us possesses internal energy, so why not take heat from the atmosphere and use that to power a car engine? The car will do work on the surrounding air while moving and thereby return energy to it, satisfying the first law of thermodynamics (Figure 5.1). If we can invent such an engine we will have limitless power at our disposal and never need to refuel our car. This wonderful engine sounds too good to be true, but there is nothing in the principle of energy conservation that forbids it.

To understand why we cannot make an engine that runs without any fuel, let us start by thinking about how to transform heat into work. Our heat source is a combustor that is continuously supplied fuel and generates heat (Figure 5.2a), maintaining a constant temperature ($T_H$). To use this heat to do work we bring the combustor into contact with a cylinder containing gas that expands as its temperature increases, receiving $Q_H$ joules of heat and doing $W$ joules of work in raising the

---

*Energy, Entropy and Engines: An Introduction to Thermodynamics*, First Edition. Sanjeev Chandra.
© 2016 John Wiley & Sons, Ltd. Published 2016 by John Wiley & Sons, Ltd.
Companion website: www.wiley.com/go/chandraSol16

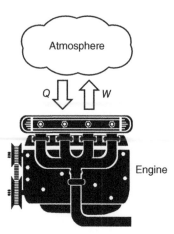

**Figure 5.1**   A proposed engine takes heat ($Q$) from the atmosphere and does an equal amount of work ($W$) on the surroundings.

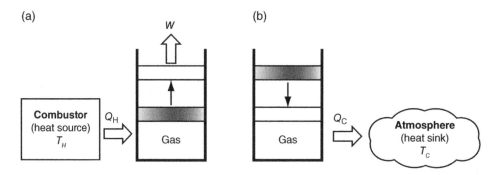

**Figure 5.2**   (a) Heat ($Q_H$) is transferred from a combustor at a high temperature ($T_H$) to gas contained in a cylinder. The gas expands, raises the piston, and does work ($W$) on the surroundings. (b) The gas in the cylinder loses heat ($Q_C$) to the surrounding atmosphere at a lower temperature ($T_C$) until the piston returns to its original position.

piston. The piston will rise until the gas in the cylinder is heated to $T_H$, reaching thermal equilibrium after which there will be no further heat transfer to the gas or movement of the piston. The system has not run out of energy – the combustor can draw as much fuel as necessary – but by definition a system at equilibrium cannot do work on its surroundings.

To use the gas in the cylinder to do more work we have to disrupt the equilibrium and restore the piston to its original position. We can do this by separating the cylinder from the combustor and letting the gas in it cool (Figure 5.2b), dissipating $Q_C$ joules of heat to the surrounding atmosphere at a lower temperature ($T_C$). By repeating this heating and cooling cycle we can continuously do work. An engine always operates between a high temperature heat source and a low temperature heat sink, alternating between the two so that it never stays at equilibrium.

If we reduce the temperature of the combustor the distance the piston moves will decrease and therefore the amount of work done will be less. Conversely, if we increase the temperature of the combustor the gas expands further and the piston rises higher, producing more work. The work output of an engine depends not only on the amount of heat added to it but also the temperature at which it is added, and in the limit that $T_H \to T_C$ the work goes to zero. The energy dissipated to the atmosphere ($Q_C$) does not disappear but since it is at the temperature of the heat sink it is useless for doing more work. Not all heat inputs are equal – the higher the temperature at which heat is supplied, the more effectively it can be used to do work. We need to describe not only the *quantity* of heat that is added to the engine, but also its *quality*.

## 5.2  A New Extensive Property: Entropy

Can we define a property that measures the quality of heat and tells us how efficiently it can be transformed to work? This property will obviously be a function of the amount of heat, but we have just shown that the temperature of the heat source is equally important, so we must combine these two quantities. Let us postulate that every system has an extensive property called *entropy* that changes when heat is added to or removed from it. The magnitude of change is

$$\text{Entropy change} = \frac{\text{Heat transferred}}{\text{Temperature}}.$$

The temperature we use to evaluate entropy is that of the system boundary across which heat transfer occurs. The higher the temperature at which the heat is added to an engine, the smaller the change in its entropy. For efficient energy conversion we would like to minimise the entropy increase of a heat engine when we add a given amount of heat to it.

Rudolf Clausius first identified entropy as a thermodynamic property in 1865, deriving the word from the Greek word for transformation. He deliberately picked a name that sounded similar to energy since the two are closely related, both being extensive properties that increase when we heat a system. When we add an infinitesimal amount of heat ($\delta Q$) to a closed system at temperature $T$, its internal energy increases by $dU = \delta Q$ (assuming no work is done) and its entropy, denoted by the symbol $S$, increases by

$$dS = \frac{\delta Q}{T}, \tag{5.1}$$

where $T$ is the absolute temperature, measured in kelvin (K). The units of entropy are those of energy / temperature, or J / K, in SI units. The change in entropy as a system goes from an initial state 1 to a final state 2 is

$$\Delta S = \int_1^2 dS = \int_1^2 \frac{\delta Q}{T}. \tag{5.2}$$

When calculating entropy changes it is convenient to use the concept of a *thermal reservoir, which is a system whose temperature remains constant and uniform in spite of heat transfer to or from it*. Examples of systems that can be treated as thermal reservoirs are very large masses such as the atmosphere or a lake. A combustor, such as that shown in Figure 5.1, is also a thermal reservoir since its temperature does not change even when it is supplying heat. To calculate the entropy change of a thermal reservoir we take $T$, which is constant, outside the integral in Equation (5.2), so that

$$\Delta S = \frac{1}{T}\int_1^2 \delta Q = \frac{Q_{12}}{T},$$

(5.3)

where $Q_{12}$ is the total amount of heat added to or removed from the system. We can also write this equation on a rate basis,

$$\dot{S} = \frac{\dot{Q}}{T},$$

(5.4)

where $\dot{Q}$, measured in watts (W), is the rate of heat transfer to or from a thermal reservoir and $\dot{S}$ is its rate of entropy change, measured in W / K.

Entropy and energy are related properties, but there is one very important difference between them: *energy is conserved but entropy can be generated*. If thermal reservoir $A$ at temperature $T + \Delta T$ loses $Q$ joules of heat to thermal reservoir $B$ at lower temperature $T$ (Figure 5.3) the entropy decrease of the hotter system $A$ is

$$\Delta S_A = \frac{Q}{T + \Delta T},$$

(5.5)

and the increase in entropy of $B$ is

$$\Delta S_B = \frac{Q}{T}.$$

(5.6)

Let us not worry about the sign of $\Delta S$, since $Q$ is negative for $A$ and positive for $B$, and consider only the magnitude of entropy changes. For this section, therefore, the symbols $Q$, $W$ and $\Delta S$ refer only to magnitudes of heat transfer, work and entropy change, without any signs attached.

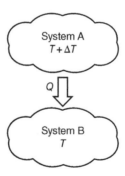

**Figure 5.3**   Heat transfer from a thermal reservoir $A$ at a higher temperature ($T + \Delta T$) to another thermal reservoir $B$ at a lower temperature ($T$).

For heat transfer to occur $\Delta T$ must be positive, so $\Delta S_A < \Delta S_B$. The entropy of system $B$ increases more than the entropy of system $A$ decreases, implying that every time we transfer heat we generate entropy:

$$S_{gen} = \Delta S_B - \Delta S_A. \qquad (5.7)$$

In the limit that $\Delta T \to 0$, $S_{gen} \to 0$. *There is no entropy generated when heat transfer takes place between two thermal reservoirs whose temperatures differ by an infinitesimal amount.* An infinitesimally small temperature difference implies infinitely slow transport of energy so all real heat transfer processes result in the generation of entropy.

For $S_{gen}$ to be negative we need $\Delta T < 0$. Destruction of entropy requires that heat move from a cold to a hot region, which cannot happen spontaneously. *Entropy can be generated but not destroyed.*

If an isolated system is not at equilibrium, so that part of it is hotter than the rest, heat diffuses from the hot region to surrounding areas (Figure 5.4). Entropy is generated due to this heat transfer, and since it cannot leave the system (which is isolated) the system entropy increases. Entropy production continues as long as temperature gradients exist in the system and stops only when equilibrium is reached. *The entropy of an isolated system increases until it reaches equilibrium and then remains constant.*

Another important difference between entropy and energy is that, according to our definition [Equation (5.1)], when we do work on a system its energy increases but its entropy remains the same because entropy is a function of heat, not work. *Doing work on a system does not increase its entropy.* When a mass is pushed up a frictionless ramp its entropy remains constant while its potential energy increases. If, though, there is friction between the mass and the ramp part of the work done on the system is converted to heat, which when dissipated to the surroundings creates entropy.

Let us summarise what we know so far about entropy:

- Entropy is an extensive property.
- The change in entropy of a thermal reservoir being heated or cooled is $\Delta S = Q / T$.
- Entropy is created when heat is transferred through a finite temperature difference.
- Entropy cannot be destroyed.

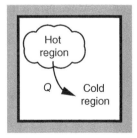

**Figure 5.4**  Isolated system that is not at equilibrium since a portion of it is hotter than the rest. Heat ($Q$) is transferred from the hot region to the cold region and the entropy of the system increases until it reaches equilibrium and its temperature is uniform everywhere.

- The entropy of an isolated system increases until it reaches equilibrium and then remains constant.
- Transferring energy in the form of work does not change the entropy of a system.

**Example 5.1**
**Problem:** A 10 kg mass of lead heated to 200 °C is dropped into a lake at a temperature of 18 °C. Find the increase in entropy of the lake.
**Find:** Increase in entropy of the lake $\Delta S$.
**Known:** Mass of lead $m = 10$ kg, initial temperature of lead $T_1 = 200$ °C, temperature of lake $T_C = 18$ °C.
**Assumptions:** Lead cools down to the temperature of lake, so final temperature of lead $T_2 = 18$ °C, temperature of lake remains constant.
**Properties:** Specific heat of lead $c = 0.128$ kJ / kg°C (Appendix 2).

**Solution:**
Heat lost by lead while cooling is

$$Q_C = mc(T_2 - T_1) = 10\,\text{kg} \times 0.128\,\text{kJ/kg°C}(18\ °\text{C} - 200\ °\text{C}) = -232.96\,\text{kJ}.$$

Heat gained by lake is

$$Q_H = -Q_C = 232.96\,\text{kJ}.$$

Entropy increase of lake is

$$\Delta S = \frac{Q_H}{T} = \frac{232.96\,\text{kJ}}{(273.15 + 18)\text{K}} = 0.800\ 14\,\text{kJ/K}.$$

**Answer:** The entropy of the lake increases by 0.800 kJ / kg.                                    ■

## 5.3   Second Law of Thermodynamics

The fact that entropy increases cannot be derived from first principles: it is a consequence of the observation that the direction of spontaneous heat transfer is from higher temperatures to lower temperatures, but not the other way around. We therefore accept it as a postulate that cannot be proven but we assume to be true. Thus, the second law of thermodynamics states that:

*The entropy of an isolated system will increase until the system reaches a state of equilibrium. The entropy of an isolated system in equilibrium remains constant.*

Mathematically we can express this statement as:

$$dS_{\text{isolated}} > 0 \text{ for an isolated system not in equilibrium,}$$
$$dS_{\text{isolated}} = 0 \text{ for an isolated system in equilibrium,} \qquad (5.8)$$
$$dS_{\text{isolated}} < 0 \text{ not possible for an isolated system.}$$

## 5.4   Reversible and Irreversible Processes

The universe is not always symmetric. We can move forward or backward in space, but not in time, which only advances. Heat flows from high to low temperatures and water from high to low elevations, never the other way around. How do we differentiate between processes that occur spontaneously in nature, such as heat transmission from a hot to a cold body, and those that we know cannot happen, such as the movement of heat from cold to hot? The first law of thermodynamics does not prohibit the movement of heat in either direction: it only requires that total energy be conserved. According to the second law, however, real processes are those that produce an increase in entropy while those that require entropy to be destroyed, such as heat transfer from a low temperature to a high temperature, cannot occur. The second law complements the first and adds to our understanding of how the universe functions.

When modelling physical systems it is often useful to define idealised processes that do not produce any entropy. Assume that the pulley in Figure 5.5 is perfectly frictionless. If a force $F = mg$ pulls on the rope the mass $(m)$ will be in equilibrium and remain stationary. If the force is increased to $F + \Delta F$, where $\Delta F > 0$, the mass will accelerate as it moves upwards since there is an unbalanced force acting on it. The force does work in lifting the weight at the end of the rope, which is stored partly as an increase in potential energy of the object and partly as kinetic energy since it has a finite velocity. If we want to recover this energy we first have to bring the weight to rest and then lower it to its original position. If we stop pulling on the rope the object will continue to rise until it is brought to rest by gravity and then fall until arrested with a jerk by the rope. The kinetic energy of the mass is lost as heat to the surrounding atmosphere, generating entropy. We therefore cannot recover all the energy that was used initially to lift the weight.

As $\Delta F$ is reduced the acceleration of the mass decreases and, within the limit that $\Delta F \to 0$, it will rise infinitely slowly, in a quasi-equilibrium process, so that it can be brought to rest with zero heat lost or entropy generated. If the force on the rope is reduced from $F + \Delta F$ to $F - \Delta F$, the weight will be gradually lowered to its initial position so that the system does work on the surroundings, and all the energy is recovered. *A reversible process is one that can be*

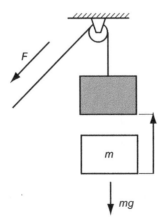

**Figure 5.5**   Mass $(m)$ raised vertically by a force $(F)$.

*reversed by an infinitesimal change in the surroundings so that both the system and surroundings are restored to their initial conditions.* Quasi-equilibrium processes, without any dissipation of energy in the form of heat, are reversible. Calculating the work done during a reversible processes allows us to determine what is the most efficient way of carrying out an energy transfer, with no wasteful conversion of work to heat.

Once the weight was raised and lowered to its initial position the energy of both system and surroundings were brought back to their initial values. There was no change in entropy of either the system or surroundings since there was no heat transfer (i.e., it was an adiabatic process). *Processes that are both reversible and adiabatic produce no change in entropy and are called isentropic processes.*

A more realistic analysis of the process of raising a weight with a rope and pulley has to account for friction in the pulley. When the rope is pulled to hoist the weight, some work will have to be done to overcome friction. The work done against friction is dissipated to the surroundings as heat. If we define the system as including both the pulley and object being raised, as in Figure 5.6a, then there is heat transfer ($Q_1$) from the pulley to the surroundings, while work ($W_1$) is done by the surroundings on the system. In the reverse process (Figure 5.6b) there is further heat transfer ($Q_2$) from the pulley to the surroundings, while work ($W_2$) is done by the system on the surroundings.

An energy balance for the complete raising and lowering process gives that

$$W_1 - W_2 = Q_1 + Q_2 \tag{5.9}$$

and, therefore, $W_1 > W_2$. The system returned less work to the surroundings than it received, converting the difference to heat that was lost to the surroundings. The entropy of the surroundings increased because of heat transfer from the system. This is an example of an *irreversible process*, since the system and surroundings were not restored to their initial state: reversing the process led to an increase in the entropy of the surroundings.

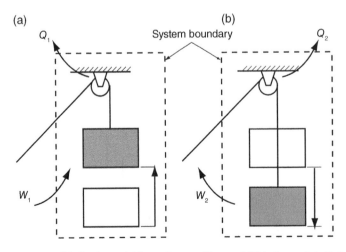

**Figure 5.6**   A pulley with friction is used (a) to raise an object and (b) to lower it to its initial position.

*A reversible process generates no entropy while an irreversible process generates entropy.* If the entropy generated during a process is denoted by $S_{gen}$,

$$S_{gen} = 0 \text{ for a reversible process,}$$
$$S_{gen} > 0 \text{ for an irreversible process.}$$

(5.10)

Entropy gives us a scale with which to measure the magnitude of irreversibility. How do we quantify the amount of friction in a pulley? By the amount of entropy created as it turns. If we oil the pulley and reduce friction the entropy generated will also decrease. The concept of irreversibility serves as a bridge between the idealized world we model in thermodynamics and the real world in which frictionless pulleys and pistons do not exist and all physical processes are irreversible.

During an irreversible process entropy may be generated in the system, the surroundings, or both. If no entropy is generated within the system it is known as an *internally reversible process*. Such a process may still produce entropy but it is created outside the system and is known as an *external irreversibility*. Sources of irreversibility, such as friction in a pulley, are typically difficult to measure and it is usually easier to exclude them from the analysis if possible. Since we are free to define the system boundary as we like it is often convenient to locate it such that the source of irreversibility lies outside the system. For example, Figure 5.7a shows a system consisting of a weight being lifted by a pulley with friction. The system boundary encompasses both the weight and the pulley so that the entropy created by friction in the pulley is internal to the system. If the system boundary excludes the pulley, as in Figure 5.7b, and the weight is raised in a quasi-equilibrium manner, entropy is created only outside the system and the process is internally reversible. Such an idealisation allows us to calculate the minimum amount of work required for raising the weight and the energy required to compensate for frictional losses can be accounted for separately.

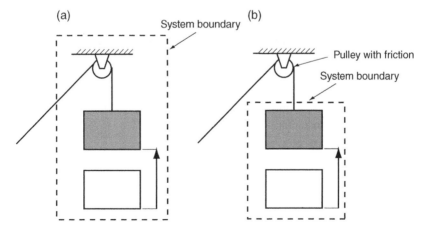

**Figure 5.7**   A weight being lifted using a pulley that has friction and creates entropy. (a) If the system boundary includes the pulley the process is internally irreversible. (b) If the system boundary excludes the pulley the process is internally reversible.

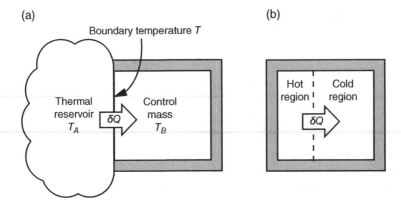

**Figure 5.8**   A control mass at an initial temperature $(T_B)$ is (a) brought into thermal contact with a thermal reservoir at a higher temperature $(T_A)$ and then (b) separated and kept isolated. An infinitesimal amount of heat $(\delta Q)$ is transferred when the two systems are in contact. Entropy $dS = \delta Q/T$ is transferred to the control mass from the reservoir. More entropy $(dS_{gen})$ is generated as the heat dissipates in the control mass.

Heat transfer may also be reversible or irreversible. In Figure 5.3, when heat transfer occurs from a thermal reservoir at temperature $T + \Delta T$ to one at lower temperature $T$, entropy is generated. Since this is created outside the boundaries of the two systems it is an external irreversibility. As $\Delta T \rightarrow 0$ the process becomes reversible since $S_{gen} \rightarrow 0$. An infinitesimal decrease in the thermal reservoir temperature to $T - \Delta T$ will restore the system and surroundings to their original state.

We can also have internal irreversibilities during heat transfer. Figure 5.8 shows a control mass that is at an initial temperature $(T_B)$ is brought into close thermal contact with a thermal reservoir at a higher temperature $(T_A)$ for a short interval of time and then separated and kept isolated.

If $\delta Q$ joules of heat are transferred when the two systems are in contact, the entropy of the control mass increases by $dS = \delta Q/T$, where $T$ is the temperature of the boundary across which heat transfer occurs. This energy will initially increase the temperature of the control mass nearest the boundary. Then, as heat propagates throughout the mass, more entropy will be generated so that the total entropy change of the system is $dS = \delta Q/T + dS_{gen}$. Since entropy is generated within the system, this is an internal irreversibility. The entropy increase of a system undergoing heat transfer is:

For an internally reversible process

$$dS = \left(\frac{\delta Q}{T}\right)_{int\ rev} . \tag{5.11}$$

For an internally irreversible process

$$dS = \left(\frac{\delta Q}{T}\right)_{irr} + dS_{gen}\ ; \tag{5.12}$$

and since entropy generated is always positive

$$dS > \left(\frac{\delta Q}{T}\right)_{irr}.$$

(5.13)

## Example 5.2

**Problem:** A 30 kg mass is pushed horizontally, sliding a distance of 8 m across the floor. The force of friction resisting motion is $F_f = 0.1F_N$, where $F_N$ is the normal force between the mass and the floor. Find the entropy increase of the surrounding atmosphere, which is at 25 °C.

**Find:** Increase in entropy of the atmosphere $\Delta S$.

**Known:** Mass of body $m$ = 30 kg, distance moved $\Delta x$ = 8 m, temperature of atmosphere $T_{atm}$ = 25 °C = 298.15 K.

**Assumptions:** The work done against friction is lost as heat to the surrounding air.

**Solution:**

Frictional force is

$$F_f = 0.1F_N = 0.1mg = 0.1 \times 30\,\text{kg} \times 9.81\,\text{m}/\text{s}^2 = 29.43\,\text{N}.$$

Work done against friction is

$$W = F_f \Delta x = 29.43\,\text{N} \times 8\,\text{m} = 235.4\,\text{J}.$$

All work done against friction is heat transferred to the atmosphere:

$$Q = W.$$

Entropy increase of the atmosphere is

$$\Delta S = \frac{Q}{T_{atm}} = \frac{235.4\,\text{J}}{298.15\,\text{K}} = 0.7895\,\text{J}/\text{K}.$$

**Answer:** The entropy of the atmosphere increases by 0.79 J / K.                                        ∎

## 5.5  State Postulate

What information do we need about a system to calculate its entropy? To fully specify the thermodynamic state of a system we must know its mass and energy. Energy is always measured relative to some arbitrary reference state in which the energy of the system is set to zero. We must specify one property for each possible way of doing work (e.g., boundary work, shaft work, electrical work, etc.) to determine the change in energy from the datum state by work. Heat transfer can also change system energy, and we can quantify this by specifying the internal energy of the system. Therefore, to fix the state of a system we need to know its mass, internal energy, and one property for each possible work mode. We are most interested in simple compressible systems, whose only work mode is expansion or compression, for which the amount of work done is determined by its volume. This conclusion is summarised by the *state postulate*:

*The equilibrium state of a pure, simple compressible substance is completely described by its mass (m), volume (V) and internal energy (U).*

Given $m$, $V$ and $U$ we can, in principle, calculate any other property. Since entropy is an extensive property,

$$S = S(U,V,m). \tag{5.14}$$

Equation (5.14) is known as the *fundamental equation*, since it contains all the information necessary to calculate the thermodynamic properties of a system. Unfortunately, we usually do not have enough information to write the fundamental equation for most systems. Instead, we try to deduce information about thermodynamic properties from equations of state, linking entropy to changes in temperature, pressure and volume.

We can define an intensive property, specific entropy, which is the entropy per unit mass of the system with units of J / kgK:

$$s = \frac{S}{m}, \tag{5.15}$$

so that

$$s = s(u,v). \tag{5.16}$$

We can rearrange this relation to write $u = u(s,v)$ or $v = v(u,s)$. In general, *the state of a simple compressible system is fixed if we know any two independent, intensive properties.*

## 5.6   Equilibrium in a Gas

A rigid vessel is evacuated and connected to a large gas reservoir (Figure 5.9). The valve regulating flow through the connecting pipeline is opened and gas begins to flow into the empty tank. How does pressure in the vessel change with time? This deceptively simple question is impossible to answer. In practice, the pressure would depend on where it is measured within the vessel. If a pressure transducer is placed at $P_1$, it will read a higher pressure than one at $P_2$. Almost all gas molecules entering the vessel through the pipe impinge directly onto $P_1$ but only a few of them will rebound and hit $P_2$ while the vessel is relatively empty. It is meaningless to talk about the pressure or temperature of a system that is not at equilibrium since these properties may vary from one place to another within the system. Only when gas flow is shut off and the system is isolated from the surroundings will readings from the two pressure sensors begin to converge to the same value; then after a sufficiently long time, when the system is at equilibrium, we can define the system pressure. The equilibrium gas pressure $(P_{eq})$ is a system property that depends only on the mass, volume and temperature of the gas and can be calculated from an equation of state such as the ideal gas equation. Once equilibrium is reached it no longer matters how rapidly the tank was filled or what the supply pressure was.

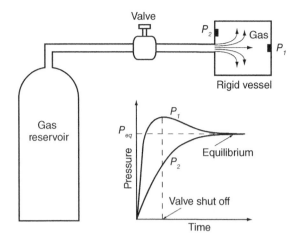

**Figure 5.9**   A rigid vessel is evacuated and then filled with gas by connecting it to a large reservoir. $P_1$ and $P_2$ are two pressure sensors that initially give different values, but converge when equilibrium is reached.

How do we know that a system is at equilibrium? In this example we monitored the pressure and determined when it no longer changed with time. But did this really establish that the system was at equilibrium? A system is not necessarily at equilibrium simply because a single property is constant: the temperature in the tank may have been constant while its pressure was still changing.

The second law of thermodynamics gives us a criterion to determine if an isolated system is at equilibrium: entropy increases until the system reaches equilibrium and then remains constant. Given enough information about the system we can use the state principle to calculate what the final entropy will be and by tracking $S$ we can determine when equilibrium is reached.

What does it mean, at a molecular level, for a gas to be in equilibrium? A gas is made up of molecules that are in incessant motion, changing energy and velocity every time they collide. Yet these random interactions produce constant values of macroscopic properties such as temperature, pressure and internal energy. If the properties of its molecules are constantly changing, how can we say the system is in equilibrium? What changes occur between the instant the valve is shut off in Figure 5.9 and the time when we decide gas properties have reached equilibrium?

A gas sealed in a rigid, insulated container is an isolated system whose total mass and energy are fixed. However, the individual molecules that make up the gas are constantly colliding with each other and the walls of the container and their velocities and energies are changing. Quantum mechanics tells us that the energies of individual molecules do not vary continuously, but can only occupy certain discrete energy levels. For this discussion we do not need to know the exact values of these energy levels – we would have to solve the Schrödinger equation to obtain those – but let us denote them by $\varepsilon_1$, $\varepsilon_2$, $\varepsilon_3$ and so on.

The molecules of gas are distributed among available energy levels subject to the constraints that the total number of molecules equals $n$ and the sum of their energies remains constant and equal to the internal energy, $U$. Therefore, if the number of molecules in any energy level $\varepsilon_i$ is $n_i$ then

$$n = \sum_i n_i \qquad (5.17)$$

and

$$U = \sum_i \varepsilon_i n_i. \qquad (5.18)$$

There are many different arrangements of molecules within the energy levels that will satisfy Equations (5.17) and (5.18). Each individual distribution of molecules among the available energy states, which satisfies the constraints imposed, will be known as a *microstate* of the system. If, as a simple example, a system of five molecules has a total energy of 10 units, it can be divided between the five available energy states shown in Figure 5.10 in five different ways. The number of microstates is $\Omega = 5$ for this system. We cannot directly observe changes in microstates but only measure the total energy of the system, which remains constant at equilibrium. Therefore the *macrostate* of the system, defined by macroscopic properties such as the system energy and volume, remains unchanged even as molecules move from one microstate to another.

How does the system shown in Figure 5.9 reaches equilibrium once the gas supply line is shut off? If we assume that gas molecules in the pipeline all have velocity and temperature lying within a narrow range, they must have similar energies. To keep the discussion simple let us idealise the system and assume that all molecules entering the vessel have identical energy. A histogram of the energy distribution is shown in Figure 5.11a where initially all the molecules occupy the same energy level ($\varepsilon_s$). Once the gas is sealed in a rigid container the velocities of the molecules will begin to change as they collide with each other and the surrounding surfaces. Some molecules move faster and others slower after collisions, depending on the impact conditions. The number of energy levels occupied by the molecules will increase, so that the distribution will become like that shown in Figure 5.11b. Finally, after sufficient time, the system will reach equilibrium and the distribution of molecules will look like that in Figure 5.11c.

What will the equilibrium distribution be? The English physicist James Clerk Maxwell used a novel approach to calculate the equilibrium energy distribution in an ideal gas. Since it is

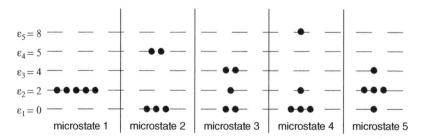

**Figure 5.10**   Different distributions of five molecules that give a total energy of 10 units.

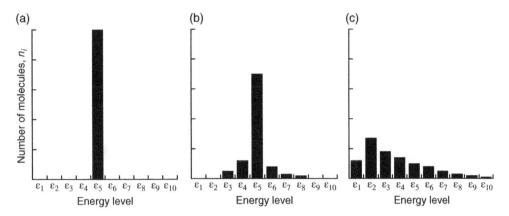

**Figure 5.11** Distribution of gas molecules among energy levels: (a) initial state with all molecules having the same energy, (b) intermediate state and (c) equilibrium distribution.

impossible to gather enough information about the state of an individual molecule to calculate its energy, he solved the problem using probability theory. If we flip a coin 1000 times we have no way of knowing for certain whether any particular toss will produce heads or tails, but we can predict with great confidence that in total we will obtain 500 heads and 500 tails. Similarly, though we cannot calculate the energy of any individual molecule, if we assume that all microstates are equally probable we can estimate the number of molecules in each energy level. Maxwell calculated that for a total of $n$ molecules in the gas the fraction $\Delta n_\varepsilon / n$ with energies lying between $\varepsilon$ and $\varepsilon + \Delta\varepsilon$ is given by

$$\frac{\Delta n_\varepsilon}{n} = \frac{2}{\sqrt{\pi}} (kT)^{-3/2} \varepsilon^{1/2} \exp\left(-\frac{\varepsilon}{kT}\right) \Delta\varepsilon. \tag{5.19}$$

This is the famous Maxwell–Boltzmann distribution, which describes how the internal energy of an ideal gas at equilibrium is distributed among its molecules. Once a gas has reached equilibrium the energy of individual molecules fluctuates constantly as they collide and rebound. In spite of this incessant turmoil the number of molecules in each energy level at any instant remains virtually constant, with fluctuations that are too small to detect. Figure 5.12 shows plots of Equation (5.19), evaluated at both $T = 200$ K and $T = 400$ K. The distribution is not symmetric, because molecular energies have a lower bound of zero (they cannot be negative), but the upper bound is very large – in principle one molecule could have all the energy of the gas while all other molecules had none. The population of molecules in the highest energy levels is very low, because if a few molecules have very high energy, the others are forced to occupy low energy states to keep the total system energy constant. There are very few microstates corresponding to such an unequal arrangement. Conversely, there are a lot of ways of arranging molecules if they share energy more equitably. If gas temperature is raised the average energy of molecules becomes greater and more molecules occupy higher energy levels. The number of system microstates increases with temperature.

Maxwell's results were rather difficult for nineteenth century physicists to accept, since they were more used to solving equations than playing games of chance. Fortunately the

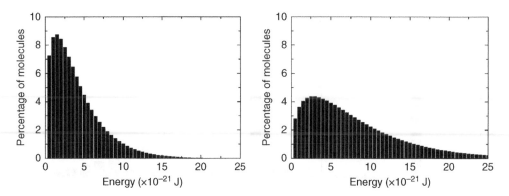

**Figure 5.12**  Equilibrium energy distributions in a gas at (a) 200 K and (b) 400 K.

theory was easy to test. Equation (5.19) does not have the mass of the molecules in it – it predicts that the energy distribution of all gases will be the same, depending only on temperature. If this is true molecules of a lighter gas must move faster than those of a heavier gas for them to have the same kinetic energy. For example, if one mole of oxygen and one mole of hydrogen are kept in identical containers at the same temperature, their average molecular kinetic energies must be the same. Therefore,

$$\frac{1}{2} m_{H_2} \left( \mathbf{V}_{H_2} \right)^2 = \frac{1}{2} m_{O_2} \left( \mathbf{V}_{O_2} \right)^2 ,$$

$$\left( \frac{\mathbf{V}_{H_2}}{\mathbf{V}_{O_2}} \right)^2 = \frac{m_{O_2}}{m_{H_2}} = \frac{32}{2} = 16,$$

$$\frac{\mathbf{V}_{H_2}}{\mathbf{V}_{O_2}} = 4.$$

The average velocity of hydrogen molecules is four times that of oxygen molecules at the same temperature. If we drill equal-sized holes through the walls of two gas containers containing hydrogen and oxygen respectively, hydrogen will leak out four times as fast as oxygen so the pressure in the hydrogen container will fall proportionately faster. This was a relatively simple experiment to do, and the results proved to agree exactly with Maxwell's predictions, convincing many of his critics that statistical analysis was a reasonable way of doing physics.

We are still left with many questions. Why do gas molecules remain in the Maxwell–Boltzmann distribution? If all microstates are equally probable, why do we not see the gas ever revert to the original state where the total energy was distributed equally between all molecules? Why do changes in the energy distribution of molecules always proceed in one direction so that irrespective of the starting condition, they always reach the same equilibrium state? If we can answer these questions we will be closer to our goal of understanding how a gas reaches equilibrium.

## 5.7    Equilibrium – A Simple Example

A state of equilibrium is defined only at a macroscopic level, by averaging the properties of a large number of molecules. A system is said to be at equilibrium when these averages – represented by properties such as pressure, temperature, internal energy and entropy – are constant. To relate macroscopic to microscopic properties we need to apply statistical analysis, which the field of statistical thermodynamics does rigorously, but that level of mathematical analysis is not necessary for our purposes since we can explain the necessary concepts using a much simpler analogy. We will analyse a much simpler system consisting of ten dice, each initially displaying the number 1 as shown in Figure 5.13. The state of the

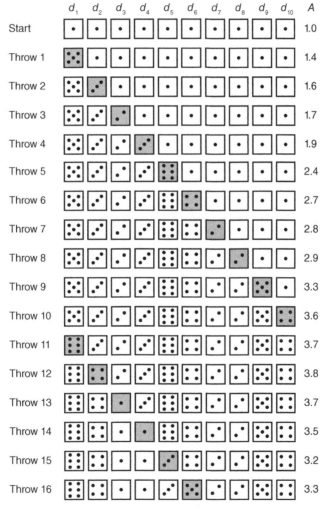

**Figure 5.13**   A system of 10 dice, all having a value of 1 initially. The shaded die is thrown in each turn to give a new value. $A$ is the average value of the dice which increases until all the dice are thrown and then remains approximately constant with a value fluctuating about 3.5.

system is defined by the numbers on the upper faces of the dice and changes as each die is picked up in turn and thrown to get a new number.

How does this set of dice relate to a gas? In our analogy dice represent molecules. A molecule in a gas occupies one of a number of discrete energy levels, while each die can be in one of six different states, given by the number on its upper face. Gas molecules jump from one energy state to another as they collide with each other and with the walls of the container; the state of the dice changes as they are thrown, one after another. A gas in a vessel can be described in two different ways: by listing the position and energy of each individual molecule, an impossible task, or by looking at macroscopic properties, such as pressure or temperature, which are averages of the properties of the molecules in a gas. We can describe our system of dice by either listing all ten values $(d_1, d_2, ..., d_{10})$ or, if we want to represent the system by a single number, by averaging their value and calculating the arithmetic mean $(A)$.

Figure 5.13 shows how the state of the system evolves as the dice are thrown one by one. Each row shows the set of dice with their values indicated by $d_1$, $d_2$ and so on. We specify the initial state of the system, shown in the first row of Figure 5.13, as all the dice having value 1 so that $A = 1$. The first die, shown shaded in the second row, is picked up and thrown. For purposes of illustration we assume $d_1 = 5$, so that the average value $A = 1.4$. The system may now be described in two different ways: we can specify the microstate, which is a list of the individual dice values (5, 1, 1, 1, 1, 1, 1, 1, 1, 1) or, more concisely, we can describe the macrostate, which is 1.4. Next, the second die is thrown and shows a value $d_2 = 3$ and the average value rises to $A = 1.6$. The third throw gives $d_3 = 2$ and $A = 1.7$.

Figure 5.14 plots the variation of $A$ as the game progresses. By the time we reach the 10th (and last) die, $A$ has increased to 3.6. After all the dice have been thrown, on the 11th throw we go back to the first die and toss it again. The system is now in a state of equilibrium in that the average fluctuates about a value of 3.5, even though values of individual elements are constantly altering.

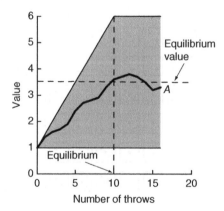

**Figure 5.14**  Variation of the average value of the system of 10 dice as they are thrown one after another. The shaded area is the range of possible values of $A$.

When we threw the first die its value could vary randomly from 1 and 6 and $A$ could lie anywhere between 1 and 1.5. After the second throw $A$ could be between 1 and 2: the shaded area in Figure 5.14 demarcates this range of possible values of $A$. Once all 10 dice have been thrown $A$ may lie, in theory, anywhere between 1 and 6, so why does the average always remain close to 3.5? Why does it not suddenly jump to 5.5 or fall to 1.5?

Not all values of $A$ are equally probable. The total number of ways we can arrange 10 dice is $\Omega_{tot} = 6^{10} = 60\ 466\ 176$. Of these more than 60 million possible microstates of the system, only one corresponds to an average of 1. Similarly, only one microstate gives an average of 6. The odds of getting either of these values, as any gambler will tell you, are negligibly small. Most of the other possible combinations of dice will give averages that are close to the mean value of $A = 3.5$. If the number of microstates that give a particular average $A$ is $\Omega_A$, the probability of getting $A$ is

$$\text{Probability of getting } A = \frac{\text{number of microstates giving average } A}{\text{total number of possible microstates}} = \frac{\Omega_A}{\Omega_{tot}}. \qquad (5.20)$$

Table 5.1 shows the number of microstates corresponding to different averages when ten dice are thrown, and the probability of obtaining that average. We will not delve into the details of how to obtain these numbers, but it is not very difficult to write a computer program that lists all the possible ways in which 10 dice can be arranged and calculate their average value.

Only one microstate corresponds to $A = 1$, and the probability of getting this in a random throw is only $1.65 \times 10^{-6}\%$. The odds of getting $A = 3$ are almost three million times higher, at 4.85%, and the highest probability, 7.27%, is of getting $A = 3.5$. It is not surprising that the average values of the dice cluster about this value at equilibrium: there is about a 70% chance of getting an average value between 3 and 4.

As the number of dice in the system increases our ability to predict the average value gets better. Figure 5.15 shows the distribution of probabilities for the average value obtained after

**Table 5.1**   Number of microstates ($\Omega_A$) corresponding to different averages ($A$) when 10 dice are thrown and the probability of obtaining that average.

| Average, $A$ | Number of microstates, $\Omega_A$ | Probability, $(\Omega_A / \Omega_{tot})\%$ |
|---|---|---|
| 1.0 | 1 | $1.65 \times 10^{-6}$ |
| 1.5 | 2002 | $3.31 \times 10^{-3}$ |
| 2.0 | 85 228 | 0.141 |
| 2.5 | 831 204 | 1.37 |
| 3.0 | 2 930 455 | 4.85 |
| 3.5 | 4 395 456 | 7.27 |
| 4.0 | 2 930 455 | 4.85 |
| 4.5 | 831 204 | 1.37 |
| 5.0 | 85 228 | 0.141 |
| 5.5 | 2002 | $3.31 \times 10^{-3}$ |
| 6.0 | 1 | $1.65 \times 10^{-6}$ |

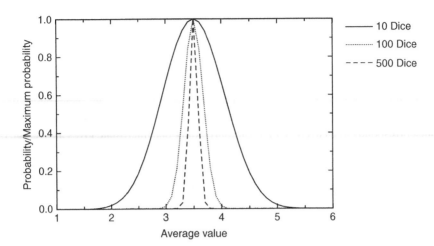

**Figure 5.15** Probability of obtaining a particular average value when throwing sets of 10, 100 and 500 dice.

throwing 10, 100 and 500 dice. To show all the curves in one graph we normalise them by dividing probability values by the maximum probability, so that the peaks of all the curves coincide at a value of 1. As the number of dice increases the probabilities cluster closer around the central value of 3.5. For 100 dice the probability of getting a total between 3 and 4 is over 99.7%. For a system with one million dice we could predict with near certainty that once all of them have been thrown, their average will be 3.5. Why is this? For the average to be significantly different from 3.5 when we throw one million dice a very large proportion of the dice must land with faces showing extreme values such as 1 or 6. This is so implausible that we can safely disregard the possibility of it happening. Suppose that when a set of 10 dice are thrown obtaining three sixes significantly skews the average. The probability of this happening is $(1/6)^3 = 0.5\%$: small, but not impossible. With a million dice, the probability that 30% of them show sixes would be $(1/6)^{300\,000}$, a number so small that we cannot use an ordinary calculator to evaluate it. If the number of microstates corresponding to the equilibrium state is $\Omega_{eq}$, we can quite reasonably assume that for large systems $\Omega_{eq} \approx \Omega_{tot}$, meaning that almost all the states of the system correspond to an average of 3.5. Other states such as one in which a million dice all show sixes, though possible, are so uncommon as to be of little interest.

*The equilibrium macrostate of any system is the one that has the largest number of corresponding microstates.* Though the individual values of dice in our simple example change each time they are thrown, so many of the combinations give an average value of 3.5 that it would be very rare for any other number to appear. It does not matter what the initial state of the system was, it will always reach the same equilibrium. If we had started with the dice all showing sixes, they would still eventually reach an average of 3.5.

How do we define a macroscopic property, $S$, that can be used to determine when a system has reached equilibrium? At equilibrium $\Omega$ is at its maximum, so we could simply assume that

$$S = \Omega. \tag{5.21}$$

In our example of ten dice we start with $S = 1$ since there is only one possible initial microstate. As the system evolves and passes through states with more microstates, $S$ will increase. Finally, at equilibrium $S_{eq} = \Omega_{eq}$. $S$ is now at a maximum value and does not change further. If we are dealing with systems that have a large number of elements, $\Omega_{eq} \approx \Omega_{tot}$, so at equilibrium

$$S_{eq} = \Omega_{tot}. \tag{5.22}$$

The advantage of this assumption is that we can calculate $\Omega_{tot}$ without detailed information about the system at every instant: if we know the number of elements in it (10 dice in our example) and the number of states for each element (6 for each of the dice) we can calculate $S$ at equilibrium ($S_{eq} = 6^{10}$).

We have succeeded in finding a property that we can use to determine equilibrium. There is, though, still one problem: $S$, according to this definition, is not an extensive property since extensive properties need to be additive. If we have two systems, $A$ and $B$, each consisting of 10 dice, then at equilibrium

$$S_A = S_B = 6^{10}.$$

But if we combine the two systems, to get a system $C$ with 20 dice,

$$S_C = (S_A)(S_B) = 6^{20} \neq S_A + S_B.$$

To calculate $S_C$ we cannot simply add $S_A$ and $S_B$, as we would like. We can remedy this shortcoming by adopting a slightly different definition of $S$ that would make it an extensive property:

$$S = \ln \Omega \tag{5.23}$$

This expression for $S$ makes it an extensive property so that

$$S_C = S_A + S_B. \tag{5.24}$$

We can now use this definition of $S$ to determine when our system of dice reaches equilibrium. Table 5.2 lists the values of $A$ as the dice are thrown. Also listed are the corresponding values of $\Omega$ and $S$. We know that the maximum number of microstates, corresponding to the equilibrium value $A = 3.5$, is $\Omega_{eq} = 4\,395\,456$ (see Table 5.1). Therefore at equilibrium $S_{eq} = \ln \Omega_{eq} = 15.296$. If the number of elements in our system had been larger we could have assumed that $S_{eq} = \ln \Omega_{tot} = 17.918$. Ten dice is too few for this assumption to be very good, but with larger numbers it starts getting much better.

We track the evolution of our system starting with $A = 1$, when $S$ has a value of 0 since there is only a single possible microstate. $S$ grows larger as more dice are thrown and $\Omega$ increases, until after the 10th throw $S = 15.280$. Given the small size of the system, this is close enough to $S_{eq}$ to declare that we have reached equilibrium. On subsequent throws $S$ fluctuates about this value, confirming that it is the equilibrium value.

We now know how to describe equilibrium for our simple system of dice. How do we define $S$ for the far more complicated case of a gas consisting of an immense number of molecules?

Table 5.2    Number of microstates ($\Omega$) corresponding to different averages ($A$) when 10 dice are thrown and the corresponding value of entropy ($S$).

| Throw number | Average, $A$ | Number of microstates, $\Omega$ | $S = \ln\Omega$ |
|---|---|---|---|
| Start | 1.0 | 1 | 0.000 |
| Throw 1 | 1.4 | 715 | 6.5723 |
| Throw 2 | 1.6 | 4995 | 8.5162 |
| Throw 3 | 1.7 | 11 340 | 9.3361 |
| Throw 4 | 1.9 | 46 420 | 10.745 |
| Throw 5 | 2.1 | 147 940 | 11.905 |
| Throw 6 | 2.7 | 1 535 040 | 14.244 |
| Throw 7 | 2.8 | 1 972 630 | 14.495 |
| Throw 8 | 2.9 | 2 446 300 | 14.710 |
| Throw 9 | 3.3 | 4 121 260 | 15.232 |
| Throw 10 | 3.6 | 4 325 310 | 15.280 |
| Throw 11 | 3.7 | 4 121 260 | 15.232 |
| Throw 12 | 3.8 | 3 801 535 | 15.151 |
| Throw 13 | 3.7 | 4 121 260 | 15.232 |
| Throw 14 | 3.5 | 4 395 456 | 15.296 |
| Throw 15 | 3.2 | 3 801 535 | 15.151 |
| Throw 16 | 3.3 | 4 121 260 | 15.232 |

## Example 5.3

**Problem:** A set of four identical coins is flipped repeatedly. If each coin has equal probability of showing either heads or tails, how many microstates and how many macrostates (defined by the number of heads and tails) are possible? What is the probability of each macrostate and which is the most likely macrostate? Define $S$ for this system and calculate its value at equilibrium.

**Find:** Entropy $S$ for the system and its value at equilibrium.

**Solution:** List all the possible ways in which four coins can be arranged.

| Arrangements | Macrostate | Number of microstates $\Omega$ | Probability $\Omega / \Omega_{tot}$ | $S = \ln\Omega$ |
|---|---|---|---|---|
| HHHH | 4 Heads | 1 | 0.0625 | 0 |
| HHHT HHTH HTHH THHH | 3 Heads 1 Tail | 4 | 0.25 | 1.39 |
| HHTT HTHT TTHH THTH HTTH THHT | 2 Heads 2 Tails | 6 | 0.375 | 1.79 |
| TTTH TTHT THTT HTTT | 1 Head 3 Tails | 4 | 0.25 | 1.39 |
| TTTT | 4 Tails | 1 | 0.0625 | 0 |

The total number of microstates $\Omega_{tot} = 16$. The most likely macrostate is one with two heads and two tails for which there are six microstates. The equilibrium value for this most probable microstate is $S = \ln(6) = 1.79$.

**Answer:** For this system entropy $S = \ln\Omega$. At equilibrium entropy is 1.79.    ■

## 5.8   Molecular Definition of Entropy

A gas sealed in a rigid, insulated container is an isolated system whose mass and energy are fixed. The molecules of gas are distributed among available energy levels with the constraint that their total number and energy are fixed. There are many different arrangements of molecules that satisfy these conditions, and each one is considered a microstate of the system. The microstates of the gas are analogous to the number of possible states of the set of dice in the previous example. The total number of microstates in the gas is $\Omega_{tot}$, but since the number of molecules is very large (on the order of $10^{23}$ in a few grams of gas) we can safely assume that almost all of these correspond to the equilibrium state so that $\Omega_{eq} \approx \Omega_{tot}$. We will therefore drop the subscript and simply refer to the number of microstates as $\Omega$.

Calculating the energy levels of a particle in a box by solving the Schrödinger equation is a classic problem in quantum mechanics. We can use the solution for the case where $n$ atoms of a monoatomic ideal gas (such as argon or helium) are confined in a rigid container of volume $V$. If the total internal energy of the gas is $U$ it can be shown that the number of microstates is

$$\Omega(U, V, n) = f(n) V^n U^{3n/2},\tag{5.25}$$

where $f$ is a function of $n$ only. For a fixed mass of gas the number of available microstates grows larger if we increase either the volume or the internal energy of the system.

Our definition of entropy was $S = \ln\Omega$, but this makes $S$ a dimensionless quantity. Ludwig Boltzmann, when presenting his molecular theory in 1877, had to introduce a constant to make the units of entropy the same as those proposed earlier by Clausius. The molecular definition of entropy is

$$S = k \ln \Omega,\tag{5.26}$$

where $k = 1.38 \times 10^{-23}$ J / K is the Boltzmann constant. If a system goes from initial state 1 to a final state 2 the change in entropy is

$$\Delta S = S_2 - S_1 = k \ln \Omega_2 - k \ln \Omega_1 = k \ln \frac{\Omega_2}{\Omega_1}.\tag{5.27}$$

Introducing the expression for the number of microstates from Equation (5.25) into the definition of entropy Equation (5.26) gives us the fundamental equation:

$$S(U,V, n) = nk \ln V + \frac{3}{2} nk \ln U + k \ln f(n)\tag{5.28}$$

and the change in entropy in going from state 1 to state 2 is

$$\Delta S = nk \left[ \ln \frac{V_2}{V_1} + \frac{3}{2} \ln \frac{U_2}{U_1} \right].\tag{5.29}$$

Recalling from Chapter 3 that $nk = NR_u$ and that for a monoatomic ideal gas $U = 3NR_u T / 2$,

$$\Delta S = NR_u \left[ \ln \frac{V_2}{V_1} + \frac{3}{2} \ln \frac{T_2}{T_1} \right].\tag{5.30}$$

Converting from molar to mass units,

$$\Delta S = mR \left[ \ln \frac{V_2}{V_1} + \frac{3}{2} \ln \frac{T_2}{T_1} \right].$$                     (5.31)

We can view entropy changes from either a microscopic or a macroscopic perspective. In a macroscopic description heating an ideal gas results in an increase in either its temperature (for constant $V$) or its volume (for constant $T$). Either of these will produce an increase in entropy. Alternately, in a microscopic description, more energy states become accessible if either the volume or temperature of the gas increases, so that the Maxwell–Boltzmann distribution becomes wider, as shown in Figure 5.12. Molecules in the gas will occupy these new energy states and, since $\Omega_2 > \Omega_1$, the change in entropy is $\Delta S > 0$. Eventually the system will reach equilibrium when entropy is maximum. Heat transfer through a finite temperature difference is irreversible since the system will not spontaneously go from a state that has a very high probability to a state that is much less likely.

Entropy can be understood using statistical concepts but in classical thermodynamics, where matter is assumed to be a continuum, it is treated as a fundamental property just as mass, volume and energy are. We will therefore spend a lot of time discussing the attributes of entropy to build an intuitive understanding of the concept, but we will never be able to give a succinct definition of it.

**Example 5.4**

**Problem:** The entropy of a kmol of gas is 2.3 kJ / K. On average, how many microstates are there for each molecule in the gas?

**Find:** Number of microstates per molecule of the given gas.

**Known:** Entropy per kgmole $S = 2.3$ kJ / K.

**Governing Equation:**

Entropy                                    $S = k \ln \Omega$

**Properties:** Number of molecules in a kmol $N_A = 6.022 \times 10^{26}$, Boltzmann constant $k = 1.38$ J / K.

**Solution:**

$$\Omega = \exp \left( \frac{S}{k} \right) = \exp \left( \frac{2.3 \times 10^3 \, \text{J} / \text{K}}{1.38 \times 10^{-23} \, \text{J} / \text{K}} \right) = e^{1.6667 \times 10^{26}}$$

Average number of microstates per molecule is

$$\frac{\Omega}{N_A} = \frac{e^{1.6667 \times 10^{26}}}{6.022 \times 10^{26}}.$$

**Answer:** The number of microstates per molecule is $\exp(1.67 \times 10^{26}) / (6.022 \times 10^{26})$. This is a very large number.                                                                                        ∎

## Example 5.5

**Problem:** What is the increase in entropy of 1 kg of air when it is heated from 300 K to 305 K? What is the ratio of the number of microstates at 305 K to those at 300 K?

**Find:** Increase in entropy and the ratio of microstates.

**Known:** Mass of air $m = 1$ kg, initial temperature $T_1 = 300$ K, final temperature $T_2 = 305$ K.

**Assumptions:** Since the temperature increase is small we will assume that the air temperature remains approximately constant during heating at $T_{avg} = (T_1 + T_2)/2 = (300\,K + 305\,K)/2 = 302.5\,K$.

**Properties:** For air at 300 K $c_p = 1.005$ kJ / kgK (Appendix 4).

**Solution:**

Heat added to air,

$$Q = mc_p\left(T_2 - T_1\right) = 1\,kg \times 1.005\,kJ / kgK \times \left(305\,K - 300\,K\right) = 5.025\,kJ.$$

Change in entropy,

$$\Delta S = \frac{Q}{T_{avg}} = \frac{5.025\,kJ}{302.5\,K} = 0.016\ 612\ kJ / K.$$

Also,

$$\Delta S = kln\frac{\Omega_f}{\Omega_i}$$

$$\frac{\Omega_f}{\Omega_i} = \exp\left(\frac{\Delta S}{k}\right) = \exp\left(\frac{16.612\,J / K}{1.38\times10^{-23}\,J / K}\right) = \exp\left(1.2043 \times 10^{24}\right).$$

**Answer:** The increase in entropy is 0.0166 kJ / K. The ratio of the number of microstates is $\exp\left(1.20\times10^{24}\right)$. ∎

## 5.9   Third Law of Thermodynamics

Entropy, unlike energy, can have an absolute value of zero. Energy is always measured relative to some datum, where it is arbitrarily assumed to be zero. Zero entropy, however, has a physical meaning: it implies that there is only one possible microstate. If a pure substance is cooled to 0 K all its molecules occupy their lowest energy states so that $\Omega = 1$ and $S = 0$. This is stated as the third law of thermodynamics:

*The entropy of a pure substance in thermodynamic equilibrium is zero at a temperature of absolute zero.*

## 5.10   Production of Entropy

An isolated system consists of an insulated, rigid container subdivided into two compartments separated by a rigid wall (Figure 5.16). One compartment is filled with gas while the other is evacuated. The entropy of the gas is constant, since it is in equilibrium

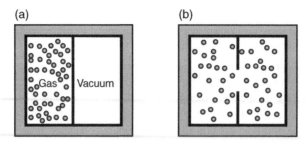

**Figure 5.16**   Gas (a) initially occupying half of a rigid vessel and (b) filling the entire vessel.

and its macroscopic properties do not change with time. The wall separating the two chambers is punctured so that gas fills the entire container.

The gas does not do any work while expanding since it does not meet any resistance when it flows into a vacuum. Therefore the internal energy does not change ($\Delta U = 0$) and, assuming ideal gas behaviour, there is no change in temperature ($T_2 = T_1$). The number of microstates increases because of the change in volume and from Equation (5.25) $\Omega \sim V^n$. There are many more microstates available to the gas in the entire container than in a part of it so if the volume increases from $V_1$ to $V_2$ the number of microstates increases from $\Omega_1$ to $\Omega_2$ where

$$\frac{\Omega_2}{\Omega_1} = \left(\frac{V_2}{V_1}\right)^n \tag{5.32}$$

and the number of molecules is typically $n \sim 10^{23}$. The expansion is an irreversible process. The gas will never spontaneously vacate one chamber and return to its initial state because the probability of all the molecules in the gas moving by themselves to one side of the vessel is so low that to see it happen we would have to wait for a period much longer than the anticipated life of the universe.

The entropy increases when the gas expands as

$$\Delta S = k \ln \frac{\Omega_2}{\Omega_1} = NR_u \ln \frac{V_2}{V_1}. \tag{5.33}$$

The system was isolated, so the increased entropy was not transferred from the surroundings – it was generated within the system itself due to molecules occupying new microstates. Once entropy has been created in an isolated system it cannot disappear. The only way to reduce the entropy of a system is to transfer it to the surroundings.

The entire universe can be considered an isolated system since it does not, by definition, interact with any other system. The total entropy of the universe is constantly increasing as mass and energy spread out from regions where they are concentrated to other areas where they are more thinly dispersed. The initial state of the universe was one of very low entropy, with all mass and energy concentrated in an infinitesimally small region of space. Ever since the Big Bang that gave birth to the universe both mass and energy have been spreading out, occupying space available to them. Stars and planets radiate heat out into space, which is at a much lower temperature, and lose mass that diffuses away into the surrounding vacuum. The final, equilibrium state of the universe will be one of maximum entropy with all mass and energy spread perfectly uniformly through all space. At this time all thermodynamic processes will halt, since there will be no gradients in mass concentration, temperature or pressure to drive the spontaneous movement of mass or energy.

## 5.11 Heat and Work: A Microscopic View

If a gas contains molecules such that each energy level, $\varepsilon_i$, contains $n_i$ molecules, then the total internal energy of the gas is

$$U = \sum_i \varepsilon_i n_i. \tag{5.34}$$

A change in energy is given by

$$dU = \sum_i \varepsilon_i dn_i + \sum_i n_i d\varepsilon_i. \tag{5.35}$$

There are two ways in which the energy of the system may be changed: either by varying the number of molecules allotted to each energy state $(dn_i)$, or by altering the amount of energy associated with each level $(d\varepsilon_i)$. Both of these changes modify the total energy of the system, but their impact on entropy is very different. If some molecules are moved from one energy level to another $(dn_i)$, then the number of microstates of the system changes and therefore its entropy does as well. If the energy levels themselves are altered $(d\varepsilon_i)$, but there is no redistribution of molecules within them, there is no change in the number of microstates or the entropy. The first term in Equation (5.35) is therefore associated with entropy change, whereas the second term is not.

If we bring two identical masses that are at different temperatures in thermal contact, the hotter body will cool by $\Delta T$ while the temperature of the colder body will increase by $\Delta T$. Entropy is a logarithmic function of temperature, so the slope of $S$ decreases as $T$ increases (Figure 5.17). At low temperatures, when there are few available microstates, adding energy

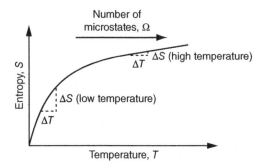

**Figure 5.17**   Variation of entropy ($S$) with temperature ($T$). At low temperature a given change $\Delta T$ produces a much larger change in $S$ than it does at high temperature.

produces a large increase in entropy $(S)$; at high temperatures the same amount of energy results in a much smaller change in $S$. Heat transfer therefore not only moves entropy from the hot body to the cold one, but also generates additional entropy.

Since adding or removing heat always results in shifting of molecules between energy levels, at a microscopic level,

$$\delta Q = \sum_i \varepsilon_i dn_i. \qquad (5.36)$$

Doing work on a system has quite a different effect on its entropy. A rigid, insulated container filled with gas is tied to a rope and hoisted up using a pulley so that the potential energy of the system increases (Figure 5.18a). The energy of each gas molecule increases, but by exactly the same amount as every other molecule in the system, since they have all been raised by the same amount. There is no movement of molecules from one energy state to another, so $dn_i = 0$ and there is no entropy change.

Similarly, if the system is accelerated from rest to velocity $V$ the kinetic energy of the system increases, but this added energy is shared equally between the molecules of the system (Figure 5.18b). There is no change in $\Omega$ and the entropy of the system is unchanged. Changes in kinetic or potential energy of a system do not produce changes in entropy.

Work can be done on a gas by compressing it in a piston–cylinder system. If the piston moves forward rapidly gas adjoining it will be compressed and molecules near the piston face will cluster together, while those far away will not experience the effect of the piston until later and remain relatively far apart (Figure 5.19a). Confining gas molecules to a smaller volume requires work, so molecules near the piston will be shifted to higher energy levels than those that are distant. The total number of microstates of the system will be increased, as will the entropy.

If the piston is moved slowly, rather than rapidly, when compressing the gas, less entropy will be generated. Gas molecules will have more time to move away from the advancing piston and experience less compression so fewer molecules will be transferred to higher energy levels.

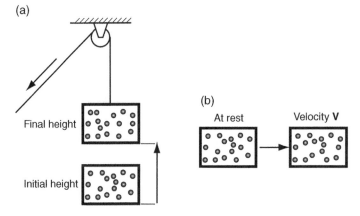

**Figure 5.18** Rigid insulated container filled with gas being (a) raised vertically under the force of gravity and (b) accelerated from rest.

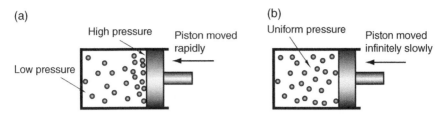

**Figure 5.19** Compression of a gas in a piston–cylinder system in (a) an irreversible process where the piston is moved rapidly and (b) a reversible process in which the piston is moved infinitely slowly.

In the limiting case that the velocity of the piston approaches zero, which is a quasi-equilibrium process, all molecules in the system will be compressed by the same amount, and all their energy levels will increase by the same amount (Figure 5.19b). Since there is no redistribution of molecules amongst energy levels, $dn_i = 0$, the process is isentropic.

*Energy transfers to or from a system that do not alter its entropy are classified as reversible work.* This corresponds to the second term in Equation (5.35), so that

$$\delta W = \sum_i n_i d\varepsilon_i. \tag{5.37}$$

By substituting Equations (5.36) and (5.37) into Equation (5.35) we obtain the first law of thermodynamics:

$$dU = \delta Q + \delta W. \tag{5.38}$$

## 5.12 Order and Uncertainty

Entropy is sometimes called a measure of disorder in a system. Highly ordered systems, if allowed to change randomly, become more disordered. People perceive some systems as being ordered, if they can recognise a pattern in them. This could be simple, such as a set of dice with all showing the same value, or it may be a much more complex, subjectively defined pattern. A sequence of numbers that corresponds to your date of birth may hold meaning for you but not necessarily for someone else. However, the vast majority of combinations of dice will have no discernible pattern and we will view them as disordered. Starting from one of the few combinations that appears ordered, as the system evolves it is highly probable that the combinations that appear will have no visible pattern, giving an appearance of disorder. The reverse case, where we start from a disordered system and an ordered system spontaneously appears is much less likely, though in principle not impossible.

It is tempting to read too much significance into these ideas about disorder and leap to unjustified conclusions. An example often cited to illustrate increasing entropy is that of a room that initially has every object in its proper place and then becomes increasingly messy. Though an appealing analogy, it is important to realise that the concept of entropy is statistical and only meaningful for a system that has a very large number of rapidly changing micro-states, which is not a description typically applicable to a room. Unless we can apply the laws of probability to a system it makes no sense to talk about its entropy. "Order" and "disorder" are also very difficult to quantify. Is an ice cube more orderly than a cup filled with water? What if we crush the cube? The ice cube and crushed ice chips have the same entropy while the water has higher entropy, but it is difficult to conclude this purely by conjecturing how "orderly" the system seems.

Entropy is also described as a measure of uncertainty about a system. If we know the macrostate of a system, how accurately can we guess its microstate? If $S = 0$, there can be only one possible microstate, so we know it with absolute certainty. Equilibrium corresponds to the largest number of microstates, where entropy is maximum and so is our uncertainty about the microstate of the system.

## Further Reading

1. W. C. Reynolds, H. C. Perkins (**1977**) *Engineering Thermodynamics*, McGraw Hill, New York.
2. B. Poirier (**2014**) *A Conceptual Guide to Thermodynamics*, John Wiley & Sons, Ltd, London.
3. D. Schroeder (**2000**) *An Introduction to Thermal Physics*, Addison Wesley, London.
4. M. A. Saad (**1997**) *Thermodynamics – Principles and Practice*, Prentice Hall, New York.

## Summary

*Entropy* is an extensive property of all thermodynamics systems. When a system at temperature ($T$) is heated or cooled by an infinitesimal amount ($\delta Q$) its entropy changes by

$$dS = \frac{\delta Q}{T}.$$

A *thermal reservoir* is a system whose temperature remains constant and uniform in spite of heat transfer to or from it. The entropy change of a thermal reservoir is

$$\Delta S = \frac{1}{T}\int_1^2 \delta Q = \frac{Q_{12}}{T}.$$

The *second law of thermodynamics* states that the entropy of an isolated system will increase until the system reaches a state of equilibrium. The entropy of an isolated system in equilibrium remains constant.

A *reversible process* is one that can be reversed by an infinitesimal change in the surroundings and both the system and surroundings restored to their initial conditions. A reversible process generates no entropy while an irreversible process generates entropy. Processes that are both reversible and adiabatic produce no changes in entropy and are called *isentropic* processes:

$$S_{gen} = 0 \text{ for a reversible process,}$$

$$S_{gen} > 0 \text{ for an irreversible process.}$$

*Entropy is always transferred with heat.* Changes in kinetic or potential energy of a system do not produce changes in entropy. Entropy cannot be destroyed once it is generated.

The *state principle* postulates that the equilibrium state of a pure, simple compressible substance is completely described by its mass (*m*), volume (*V*) and internal energy (*U*). The *fundamental equation* is

$$S = S(U, V, m).$$

The *third law of thermodynamics* states that the entropy of a pure substance in thermodynamic equilibrium is zero at a temperature of absolute zero.

Each individual distribution of molecules among available energy states, which satisfies the constraints imposed, is a *microstate* of the system. The *macrostate* of the system is defined by macroscopic properties such as energy and volume. The equilibrium macrostate of any system is the one that has the largest number of corresponding microstates. The molecular definition of entropy is

$$S = k \ln \Omega.$$

If a system goes from an initial state with $\Omega_1$ microstates to a final state with $\Omega_2$ microstates, the change in entropy is

$$\Delta S = k \ln \frac{\Omega_2}{\Omega_1}.$$

The entropy change for a monoatomic ideal gas is

$$\Delta S = NR_u \left[ \ln \frac{V_2}{V_1} + \frac{3}{2} \ln \frac{T_2}{T_1} \right] = mR \left[ \ln \frac{V_2}{V_1} + \frac{3}{2} \ln \frac{T_2}{T_1} \right].$$

## Problems

5.1 A 0.1 A current is passed through a 1 k$\Omega$ resistor for 5 min. The heat generated is lost to the surrounding air at 20 °C. What is the increase in entropy of the atmosphere?

5.2 A fireplace burns 6 kg of coal which supplies 15 MJ of heat for every kilogram consumed. Find the increase in the entropy of the surrounding air at 24 °C.

5.3 Two large bodies, one at $T_A = 400$ K and the other at $T_B > T_A$ are briefly brought into contact so that 1 kJ of heat is transferred from the hotter to the colder mass. The entropy generated during this process is $S_{gen} = 0.5$ J / K. Determine $T_B$, assuming that $T_A$ and $T_B$ remain constant.

5.4 A lake with a water temperature of 15 °C receives heat from the air above it, which is at 25 °C, at the rate of 12 W for each square meter of surface area. The total surface area of the lake is 4 km². Find the entropy generated in one hour due to this heat exchange.

5.5 A 6 kg aluminium plate is removed from a furnace at 400 °C and allowed to cool in air at 15 °C until it reaches equilibrium. What is the increase in entropy of the atmosphere during this process?

5.6    A bucket with 5 kg of water at 30 °C is exposed to air at 25 °C until it cools to the surrounding temperature. Estimate the entropy change of the air.

5.7    A rigid tank with a volume of 0.5 m³ initially contains air at 300 kPa and 400 K. It loses heat to the surrounding air at 300 K until it reaches equilibrium. Find the increase in entropy of the atmosphere.

5.8    An ice cube with a mass of 200 g is dropped into a glass of water at 0 °C placed in a room maintained at 23 °C. Calculate the entropy change of (a) the air, (b) the ice and (c) the total entropy generated when the ice has melted completely. Assume that the heat transfer required to melt ice is 334 kJ / kg.

5.9    An ideal gas contained in a well-insulated piston–cylinder device is at equilibrium with the surrounding atmosphere. The piston is first pushed to compress the gas from its initial volume ($V_1$) to final volume ($V_2$). The piston is then released and the gas allowed to expand until it reaches equilibrium. Draw the process on a $P$-$V$ diagram assuming (a) that the piston is frictionless and (b) that the piston has friction.

5.10   A 30 kg mass falls onto the ground from a 3 m high shelf. Calculate the increase in entropy of the surrounding air at 27 °C.

5.11   A 300 g rubber ball dropped from a height of 2 m rebounds to a height of 1.6 m. Calculate the increase in entropy of the atmosphere at a temperature of 21 °C.

5.12   A 1500 kg car travelling at a speed of 90 km / h is brought to a halt when its brakes are applied. Calculate the increase in entropy of the atmosphere at 17 °C.

5.13   A 10 kg mass slides down a 2 m long ramp inclined at 30° to the horizontal. The force of friction resisting motion $F_f = 0.2F_N$, where $F_N$ is the normal force between the mass and the ramp. Find the velocity of the mass when it reaches the bottom of the ramp and the entropy increase of the surrounding atmosphere, which is at 25 °C.

5.14   A 0.5 kg ball travelling at 10 m / s hits a wall and rebounds with 80% of its impact velocity. Find the entropy generated in the surrounding air at 15 °C.

5.15   A 100 W light bulb is switched on for 4 h in a house that is maintained at 24 °C. The house loses heat through its walls to the exterior air which is at a temperature of −10 °C. What is the entropy change of (a) the air in the house and (b) the air outside the house?

5.16   A 1 kW electrical heater is placed inside a rigid tank filled with water. The temperature of the water remains constant at 80 °C as it loses heat continuously to the surrounding air at 25 °C. Calculate the rate of entropy change of (a) the water and (b) the air.

5.17   Water enters a 2 kW pump at a rate of 2 kg / s with a pressure of 100 kPa and leaves at 1 MPa. Due to frictional losses in the pump part of the energy supplied to it is lost as heat to the surrounding air. Assuming that the temperature change of water is negligible and that its specific volume is constant at 0.001 m³ / kg, calculate the rate of entropy increase of the surrounding air at 17 °C.

5.18   Water flows through a straight length of pipe at the rate of 0.5 kg / s. The pressure drop from the inlet to the outlet of the pipe is 10 kPa. When a bend is introduced into the middle of the pipe the pressure drop increases to 20 kPa. Calculate the rate of entropy increase of the surrounding air at 20 °C for the pipe both with and without the bend. Assume that the temperature change of water is negligible and that its specific volume is 0.001 m³ / kg.

5.19   Air enters a gas turbine at 800 kPa and 900 K and leaves at 100 kPa and 480 K. If the work output of the turbine is 400 kJ per kg of air flowing through it find the entropy increase (per kg of airflow) of the surrounding atmosphere at 20 °C due to heat transfer from the casing of the turbine.

5.20   The interior of a house is at 23 °C on a night when the external temperature is 0 °C (Figure P5.20). The temperature in the wall of the house, which is 20 cm thick, varies linearly as shown in the figure. If 300 kJ of heat is conducted through the wall find (a) the entropy change of the interior (b) the entropy change of the exterior, (c) the entropy generated in the wall and (d) the entropy that crossed the mid-plane of the wall.

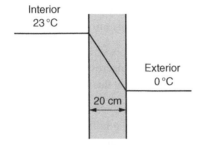

**Figure P5.20**   The temperature distribution in the wall of a house whose interior is at 23 °C when the external temperature is 0 °C.

5.21   A pot of water is placed on a 500 W electric heater that is switched on for 5 min (Figure P5.21). The bottom of the pot is at a temperature of 104 °C. We can define our

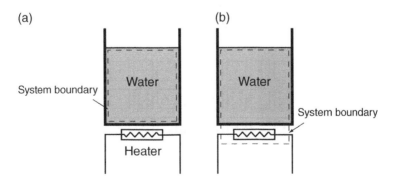

**Figure P5.21**   A pot of water is placed on a 500 W electric heater that is switched on for 5 min. (a) System includes only the water. (b) System includes both the heater and the water.

system as (a) including only the water or (b) including both the heater and water. In each of these cases state whether the system is internally reversible. Calculate the change in entropy of the system where possible.

5.22    A system consists of three molecules, which may occupy any one of five energy levels, equal to 0, 1, 2, 3 or 4 energy units respectively. If the total energy of the system is six units, how many microstates does the system have?

5.23    A system of three particles, $A$, $B$ and $C$, which may occupy four energy levels, having 1, 2, 3 or 4 units of energy, has a total energy of 9 units. How many different microstates are possible if the particles may occupy any of the energy states?

5.24    A box contains $N$ identical particles that move randomly. If it is equally likely that any given particle is in the left or right half of the box at any instant, what is the probability that all of the particles are simultaneously in the left half if (a) $N = 2$ and (b) $N = 100$?

5.25    A box contains four particles labelled $A$, $B$, $C$ and $D$. If each particle is equally likely to occupy either half of a box, list all the possible arrangements. How many microstates and how many macrostates (defined by the number of particles in each half) does the system have? What is the equilibrium entropy of the system?

5.26    A pair of dice is rolled. List all the possible outcomes. If we define the macrostate by the sum of the two dice, how many macrostates are possible for this system and which is the most likely? Define the entropy of this system. Calculate the entropy corresponding to each macrostate.

5.27    A deck of 52 playing cards, half red and half black, is shuffled randomly. How many microstates, corresponding to unique arrangements of the cards, are there? We define a macrostate $A$ as being equal to the number of red cards in the upper half of the deck. What is the probability of $A = 0$? What would you expect the equilibrium value of $A$ to be?

5.28    Three sets of dice are shown in Figure P5.28. Which one would you call the most ordered and which one the most disordered? Use the data from Table 5.1 to calculate the macrostate, the number of microstates and the entropy for each set. Is "degree of disorder" an accurate description of entropy?

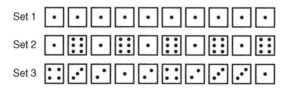

**Figure P5.28**    Three sets of dice.

5.29    A volume of gas has $10^{24}$ molecules and the number of microstates equals $10^{23}$ times the number of molecules. What is the entropy of the gas?

5.30   The entropy of 1 kg of liquid water at 100 °C and atmospheric pressure is 1.3072 kJ / K, while that of steam at the same temperature and pressure is 7.3541 kJ / K. How many times more microstates does steam have than water?

5.31   A 1-kg mass of copper at 300 K is heated by 1 K. Estimate the ratio of the number of microstates in the final state to those in the initial state.

5.32   A rigid tank containing 0.5 kg of helium at 300 °C cools down in a room kept at 17°C until it reaches equilibrium with its surroundings. Find (a) the change in entropy of the helium, (b) the change in entropy of the surrounding air and (c) the entropy generated.

5.33   A rigid, insulated container is divided into two equal compartments by a partition. One half contains 1 kmol of argon while the other half is evacuated. The partition ruptures so that the gas fills the entire container. What is the entropy generated?

5.34   One kmol of helium is expanded isothermally from 0.2 to 0.6 m³. Find the ratio of the final number of microstates to the initial number of microstates. What is the increase in entropy?

5.35   A monoatomic ideal gas is compressed isentropically from a volume of 0.3 m³ and temperature of 27 °C to a volume of 0.1 m³. Find the final temperature.

5.36   A rigid container filled with 0.5 kg of argon is heated from 300 to 400 K. Find the increase in entropy of the gas.

5.37   A cylinder containing 0.3 kg of argon at 200 kPa and 350 K is heated in a constant pressure process in which 30 kJ of heat is transferred to it. Find the increase in entropy of the gas.

5.38   A cylinder is filled with 15 g of helium with volume 0.1 m³ and pressure 100 kPa. A piston is used to compress the gas until it is at 0.05 m³ and 300 kPa. Find the change in entropy of the gas.

5.39   Show that for a monoatomic ideal gas undergoing an isentropic process from state 1 to state 2 that

$$\frac{T_2}{T_1} = \left(\frac{V_1}{V_2}\right)^{\gamma-1},$$

where the specific heat ratio is $\gamma = c_p / c_v$. Hint: for a monoatomic ideal gas $c_v = 3R/2$.

5.40   Show that for a monoatomic ideal gas undergoing an isentropic process from state 1 to state 2 that

$$PV^{\gamma} = \text{constant},$$

where the specific heat ratio $\gamma = c_p / c_v$.

# 6

# The Second Law of Thermodynamics

---

**In this chapter you will:**

- State the postulates of classical thermodynamics.
- Define thermodynamic temperature and pressure.
- Calculate entropy changes for gases, solids and liquids.
- Do entropy balances for open and closed systems.
- Analyse steady flow devices (turbines, compressors, pumps, nozzles) assuming isentropic flow and using isentropic efficiencies.

---

## 6.1 The Postulates of Classical Thermodynamics

If you had to describe all of physical existence using only four words, which would you choose? The most basic attribute of the universe is that it consists of matter, which we describe using the concept of *mass*. Matter occupies space whose extent is defined by its *volume*. *Energy* describes how pieces of matter move and interact with each other. *Entropy* tells us that these interactions, if not impeded, eventually lead to a state of equilibrium. These fundamental properties, which we cannot describe in simpler terms, are used to develop mathematical models of natural phenomena. Classical thermodynamics is based on these four extensive properties:

- Mass
- Volume
- Energy
- Entropy.

---

*Energy, Entropy and Engines: An Introduction to Thermodynamics*, First Edition. Sanjeev Chandra.
© 2016 John Wiley & Sons, Ltd. Published 2016 by John Wiley & Sons, Ltd.
Companion website: www.wiley.com/go/chandraSol16

Time is not included in this list, though it is a fundamental concept, since we will only be dealing with equilibrium states or quasi-equilibrium processes. Temperature and pressure are not fundamental properties and in this chapter we will define them as functions of the four listed above. All other properties such as enthalpy and specific heat can also be expressed in terms of fundamental properties.

We state four postulates to describe the workings of the physical world:

- *State Postulate:* All isolated systems reach a state of equilibrium. The equilibrium state of a pure, simple compressible substance is completely described by its mass, volume and internal energy.
- *First Law of Thermodynamics:* The change in energy of a closed system equals the net energy transferred to it in the form of work and heat.
- *Second Law of Thermodynamics:* The entropy of an isolated system will increase until the system reaches a state of equilibrium. The entropy of an isolated system in equilibrium remains constant.
- *Third Law of Thermodynamics:* The entropy of a pure substance in thermodynamic equilibrium is zero at a temperature of absolute zero.

In the statement of the first law we have used the terms "work" and "heat". We understand these to be transfers of energy across a system boundary. Heat is an energy transfer that increases the entropy of a system while reversible work does not.

In the third law statement we have used the word "temperature", which we have not defined as yet. Strictly speaking, we should have said that entropy is zero when $(\partial U / \partial S)_{m,v} = 0$ since, as we will see in the next section, this partial derivative is how we define temperature. Though necessary for logical consistency, this alternate statement is harder to understand.

These postulates are the basis of all of thermodynamics and we will not need to make any more assumptions. It is remarkable that these seemingly straightforward statements, which would fit comfortably on a single sheet of paper, make it possible to predict the behaviour of widely disparate mechanical, chemical and biological systems. The elegance and power of thermodynamics is rooted in the simplicity and clarity of its foundations.

## 6.2   Thermal Equilibrium and Temperature

How do we relate entropy to temperature? Let us see how both properties change as a system comes to thermal equilibrium. Figure 6.1 shows two insulated, rigid masses, labelled $A$ and $B$. They are initially at equilibrium when, suddenly, they are brought into close contact so that heat transfer can take place between them while remaining insulated from the surroundings. The two bodies together form a composite isolated system, labelled $C$.

According to the state postulate, $S_A = S_A(U_A, V_A, m_A)$ and $S_B = S_B(U_B, V_B, m_B)$. Since entropy is an extensive property we can add the entropy of the two components to get the entropy of the composite system,

$$S_C = S_A\left(U_A, V_A, m_A\right) + S_B\left(U_B, V_B, m_B\right).$$

(6.1)

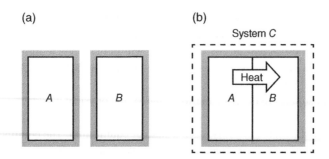

**Figure 6.1** Two rigid, insulated bodies $A$ and $B$ that are (a) held separately and (b) brought into close thermal contact to form a composite system $C$ in which $A$ and $B$ are separated by a rigid wall.

At equilibrium the entropy of the composite system $C$ does not change ($dS_C = 0$). The two sub-systems have constant volume and mass, so $m_A$, $m_B$, $V_A$ and $V_B$ are fixed. Differentiating Equation (6.1) gives

$$dS_c = \left(\frac{\partial S_A}{\partial U_A}\right)_{m_A, V_A} dU_A + \left(\frac{\partial S_B}{\partial U_B}\right)_{m_B, V_B} dU_B = 0. \tag{6.2}$$

Since $C$ is a closed system $U_C = U_A + U_B =$ constant and

$$dU_A + dU_B = 0, \tag{6.3}$$

so that

$$dU_A = -dU_B. \tag{6.4}$$

Substituting Equation (6.4) into Equation (6.2) gives

$$dS_c = \left[\left(\frac{\partial S_A}{\partial U_A}\right)_{m_A, V_A} - \left(\frac{\partial S_B}{\partial U_B}\right)_{m_B, V_B}\right] dU_A = 0, \tag{6.5}$$

or

$$\left(\frac{\partial S_A}{\partial U_A}\right)_{m_A, V_A} = \left(\frac{\partial S_B}{\partial U_B}\right)_{m_B, V_B}. \tag{6.6}$$

What does this equation mean? We know that the two systems, $A$ and $B$, are in thermal equilibrium. How do we include this observation in our analysis? We have not yet defined temperature, but from everyday usage of the term we know that it is a property that determines the direction of heat transfer (from high to low temperature) and establishes thermal equilibrium (when the temperatures of two systems are the same). A definition of temperature in terms of fundamental thermodynamic properties, that satisfies our intuitive understanding of the word, is

$$T \equiv \left(\frac{\partial U}{\partial S}\right)_{m, V}. \tag{6.7}$$

Temperature ($T$) is an intensive property since it is the ratio of entropy ($S$) and internal energy ($U$), which are both extensive properties. Substituting this definition of temperature into Equation (6.5) gives

$$dS_C = \left[ \frac{1}{T_A} - \frac{1}{T_B} \right] dU_A. \tag{6.8}$$

If $T_A = T_B$ the system will be at equilibrium and $dS_C = 0$. If initially the system is not in equilibrium, so that $T_A \neq T_B$, then heat transfer will take place from the hotter to the colder side until equilibrium is established. As a result of this irreversible process the entropy of the system must increase, so that $dS_C > 0$. Therefore,

$$\underbrace{\left[ \frac{1}{T_A} - \frac{1}{T_B} \right]}_{<0 \, \text{if} \, T_A > T_B} \underbrace{dU_A}_{<0} > 0. \tag{6.9}$$

If $T_A > T_B$ the term in parentheses is negative and therefore $dU_A < 0$ to satisfy the inequality. This implies that heat transfer takes place from $A$ to $B$, which is what we expect if $T_A > T_B$. The definition of temperature now satisfies both our expectations: equal temperatures lead to thermal equilibrium and heat transfer takes place from higher temperature to lower temperature.

Equation (6.7) gives an entirely new way of thinking about temperature: it is a measure of how sensitive entropy is to changes in energy. A given amount of energy added to a system at low temperature produces a much bigger increase in entropy than it does at high temperature.

## 6.3   Mechanical Equilibrium and Pressure

An insulated cylinder is divided into two compartments, $A$ and $B$, by a sliding piston that is initially locked in place by a pin (see Figure 6.2). The pin is removed and the piston moves until it reaches equilibrium. The two subsystems $A$ and $B$ can exchange energy in the form of both heat and work but their outer walls are insulated and rigid so the total internal energy and volume of the isolated, composite system $C$ remains constant.

In this case

$$U_C = U_A + U_B = \text{constant}, \tag{6.10}$$

and

$$V_C = V_A + V_B = \text{constant}. \tag{6.11}$$

Differentiating these two expressions gives

$$dU_A = -dU_B \tag{6.12}$$

and

$$dV_A = -dV_B. \tag{6.13}$$

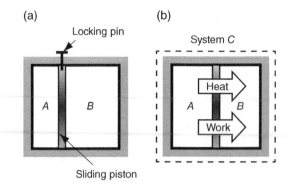

**Figure 6.2**   An insulated cylinder is divided into two sections $A$ and $B$ separated by a sliding piston. (a) The piston is locked in place. (b) The piston is allowed to move and transfer heat until the composite system $C$ comes to equilibrium.

The total entropy of the system is

$$S_C = S_A\left(U_A, V_A, m_A\right) + S_B\left(U_B, V_B, m_B\right)\tag{6.14}$$

and at equilibrium

$$dS_c = \left(\frac{\partial S_A}{\partial U_A}\right)_{m_A, V_A} dU_A + \left(\frac{\partial S_A}{\partial V_A}\right)_{m_A, U_A} dV_A + \left(\frac{\partial S_B}{\partial U_B}\right)_{m_B, V_B} dU_B + \left(\frac{\partial S_B}{\partial V_B}\right)_{m_B, U_B} dV_B = 0.$$

$$\tag{6.15}$$

Substituting in Equations (6.12) and (6.13) gives

$$dS_c = \left[\left(\frac{\partial S_A}{\partial U_A}\right)_{m_A, V_A} - \left(\frac{\partial S_B}{\partial U_B}\right)_{m_B, V_B}\right]dU_A + \left[\left(\frac{\partial S_A}{\partial V_A}\right)_{m_A, U_A} - \left(\frac{\partial S_B}{\partial V_B}\right)_{m_B, U_B}\right]dV_A = 0. \quad(6.16)$$

Substituting the definition of temperature, Equation (6.7),

$$\left[\frac{1}{T_A} - \frac{1}{T_B}\right]dU_A + \left[\left(\frac{\partial S_A}{\partial V_A}\right)_{m_A, U_A} - \left(\frac{\partial S_B}{\partial V_B}\right)_{m_B, U_B}\right]dV_A = 0.\tag{6.17}$$

At equilibrium $T_A = T_B$ so the first term in Equation (6.17) is zero. Therefore, to have equilibrium also requires

$$\left(\frac{\partial S_A}{\partial V_A}\right)_{m_A, U_A} = \left(\frac{\partial S_B}{\partial V_B}\right)_{m_B, U_B}.\tag{6.18}$$

What does this equation mean? At equilibrium, we know that the temperature on both sides of the partition must be equal. However, in this case, since the partition is movable, we also need to have mechanical equilibrium – the pressure on both sides must be the same. Equation (6.18) can be related to pressure by examining the dimensions of $\partial S / \partial V$. The units of $S$ are energy / temperature and those of $V$ are length$^3$, so $S/V$ has units of

$$\frac{\text{energy / temperature}}{\text{length}^3} = \frac{\text{force} \times \text{length}}{\text{temperature} \times \text{length}^3} = \frac{\text{force / length}^2}{\text{temperature}}.$$

From our mechanical definition of pressure, we know that it has dimensions of force / length$^2$, therefore

$$\frac{\text{force / length}^2}{\text{temperature}} = \frac{\text{pressure}}{\text{temperature}}.$$

Let us define

$$\frac{P}{T} \equiv \left(\frac{\partial S}{\partial V}\right)_{m,U}. \tag{6.19}$$

Then, with this definition,

$$dS_c = \left[\frac{1}{T_A} - \frac{1}{T_B}\right]dU_A + \left[\frac{P_A}{T_A} - \frac{P_B}{T_B}\right]dV_A. \tag{6.20}$$

Once thermal equilibrium is reached $T_A = T_B$. In that case,

$$dS_c = \frac{1}{T_A}\left[P_A - P_B\right]dV_A. \tag{6.21}$$

If $P_A > P_B$, the system is not in mechanical equilibrium and the entropy will increase. For $dS_c > 0$, we require that $dV_A > 0$. The side with greater pressure will expand in volume until equilibrium is re-established.

## 6.4   Gibbs Equation

We can define temperature and pressure in terms of specific entropy ($s = S/m$) and specific internal energy ($u = U/m$), which are both intensive properties:

$$\frac{1}{T} = \left(\frac{\partial S}{\partial U}\right)_{m,V} = \left(\frac{\partial(S/m)}{\partial(U/m)}\right)_{(V/m)} = \left(\frac{\partial s}{\partial u}\right)_v, \tag{6.22}$$

$$\frac{P}{T} = \left(\frac{\partial S}{\partial V}\right)_{m,U} = \left(\frac{\partial(S/m)}{\partial(V/m)}\right)_{(U/m)} = \left(\frac{\partial s}{\partial v}\right)_u. \tag{6.23}$$

From the state postulate, the specific entropy is a function of the specific volume and specific internal energy:

$$s = s(u,v). \tag{6.24}$$

Differentiating,

$$ds = \left(\frac{\partial s}{\partial u}\right)_v du + \left(\frac{\partial s}{\partial v}\right)_u dv. \tag{6.25}$$

Substituting Equations (6.22) and (6.23) into Equation (6.25) gives

$$ds = \frac{1}{T}du + \frac{P}{T}dv. \tag{6.26}$$

Equation (6.26) is known as the *Gibbs equation*, which gives us a relationship between changes in entropy and properties that we can measure easily such as pressure, temperature, volume and internal energy (a function of temperature).

## 6.5   Entropy Changes in Solids and Liquids

The Gibbs equation, Equation (6.26), gives entropy as a function of changes in internal energy and volume. Liquids and solids can be assumed to be incompressible and therefore $dv = 0$. For incompressible substances the specific heats at constant volume and pressure are the same, so $c_p = c_v = c$. The internal energy of an incompressible substance can be related to temperature,

$$du = c(T)dT.$$

The Gibbs equation reduces to

$$ds = c(T)\frac{dT}{T}. \tag{6.27}$$

Assuming specific heat is constant, with an average value $c_{avg}$, and integrating Equation (6.27) from state 1 to state 2,

$$\Delta s = s_2 - s_1 = c_{avg} \int_{T_1}^{T_2} \frac{dT}{T} = c_{avg} \ln \frac{T_2}{T_1}. \tag{6.28}$$

For an isentropic process, for which $\Delta s = 0$, Equation (6.28) gives that $T_2 = T_1$: the temperature of an incompressible substance undergoing an isentropic process does not change.

### Example 6.1
**Problem:** A ship's anchor, made of 500 kg of steel and initially at a temperature of 20 °C, is dropped into the ocean which has a temperature of 7 °C. Find the entropy change of the anchor.
**Find:** Entropy change $\Delta S$ of the anchor.
**Known:** Mass of the anchor $m = 500$ kg, initial temperature of anchor $T_1 = 20\ °C = 293.15$ K, final temperature of anchor $T_2 = 7\ °C = 280.15$ K.
**Assumptions:** Steel has constant specific heat.

**Governing Equation:**

Entropy change (incompressible) $\qquad \Delta s = c_{avg} \ln \dfrac{T_2}{T_1}$

**Properties:** Specific heat of steel $c = 0.500$ kJ / kgK (Appendix 2).

**Solution:**

Entropy change of the anchor is

$$\Delta S = m\Delta s = mc \ln \frac{T_2}{T_1} = 500\,\text{kg} \times 0.500\,\text{kJ}/\text{kgK} \times \ln \frac{280.15\,\text{K}}{293.15\,\text{K}} = -11.340\,\text{kJ}/\text{K}.$$

**Answer:** The entropy of the anchor is reduced by 11.3 kJ / K. ∎

## 6.6 Entropy Changes in Ideal Gases

The change in specific internal energy ($u$) of an ideal gas is

$$du = c_v (T) dT,$$

and from the ideal gas equation we have

$$\frac{P}{T} = \frac{R}{v}.$$

We can substitute both of these equations into the Gibbs equation, Equation (6.26), to give

$$ds = \frac{c_v(T)}{T} dT + R\frac{dv}{v}. \qquad (6.29)$$

Integrating from state 1 to state 2,

$$\Delta s = s_2 - s_1 = \int_{T_1}^{T_2} \frac{c_v(T)}{T} dT + \int_{v_1}^{v_2} R\frac{dv}{v}. \qquad (6.30)$$

### 6.6.1 Constant Specific Heats

Assuming that specific heat ($c_v$) is constant we can integrate Equation (6.30) and get

$$\Delta s = s_2 - s_1 = c_v \ln \frac{T_2}{T_1} + R \ln \frac{v_2}{v_1}. \qquad (6.31)$$

Equation (6.31) gives a way of calculating entropy changes when the initial and final temperatures and specific volumes of a system are known. If we prefer to have pressure as an independent variable, noting that for an ideal gas

$$\frac{T_2}{T_1} = \frac{P_2}{P_1}\frac{v_2}{v_1}, \qquad (6.32)$$

and substituting into Equation (6.31) gives

$$\Delta s = s_2 - s_1 = c_v \ln \frac{P_2}{P_1} + (c_v + R)\ln \frac{v_2}{v_1}. \tag{6.33}$$

For an ideal gas, $c_p = c_v + R$, so

$$\Delta s = s_2 - s_1 = c_v \ln \frac{P_2}{P_1} + c_p \ln \frac{v_2}{v_1}. \tag{6.34}$$

If we want temperature ($T$) and pressure ($P$) as independent variables, we can write the Gibbs equation, Equation (6.26), as

$$Tds = du + Pdv. \tag{6.35}$$

Then, combining it with the definition of specific enthalpy, $h = u + Pv$, differentiated to give

$$dh = du + Pdv + vdP, \tag{6.36}$$

and substituting Equation (6.36) into Equation (6.35) gives

$$Tds = dh - vdP. \tag{6.37}$$

This is an alternate form of the Gibbs equation. As before, we substitute in the ideal gas equation and the change in enthalpy ($h$) of an ideal gas,

$$dh = c_p (T)dT,$$

to give

$$ds = \frac{c_p (T)}{T} dT - R \frac{dP}{P}. \tag{6.38}$$

Integrating Equation (6.38) from state 1 to state 2 gives the change in specific entropy:

$$\Delta s = s_2 - s_1 = \int_{T_1}^{T_2} \frac{c_p (T)}{T} dT - \int_{P_1}^{P_2} R \frac{dP}{P}. \tag{6.39}$$

Assuming constant specific heat ($c_p$) allows integration:

$$\Delta s = s_2 - s_1 = c_p \ln \frac{T_2}{T_1} - R \ln \frac{P_2}{P_1}. \tag{6.40}$$

Equations (6.31), (6.34) and (6.40) can be used to calculate changes in entropy with changes in temperature, pressure and volume for an ideal gas.

**Example 6.2**

**Problem:** A cylinder contains 0.2 kg of hydrogen gas with specific volume 1.0 m³ / kg and temperature 350 K. The gas is compressed by a piston to a final state of 0.2 m³ / kg and 650 K. Find the entropy change of the gas during this process.

**Find:** Entropy change $\Delta S$ of the gas.

**Known:** Mass of hydrogen gas $m = 0.2$ kg, initial temperature $T_1 = 350$ K, initial specific volume $v_1 = 1.0$ m³ / kg, final temperature $T_2 = 650$ K, final specific volume $v_2 = 0.2$ m³ / kg.

**Assumptions:** Hydrogen gas is ideal and has constant specific heats.

**Governing Equation:**

Entropy change (ideal gas, constant specific heats)
$$\Delta s = s_2 - s_1 = c_v \ln \frac{T_2}{T_1} + R \ln \frac{v_2}{v_1}$$

**Properties:** The average temperature is $T_{avg} = (T1 + T2)/2 = (650\text{ K} + 350\text{ K})/2 = 500$ K, gas constant of hydrogen $R = 4.124$ kJ / kgK (Appendix 1), at 500 K hydrogen specific heat $c_{v,avg} = 10.389$ kJ / kgK (Appendix 4).

**Solution:**

$$\Delta s = s_2 - s_1 = 10.389\,\text{kJ}/\text{kgK} \times \ln\frac{650\,\text{K}}{350\,\text{K}} + 4.1242\,\text{kJ}/\text{kgK} \times \ln\frac{0.2\,\text{m}^3/\text{kg}}{1.0\,\text{m}^3/\text{kg}} = -0.20645\,\text{kJ}/\text{kgK}$$

$$\Delta S = m\Delta s = 0.2\,\text{kg} \times (-0.20645\,\text{kJ}/\text{kgK}) = -0.041289\,\text{kJ}/\text{K}$$

**Answer:** The entropy of the hydrogen decreases by 0.0413 kJ / K. ∎

## 6.6.2  Ideal Gas Tables

What if the temperature interval over which we are calculating entropy changes is so large that we cannot assume specific heats to be constant? In that case we can use the ideal gas tables. We define a variable

$$s^o = \int_{T_{ref}}^{T} \frac{c_p(T)}{T}\,dT, \tag{6.41}$$

where $T_{ref}$ is a reference temperature. We can rewrite Equation (6.39) as

$$\Delta s = \int_{T_{ref}}^{T_2} \frac{c_p(T)}{T}\,dT - \int_{T_{ref}}^{T_1} \frac{c_p(T)}{T}\,dT - \int_{P_1}^{P_2} R\frac{dP}{P}, \tag{6.42}$$

or, substituting in Equation (6.41),

$$\Delta s = s^o(T_2) - s^o(T_1) - R\ln\frac{P_2}{P_1}. \tag{6.43}$$

Values of $s^o$ have been calculated for air and listed in tables (Appendix 7) as a function of temperature, assuming $s = 0$ at $T_{ref} = 0$ K.

## Example 6.3

**Problem:** Air is compressed from 100 kPa and 27 °C to 300 kPa and 167 °C. Find the change in specific entropy of the gas (a) assuming constant specific heat and (b) using air tables.

**Find:** Change in entropy $\Delta S$.

**Known:** Mass of the air $m = 10.0$ kg, initial temperature $T_1 = 27$ °C $= 300.15$ K, initial pressure $P_1 = 100$ kPa, final temperature $T_2 = 167$ °C $= 440.15$ K, final pressure $P_2 = 300$ kPa.

**Assumptions:** Air is an ideal gas with constant specific heats.

**Governing Equations:**

Entropy change (ideal gas, constant specific heats)
$$\Delta s = s_2 - s_1 = c_p \ln\frac{T_2}{T_1} - R\ln\frac{P_2}{P_1}$$

Entropy change (ideal gas)
$$\Delta s = s^o\left(T_2\right) - s^o\left(T_1\right) - R\ln\frac{P_2}{P_1}$$

**Properties:** The average temperature is $T_{avg} = (T1 + T2)/2 = (300.15\ \text{K} + 440.15\ \text{K})/2 = 370.15$ K, gas constant of air $R = 0.287$ kJ / kgK (Appendix 1), at 370 K (interpolation) air specific heat $c_{p,avg} = 1.010$ kJ / kgK (Appendix 4).

## Solution:

(a) With constant specific heat:

$$\Delta s = c_p \ln\frac{T_2}{T_1} - R\ln\frac{P_2}{P_1}$$

$$\Delta s = 1.010\,\text{kJ}/\text{kgK} \times \ln\frac{440\,\text{K}}{300\,\text{K}} - 0.287\,\text{kJ}/\text{kgK} \times \ln\frac{300\,\text{kPa}}{100\,\text{kPa}} = 0.071360\,\text{kJ}/\text{kgK}$$

(b) From air tables: At 300 K, $s_1^o = 1.70203$ kJ / kgK and, at 440 K, $s_2^o = 2.08870$ kJ / kgK.

$$\Delta s = s^o\left(T_2\right) - s^o\left(T_1\right) - R\ln\frac{P_2}{P_1}$$

$$\Delta s = 2.08870\,\text{kJ}/\text{kgK} - 1.70203\,\text{kJ}/\text{kgK} - 0.287\,\text{kJ}/\text{kgK} \times \ln\frac{300\,\text{kPa}}{100\,\text{kPa}} = 0.071368\,\text{kJ}/\text{kgK}$$

**Answer:** The change in specific entropy (a) assuming constant specific heat is 0.0713 kJ / kgK and (b) using air tables is 0.0713 kJ / kgK. These values are the same up until the 5th significant digit using the given constants. ∎

## Example 6.4

**Problem:** A cylinder contains 200 l of air at 150 kPa and 32 °C. The pressure of the gas is constant while 70 kJ of heat are added to the gas. Find its entropy change (a) assuming constant specific heats and (b) using air tables.

**Find:** Entropy increase of gas $\Delta S$ due to heating.

**Known:** Initial volume $V_1 = 200\ l = 0.2\ \text{m}^3$, initial pressure $P_1 = 150$ kPa, initial temperature $T_1 = 32$ °C $= 305.15$ K, final pressure $P_2 = P_1$, heat added $Q_{12} = 70$ kJ.

**Assumptions:** Air is an ideal gas with constant specific heats.

**Governing Equations:**

First law (constant pressure) $\qquad\qquad\qquad\qquad\qquad\qquad Q_{12} = m\left(h_2 - h_1\right)$

Entropy change (ideal gas, constant specific heats) $\qquad\quad \Delta s = s_2 - s_1 = c_p \ln\dfrac{T_2}{T_1} - R\ln\dfrac{P_2}{P_1}$

**Properties:** Air has gas constant $R = 0.287$ kJ / kgK (Appendix 1) and at 305 K (interpolation) air specific heat $c_p = 1.0053$ kJ / kgK (Appendix 4).

**Solution:**

Mass of gas,

$$m = \frac{P_1 V_1}{RT_1} = \frac{150 \text{ kPa} \times 0.2 \text{ m}^3}{0.2870 \text{ kJ.kgK} \times 305.15 \text{ K}} = 0.34255 \text{ kg.}$$

(a) Assuming constant specific heat:

$$Q_{12} = m\left(h_2 - h_1\right) = mc_p\left(T_2 - T_1\right)$$

$$70 \text{ kJ} = 0.34255 \text{ kg} \times 1.0053 \text{ kJ / kgK}\left(T_2 - 305.15 \text{ K}\right)$$

$$T_2 = 508.69 \text{ K}$$

The change in entropy is

$$\Delta S = m\Delta s = m\left(c_p \ln\frac{T_2}{T_1} - \underbrace{R\ln\frac{P_2}{P_1}}_{=\,0}\right) = mc_p \ln\frac{T_2}{T_1}.$$

Since $P_2 = P_1$, the second term is zero, then

$$\Delta S = m\Delta s = 0.34255 \text{ kg}\left(1.0053 \text{ kJ / kgK} \times \ln\frac{508.69 \text{ K}}{305.15 \text{ K}}\right) = 0.17598 \text{ kJ / K.}$$

(b) Using gas tables:

From the tables, specific enthalpy $h_1(305 \text{ K}) = 305.22$ kJ / kg, specific entropy $s_1(305 \text{ K}) = 1.71865$ kJ / kgK. Using the first law,

$$h_2 = \frac{Q}{m} + h_1 = 509.57 \text{ kJ / kg.}$$

Looking at the air tables (Appendix 7) the value of $h_2$ lies between $h(500 \text{ K})$ and $h(510 \text{ K})$. The value of $s_2$ must also lie in this temperature range.

| $T$ (K) | $h$ (kJ / kg) | $p_r$ | $u$ (kJ / kg) | $v_r$ | $s^o$ (kJ / kgK) |
|---|---|---|---|---|---|
| 500 | 503.02 | 8.411 | 359.49 | 170.6 | 2.21952 |
| $T_2$ | 509.57 | | | | $s_2$ |
| 510 | 513.32 | 9.031 | 366.92 | 162.1 | 2.23993 |

We can find the exact value of $s_2$ by linear interpolation (see Figure E6.4):

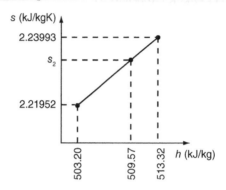

**Figure E6.4** Linear interpolation to find the value of $s$ corresponding to $h = 509.57$ kJ/kg.

$$\frac{s_2 - 2.21952}{2.23993 - 2.21952} = \frac{509.57 - 503.20}{513.32 - 503.20},$$

$$s_2 = 2.2324 \text{ kJ / kgK}.$$

$$\Delta S = m\left(s_2 - s_1\right) = 0.34255 \text{ kg} \times \left(2.2324 \text{ kJ / kgK} - 1.71865 \text{ kJ / kgK}\right) = 0.17597 \text{ kJ / K}.$$

**Answer:** The increase in entropy (a) assuming constant specific heat is 0.178 kJ / K and (b) using air tables is 0.176 kJ / K. ∎

## 6.7 Isentropic Processes in Ideal Gases

It is often convenient to idealise a process as being adiabatic and reversible, in which case it is isentropic. If $\Delta S = 0$ we can derive relationships between the temperature, pressure and volume and the start and end of the process.

### 6.7.1 Constant Specific Heats

For an isentropic process, $\Delta s = 0$. Substituting this into Equation (6.31),

$$c_v \ln \frac{T_2}{T_1} = -R \ln \frac{v_2}{v_1}. \tag{6.44}$$

Rearranging terms,

$$\ln\frac{T_2}{T_1} = \ln\left(\frac{v_1}{v_2}\right)^{R/c_v}.$$

(6.45)

For an ideal gas we have shown that

$$c_p - c_v = R.$$

Also, we defined the specific heat ratio:

$$\gamma = \frac{c_p}{c_v}.$$

Combining these two equations,

$$\frac{R}{c_v} = \frac{c_p}{c_v} - 1 = \gamma - 1.$$

(6.46)

Substituting into Equation (6.45), for an ideal gas undergoing an isentropic process

$$\frac{T_2}{T_1} = \left(\frac{v_1}{v_2}\right)^{(\gamma-1)}.$$

(6.47)

Similarly, starting with Equation (6.40), we can show that

$$\frac{T_2}{T_1} = \left(\frac{P_2}{P_1}\right)^{(\gamma-1)/\gamma}.$$

(6.48)

Finally, combining Equations (6.47) and (6.48),

$$\frac{P_2}{P_1} = \left(\frac{v_1}{v_2}\right)^{\gamma}.$$

(6.49)

Equations (6.47), (6.48) and (6.49) give the changes in pressure, temperature and specific volume of an ideal gas undergoing an isentropic process. Equation (6.49) shows that the isentropic compression or expansion of an ideal gas is a polytropic process represented by

$$Pv^{\gamma} = \text{constant}.$$

### Example 6. 5
**Problem:** An insulated cylinder fitted with a piston contains 5 kg of air at 500 kPa and 1000 K. The air expands in an adiabatic process until its volume doubles. Calculate the work done by the air.

**Find:** Work $W$ done by air during expansion.

**Known:** Mass of air $m = 5$ kg, initial pressure $P_1 = 500$ kPa, initial temperature $T_1 = 1000$ K, final volume $v_2 = 2v_1$, adiabatic process so $Q_{12} = 0$.

**Assumptions:** Air is an ideal gas with constant specific heats, the process is reversible and adiabatic so $\Delta s = 0$

**Governing equations:**

Isentropic process (ideal gas, constant specific heats)
$$\frac{T_2}{T_1} = \left(\frac{v_1}{v_2}\right)^{(\gamma-1)}$$

First law
$$\Delta U = Q_{12} + W_{12} = Q_{12} + mc_v\left(T_2 - T_1\right)$$

**Properties:** Air at 1000 K has specific heat $c_v = 0.855$ kJ / kgK (Appendix 4), specific heat ratio of air at 1000 K $\gamma = 1.336$ (Appendix 4).

**Solution:**

Solving for the final temperature,

$$T_2 = T_1\left(\frac{v_1}{v_2}\right)^{(\gamma-1)} = 1000\,\text{K}\left(\frac{1}{2}\right)^{(1.336-1)} = 792.235\,\text{K}.$$

For an adiabatic process $Q_{12} = 0$, so the first law reduces to

$$W_{12} = mc_v\left(T_2 - T_1\right) = 5\,\text{kg} \times 0.855\,\text{kJ}/\text{kgK} \times \left(792.235\,\text{K} - 1000\,\text{K}\right) = -888.195\,\text{kJ}.$$

**Answer:** The gas does 888.2 kJ of work on the surroundings.                                    ■

**Example 6.6**

**Problem:** Air at 100 kPa and 20 °C is compressed in a continuous process in an adiabatic compressor to a pressure of 400 kPa. Determine the exit temperature and the work required per kilogram of air.

**Find:** Exit temperature $T_2$ and work required per kilogram of air $w_{12}$ for compression.

**Known:** Initial pressure $P_1 = 100$ kPa, initial temperature $T_1 = 20\,°\text{C} = 293.15$ K, final pressure $P_2 = 400$ kPa, adiabatic process so $Q_{12} = 0$.

**Assumptions:** Air is an ideal gas with constant specific heat, the process is reversible and adiabatic so $\Delta s = 0$, changes in kinetic and potential energy are negligible in the compressor.

**Governing equations:**

Isentropic process (ideal gas)
$$\frac{T_2}{T_1} = \left(\frac{P_2}{P_1}\right)^{(\gamma-1)/\gamma}$$

First law (control volume)
$$w_{12} = h_2 - h_1 = c_p\left(T_2 - T_1\right)$$

**Properties:** Air at 293.15 K (interpolation) has specific heat $c_p = 1.004$ kJ / kgK (Appendix 4), specific heat ratio of air at 293.15 K (interpolation) $\gamma = 1.400$ (Appendix 4).

**Solution:**

The exit temperature is

$$T_2 = T_1 \left( \frac{P_2}{P_1} \right)^{(\gamma-1)/\gamma},$$

$$T_2 = 293.15\,\text{K} \left( \frac{400\,\text{kPa}}{100\,\text{kPa}} \right)^{(1.400-1)/1.400} = 435.66\,\text{K}$$

The work done per kilogram of gas is

$$w_{12} = c_p \left( T_2 - T_1 \right) = 1.004\,\text{kJ/kgK} \times \left( 435.66\,\text{K} - 293.15\,\text{K} \right) = 143.19\,\text{kJ/kg}$$

**Answer:** The exit temperature is 435.7 K and the work done per unit mass of air is 143.2 kJ / kg. ∎

### 6.7.2  Ideal Gas Tables

For an ideal gas undergoing an isentropic process, Equation (6.43) reduces to

$$s^o \left( T_2 \right) - s^o \left( T_1 \right) - R \ln \frac{P_2}{P_1} = 0. \tag{6.50}$$

For an isentropic compression, if we know the initial temperature $(T_1)$ and the compression ratio $(P_2 / P_1)$, we can use calculate $s^o(T_2)$ from the equation

$$s^o \left( T_2 \right) = s^o \left( T_1 \right) + R \ln \frac{P_2}{P_1}, \tag{6.51}$$

by looking up $s^o(T_1)$ from the ideal gas tables (Appendix 7). Knowing the final value of $s^o(T_2)$, the temperature $T_2$ can be determined by interpolation in the tables.

If the initial and final temperatures are known in an isentropic process the compression ratio can be determined by writing Equation (6.51) as

$$\frac{P_2}{P_1} = \frac{\exp\left[ s^o \left( T_2 \right) / R \right]}{\exp\left[ s^o \left( T_1 \right) / R \right]}. \tag{6.52}$$

The *relative pressure* $(P_r)$ is defined as

$$P_r \left( T \right) = \exp\left[ \frac{s^o \left( T \right)}{R} \right], \tag{6.53}$$

and is a function of temperature alone. We can then write Equation (6.52) as

$$\frac{P_2}{P_1} = \frac{P_{r2}}{P_{r1}}. \tag{6.54}$$

Values of the relative pressure for air as a function of temperature are tabulated in the ideal gas tables in Appendix 7.

If we need to calculate specific volumes during an isentropic process we use the ideal gas equation $v = RT/P$ and write the ratio of the specific volumes as

$$\frac{v_2}{v_1} = \frac{RT_2}{P_2}\frac{P_1}{RT_1} = \frac{RT_2}{RT_1}\frac{P_1}{P_2}. \tag{6.55}$$

Combining with Equation (6.54) gives

$$\frac{v_2}{v_1} = \frac{RT_2}{RT_1}\frac{P_{r1}}{P_{r2}} = \frac{RT_2}{P_{r2}}\frac{P_{r1}}{RT_1}. \tag{6.56}$$

We define a *relative volume* as

$$v_r(T) = \frac{RT}{P_r}. \tag{6.57}$$

Values of $v_r(T)$ have been calculated for air and are listed in the ideal gas table in Appendix 7. We can use these to calculate changes in specific volume for isentropic processes:

$$\frac{v_2}{v_1} = \frac{v_{r2}}{v_{r1}}. \tag{6.58}$$

**Example 6.7**
**Problem:** Air is compressed in an isentropic process from a pressure of 1 bar and temperature of 340 K to a temperature of 720 K. Find the final pressure.
**Find:** Final pressure $P_2$.
**Known:** Initial temperature $T_1 = 340$ K, initial pressure $P_1 = 1$ bar, final temperature $T_2 = 720$ K, isentropic process.
**Assumptions:** Air is an ideal gas.
**Governing Equation:**
Pressure ratio (isentropic)        $\dfrac{P_2}{P_1} = \dfrac{P_{r2}}{P_{r1}}$

**Properties:** For air at 340 K relative pressure $P_{r1} = 2.149$ (Appendix 7), air at 720 K relative pressure $P_{r2} = 32.02$ (Appendix 7).

**Solution:**

$$P_2 = P_1\frac{P_{r2}}{P_{r1}} = 1\,\text{bar}\left(\frac{32.02}{2.149}\right) = 14.900\,\text{bar}$$

**Answer:** The final pressure is 14.9 bar.                                                                         ∎

**Example 6.8**

**Problem:** A fixed mass of air is compressed in an isentropic process from 12 °C and 90 kPa to a final volume that is 10% of the initial volume. Find the final temperature.

**Find:** Final temperature $T_2$.

**Known:** Initial temperature $T_1 = 12\ °C = 285.15$ K, initial pressure $P_1 = 90$ kPa, final volume $V_2 = 0.1V_1$, isentropic process.

**Assumptions:** Air is an ideal gas.

**Governing Equation:**

Volume ratio (istentropic)
$$\frac{v_2}{v_1} = \frac{v_{r2}}{v_{r1}}$$

Isentropic process (ideal gas, constant specific heats)
$$\frac{T_2}{T_1} = \left(\frac{v_1}{v_2}\right)^{(\gamma-1)}$$

**Properties:** For air at 285 K (approximation) relative volume $v_{r1} = 706.1$ (Appendix 7).

**Solution:**

The final relative volume is

$$v_{r2} = v_{r1}\left(\frac{v_2}{v_1}\right) = 706.1 \times 0.1 = 70.610.$$

Interpolating in the air tables (Appendix 7) for $v_{r2} = 70.610$ gives $T_2 = 696.96$ K. Alternatively, we can assume constant specific heats: we do not know what the final temperature is, so let us use the value of $\gamma$ at 300 K: $\gamma=1.400$ (Appendix 4),

$$T_2 = T_1\left(\frac{v_1}{v_2}\right)^{(\gamma-1)} = 285.15\,\text{K}\,(10)^{(1.400-1)} = 716.26\,\text{K}.$$

We see that the temperature range is large, so that

$$T_{\text{avg}} = \frac{(T_1 + T_2)}{2} = \frac{(285.15\,\text{K} + 716.26\,\text{K})}{2} = 500.71\,\text{K}.$$

Let us recalculate, using the value of $\gamma$ at 500 K (approximation): $\gamma = 1.387$ (Appendix 4),

$$T_2 = T_1\left(\frac{v_1}{v_2}\right)^{(\gamma-1)} = 285.15\,\text{K}\,(10)^{(1.387-1)} = 695.14\,\text{K}.$$

This second iteration agrees much better with the results from the air tables.

**Answer:** The final air temperature is 697 K from the air tables, or 695 K assuming constant specific heats. ∎

## 6.8 Reversible Heat Transfer

The first law of thermodynamics for a simple compressible substance to which energy is being added as both heat ($\delta q$) and work ($\delta w$) per unit mass is

$$\delta q + \delta w = du.$$

For a simple compressible substance undergoing reversible compression or expansion $\delta w = -Pdv$, so that

$$\delta q = du + Pdv, \tag{6.59}$$

where $P$ is the system pressure. Substituting this into the Gibbs equation, Equation (6.26), gives

$$ds = \frac{1}{T}(du + Pdv) = \left(\frac{\delta q}{T}\right)_{int\ rev}. \tag{6.60}$$

The subscript "int rev" is to remind us that this expression is true only for an internally reversible process. If there are irreversibilities in the system, such as friction or temperature differences, entropy will be created and the entropy change will be larger than that given by this equation. Multiplying both sides of Equation (6.60) by the system mass ($m$), we can write it in terms of extensive properties,

$$dS = \left(\frac{\delta Q}{T}\right)_{int\ rev}. \tag{6.61}$$

Equation (6.61) gives a relation between heat transfer to a system and its change in entropy. Suppose a closed system is initially at equilibrium, in state 1. Heat is added to the system in an internally reversible process (so that no entropy is created) until it reaches a new equilibrium state 2. To find the change in entropy of the system during this process we can integrate $dS$ from the initial state 1 to the final state 2 to give

$$\Delta S = S_2 - S_1 = \int_1^2 \left(\frac{\delta Q}{T}\right)_{int\ rev}. \tag{6.62}$$

If the system is a thermal reservoir at constant temperature we can take $T$ out of the integral so that

$$\Delta S = \frac{1}{T}\int_1^2 \delta Q_{int\ rev} = \frac{Q_{12}}{T}. \tag{6.63}$$

We can write Equation (6.61) as a rate equation:

$$\dot{S} = \frac{dS}{dt} = \frac{\dot{Q}_{int\ rev}}{T}, \tag{6.64}$$

where $\dot{S}$ is the rate of entropy transfer, with units of W / K.

If a process is irreversible we can add the entropy generated ($S_{gen}$) to Equation (6.62):

$$\Delta S = S_2 - S_1 = \int_1^2 \left(\frac{\delta Q}{T}\right)_{irr} + S_{gen}. \tag{6.65}$$

Since entropy can only be generated, not destroyed, $S_{gen} > 0$ and

$$\Delta S = S_2 - S_1 > \int_1^2 \left(\frac{\delta Q}{T}\right)_{irr}. \tag{6.66}$$

## 6.9   T-S Diagrams

Thermodynamic processes involving heat transfer can be conveniently represented on diagrams in which the axes are the system temperature ($T$) and entropy ($S$), or specific entropy ($s$). Figure 6.3 shows an example of such a process in which a system, initially at equilibrium state 1, exchanges heat and work with the surroundings in a reversible manner. The final equilibrium state of the system is state 2.

Since the process is reversible the system is in a state of quasi-equilibrium throughout the process and the system temperature and entropy can be specified at each stage. From Equation (6.61),

$$\delta Q_{\text{int rev}} = TdS.$$
(6.67)

Integrating Equation (6.67) from state 1 to state 2,

$$Q_{12} = \int_{1}^{2} TdS.$$
(6.68)

$Q_{12}$ is the total heat transfer during the internally reversible process and equals the integral in Equation (6.68), which is the shaded area under the curve in Figure 6.3. $T$-$S$ diagrams give a graphical representation of heat transfer during reversible processes. Dividing both sides of Equation (6.68) by the system mass ($m$) we get

$$q_{12} = \frac{Q_{12}}{m} = \int_{1}^{2} Tds.$$
(6.69)

We can also represent reversible processes on $T$-$S$ diagrams, in which the area under the curve gives the reversible heat transfer per unit mass of the system, $q_{12} = Q_{12} / m$.

## 6.10   Entropy Balance for a Control Mass

Entropy can be transferred to a thermodynamic system ($S_{\text{in}}$) or removed from it ($S_{\text{out}}$), and entropy can also be generated within the system ($S_{\text{gen}}$) due to irreversible processes (Figure 6.4). The change in entropy of the system,

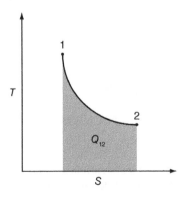

**Figure 6.3**   Thermodynamic process shown on a $T$-$S$ diagram. The shaded area under the curve represents the heat transfer $Q_{12}$ during the process in which the system goes from initial state 1 to final state 2.

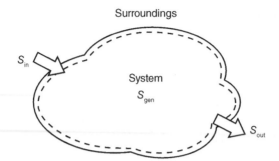

**Figure 6.4**   A system has entropy transferred to it ($S_{in}$) and loses entropy to the surroundings ($S_{out}$) while entropy is generated ($S_{gen}$) inside it.

$$\Delta S = S_{in} - S_{out} + S_{gen}.$$   (6.70)

For a system in steady state these terms must add up to zero, so that there is no net change in the entropy of the system. In general we can write an entropy balance:

$$S_{in} - S_{out} + S_{gen} = 0.$$   (6.71)

No entropy is generated in an internally reversible system so that $S_{gen} = 0$ and,

$$S_{in} = S_{out}.$$   (6.72)

We can also write the entropy balance on a rate basis,

$$\frac{dS}{dt} = \dot{S}_{in} - \dot{S}_{out} + \dot{S}_{gen},$$   (6.73)

where $\dot{S}_{in}$, $\dot{S}_{out}$ and $\dot{S}_{gen}$ are respectively the rates at which entropy enters, leaves and is generated in the system.

In a control mass the only way for entropy to enter or leave the system is with heat transfer. The rate of entropy transfer with heat ($\dot{S}_{heat}$) depends on the rate of heat transfer ($\dot{Q}$):

$$\dot{S}_{heat} = \frac{\dot{Q}}{T}.$$   (6.74)

Summing all streams of heat transfer

$$\frac{dS}{dt} = \sum_j \frac{\dot{Q}_j}{T_j} + \dot{S}_{gen},$$   (6.75)

where $T_j$ is the instantaneous temperature of the boundary across which heat transfer $\dot{Q}_j$ occurs.

**Example 6.9**
**Problem:** A rigid gas tank contains 0.5 kg of carbon dioxide at 810 K. The gas cools as it loses heat to the surrounding atmosphere at 290 K until it reaches equilibrium. Find the entropy generated.

**Find:** Entropy generated $\Delta S$.

**Known:** Mass of gas $m = 0.5$ kg, initial temperature of gas $T_1 = 810$ K, final temperature of gas $T_2 = 290$ K, temperature of atmosphere $T_2 = 290$ K, volume is constant $v_2 = v_1$.

**Assumptions:** Carbon dioxide is an ideal gas with constant specific heats.

**Governing Equations:**

First Law

$$Q + W = \Delta U$$

Rate of entropy transfer

$$\Delta S = \frac{Q}{T}$$

Entropy change (ideal gas, constant specific heats)   $\Delta s = s_2 - s_1 = c_v \ln\dfrac{T_2}{T_1} + R\ln\dfrac{v_2}{v_1}$

**Properties:** The average temperature is $T_{avg} = (T_1 + T_2)/2 = (810\ \text{K} + 290\ \text{K})/2 = 550$ K, specific heat of carbon dioxide at 550 K $c_v = 0.857$ kJ / kgK (Appendix 4).

**Solution:**

The entropy change of the gas,

$$\Delta S_{gas} = m\Delta s = m\left( c_v \ln\frac{T_2}{T_1} + \underbrace{R\ln\frac{v_2}{v_1}}_{=0} \right) = 0.5\,\text{kg} \times 0.857\,\text{kJ / kgK} \times \ln\left(\frac{290\,\text{K}}{810\,\text{K}}\right) = -0.440\ 14\,\text{kJ / K}.$$

Then, the heat lost from the $CO_2$ is

$$Q = mc_v\left(T_2 - T_1\right) = 0.5\,\text{kg} \times 0.857\,\text{kJ / kgK} \times \left(290\,\text{K} - 810\,\text{K}\right) = -222.82\,\text{kJ}.$$

The heat lost from the $CO_2$ is gained by the atmosphere, so that the entropy increase of atmosphere is

$$\Delta S_{atm} = \frac{Q}{T_{atm}} = \frac{222.82\,\text{kJ}}{290\,\text{K}} = 0.768\ 34\,\text{kJ / K}.$$

Then the amount of generated entropy during the cooling of the gas is

$$S_{gen} = \Delta S_{gas} + \Delta S_{atm} = 0.786\ 34\,\text{kJ / K} - 0.440\ 14\,\text{kJ / K} = 0.328\ 20\,\text{kJ / K}.$$

**Answer:** The entropy generated is 0.328 kJ / K.                                                      ■

## Example 6.10

**Problem:** Helium is compressed in a reversible process that requires a continuous power input of 5 kW. The temperature of the helium is kept at a constant 30 °C during compression by cooling it and transferring heat to the surrounding atmosphere at 20 °C. What is the rate of entropy generation?

**Find:** Rate of change of entropy for the helium $S_{He}$ and atmosphere $S_{atm}$.

**Known:** Work supplied $\dot{W} = 5\,\text{kW}$, temperature of helium $T_{\text{He}} = 30\ ^{\circ}\text{C} = 303.15\ \text{K}$, temperature of atmosphere $T_{\text{atm}} = 20\ ^{\circ}\text{C} = 293.15\ \text{K}$.

**Governing Equations:**

First Law as rate

$$\dot{Q} + \dot{W} = \frac{dE}{dt}$$

Rate of entropy transfer

$$\dot{S} = \frac{\dot{Q}_{\text{rev}}}{T}$$

**Solution:**

At steady state:

$$\dot{W} + \dot{Q} = 0$$

$$\dot{W} = -\dot{Q} = 5\,\text{kW}$$

$$\dot{S}_{\text{He}} = \frac{\dot{Q}}{T_{\text{He}}} = \frac{-5\,\text{kW}}{303.15\,\text{K}} = -16.493\,\text{W}\,/\,\text{K}$$

$$\dot{S}_{\text{atm}} = \frac{\dot{Q}}{T_{\text{atm}}} = \frac{5\,\text{kW}}{293.15\,\text{K}} = 17.056\,\text{W}\,/\,\text{K}$$

$$\dot{S}_{\text{gen}} = \dot{S}_{\text{atm}} + \dot{S}_{\text{He}} = 17.056\,\text{W}\,/\,\text{K} - 16.493\,\text{W}\,/\,\text{K} = 0.563\ 00\,\text{W}\,/\,\text{K}$$

**Answer:** Entropy is produced at a rate of 0.563 W / K.                                     ■

## 6.11   Entropy Balance for a Control Volume

In a control volume there is a second mechanism for entropy transfer: entropy may enter and leave the system along with mass. If the mass flow rate and specific entropy of the fluid at the inlet to the control volume are $\dot{m}_1$ and $s_1$ respectively, and those at the exit are $\dot{m}_2$ and $s_2$, then

$$\dot{S}_{\text{mass,in}} = \dot{m}_1 s_1, \tag{6.76}$$

$$\dot{S}_{\text{mass,out}} = \dot{m}_2 s_2. \tag{6.77}$$

Summing over all the streams of fluid flow and heat transfer (see Figure 6.5), Equation (6.73) becomes

$$\frac{dS}{dt} = \sum_j \frac{\dot{Q}_j}{T_j} + \sum_i \dot{m}_i s_i - \sum_e \dot{m}_e s_e + \dot{S}_{\text{gen}} \tag{6.78}$$

For a reversible ($\dot{S}_{\text{gen}} = 0$), adiabatic ($\dot{Q} = 0$) system at steady state ($dS/dt = 0$), the rate at which entropy is carried into the system with the fluid equals the rate at which is carried out. If there is only one inlet and one outlet $\dot{m}_1 = \dot{m}_2$ and

$$s_2 = s_1. \tag{6.79}$$

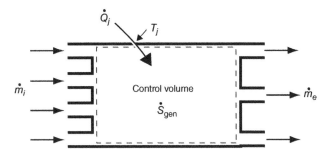

**Figure 6.5**   A control volume has entropy transferred to it from the surroundings along with mass and heat transfer, while entropy is generated ($S_{gen}$) inside it.

### Example 6.11

**Problem:** A compressor takes air from the atmosphere, which is at 100 kPa and 17 °C, and compresses it to 500 kPa. The compressor is cooled at a rate of 100 kJ / kg of air throughput and the heat rejected to the surrounding air. Irreversibilities in the compressor increase the entropy of the air flowing through it by 0.2 kJ / kg. Calculate the work done and the entropy generated per unit mass of air.

**Find:** Work done $w$ and entropy generated $\Delta s_{tot}$ per kg of air.

**Known:** Initial temperature $T_1 = 17\,°\text{C} = 290.15$ K, initial pressure $P_1 = 100$ kPa, final pressure $P_2 = 500$ kPa, temperature of atmosphere $T_{atm} = 290.15$ K, heat transfer to compressor $q = -100$ kJ / kg, increase in entropy of air $\Delta s_{air} = 0.2$ kJ / kg.

**Assumptions:** Air is an ideal gas with constant specific heats.

**Governing Equations:**

Entropy change (ideal gas, constant specific heats)

$$\Delta s = s_2 - s_1 = c_p \ln\frac{T_2}{T_1} - R\ln\frac{P_2}{P_1}$$

Steady flow balance

$$w + q = h_2 - h_1 = c_p\left(T_2 - T_1\right)$$

**Properties:** Air has gas constant $R = 0.2870$ kJ / kgK (Appendix 1), and at 290.15 K (interpolation) has specific heat $c_p = 1.0046$ kJ / kgK (Appendix 4).

### Solution:

We can find the final temperature using the increase in entropy of the air

$$\Delta s_{air} = c_p \ln\frac{T_2}{T_1} - R\ln\frac{P_2}{P_1} = 0.2\,\text{kJ/kgK},$$

$$1.0046\,\text{kJ/kgK} \times \ln\left(\frac{T_2}{290.15\,\text{K}}\right) - 0.2870\,\text{kJ/kgK} \times \ln\left(\frac{500\,\text{kPa}}{100\,\text{kPa}}\right),$$

$$T_2 = 560.750 \text{ K}.$$

Then the work per unit mass is

$$w = c_p \left(T_2 - T_1\right) - q = 1.0046 \, \text{kJ} / \text{kgK} \times \left(560.75 \, \text{K} - 290.15 \, \text{K}\right) + 100 \, \text{kJ} / \text{kg} = 371.845 \, \text{kJ} / \text{kg}.$$

The heat lost gained by the atmosphere from the compressor is $q_{\text{atm}} = -q = 100 \, \text{kJ} / \text{kg}$, so the increase in entropy of the atmosphere is

$$\Delta s_{\text{atm}} = \frac{q_{\text{atm}}}{T_{\text{atm}}} = \frac{100 \, \text{kJ} / \text{kg}}{290.15 \, \text{K}} = 0.344 \; 65 \, \text{kJ} / \text{kgK}.$$

The total entropy generated is then

$$\Delta s_{\text{tot}} = \Delta s_{\text{air}} + \Delta s_{\text{atm}} = 0.2 \, \text{kJ} / \text{kgK} + 0.344 \; 65 \, \text{kJ} / \text{kgK} = 0.544 \; 65 \, \text{kJ} / \text{kgK}.$$

**Answer:** The work done on the air is 371.8 kJ / kg and the total entropy generated is 0.545 kJ / kgK. ∎

## 6.12   Isentropic Steady Flow Devices

In Chapter 4 we learned to analyse steady flow devices such as turbines, compressors, pumps, nozzles and diffusers. Given the inlet and outlet conditions, we can do an energy balance and calculate work and heat transfers to the devices. There is one problem, though, that we will soon discover if we try to apply our theory to the design of real machines: we usually do not know what their exit conditions are.

Think about a typical situation that a design engineer faces in selecting a turbine. What information is available? To drive a gas turbine hot gases are generated in a combustion chamber whose temperature and pressure are known, so these two properties are known at the turbine inlet (Figure 6.6). Gases leaving the turbine are exhausted to the atmosphere, so the outlet pressure is fixed, but we do not know any other properties there. If the turbine was operating we could run an experiment and measure its exhaust temperature, but if we are only at the design stage we have no way of doing this. We need one more property at the outlet, since we know from the state postulate that two properties determine the state of a simple compressible substance.

We can solve this problem by assuming that the turbine is isentropic. A turbine is well insulated to prevent heat losses, so assuming it to be adiabatic is reasonable. In addition, ignoring all frictional losses, we assume flow through the device to be reversible. A reversible, adiabatic process is isentropic, so the specific entropy of the fluid in the turbine is the same at the outlet as at the inlet. We now have two independent, intensive properties, pressure and specific entropy, at the outlet and therefore the state is fixed. The work output that we calculate assuming an isentropic turbine is the highest possible amount if the inlet gas temperature and pressure and the outlet gas pressure are fixed.

Compressors and pumps can also be analysed the same way by assuming them to be isentropic and the calculations will give the lowest possible amount of work required to drive these devices. An isentropic nozzle will give the highest possible exit velocity for a given set of inlet conditions and an isentropic diffuser will give the maximum possible exit pressure, neglecting all losses due to irreversibilities.

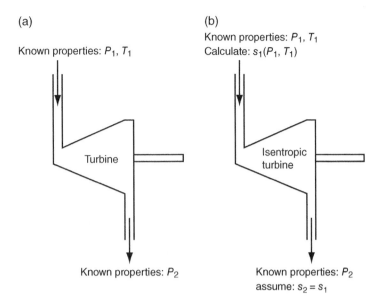

(a)

Known properties: $P_1$, $T_1$

Turbine

Known properties: $P_2$

(b)

Known properties: $P_1$, $T_1$
Calculate: $s_1(P_1, T_1)$

Isentropic
turbine

Known properties: $P_2$
assume: $s_2 = s_1$

**Figure 6.6** (a) A real turbine in which the inlet pressure ($P_1$) and temperature ($T_1$) are fixed as well as the outlet pressure ($P_2$). (b) An isentropic turbine in which the outlet and inlet specific entropy are assumed to be the same, fixing the outlet state.

**Example 6.12**

**Problem:** Air enters an adiabatic turbine at 6 MPa and 500 °C and leaves at 100 kPa with a flow rate of 2 kg / s. Determine the maximum possible power output from the turbine.

**Find:** Maximum power output $\dot{W}$.

**Known:** Mass flow rate $m = 2$ kg / s, inlet pressure $P_1 = 6$ MPa, inlet temperature $T_1 = 500$ °C $= 773.15$ K, outlet pressure $P_2 = 100$ kPa.

**Assumptions:** Air is an ideal gas with constant specific heats, the turbine is isentropic, changes in kinetic and potential energy are negligible.

**Governing Equations:**

Work output from a turbine

$$\dot{W}_{shaft} = \dot{m}\left(h_2 - h_1\right) = \dot{m}c_p\left(T_2 - T_1\right)$$

Isentropic process (ideal gas, constant specific heats)

$$\frac{T_2}{T_1} = \left(\frac{P_2}{P_1}\right)^{(\gamma-1)/\gamma}$$

**Properties:** Air at 773.15 K (interpolation) has specific heat $c_p = 1.093$ (Appendix 4), specific heat ratio of air at 773.15 K (interpolation) $\gamma = 1.357$ (Appendix 4).

**Solution:**

$$T_2 = T_1\left(\frac{P_2}{P_1}\right)^{(\gamma-1)/\gamma} = 773.15\,\text{K}\left(\frac{100\times10^3\,\text{Pa}}{6\times10^6\,\text{Pa}}\right)^{\frac{1.357-1}{1.357}} = 263.3\,\text{K}$$

$$\dot{W}_{shaft} = \dot{m}c_p\left(T_2 - T_1\right) = 2\,\text{kg}/\text{s}\times1.093\,\text{kJ}/\text{kgK}\left(263.3\,\text{K} - 773.15\,\text{K}\right) = -1115\,\text{kW}$$

**Answer:** The maximum power output is 1.1 MW. ∎

## 6.13   Isentropic Efficiencies

We can calculate the maximum work output of a turbine operating between two given pressures by assuming that the process is isentropic. But we know that entropy is generated during all real processes and that the real work output is less than that given by our calculation. How do we specify the magnitude of losses due to irreversibilities and connect our idealised analysis to the real world?

### 6.13.1   Isentropic Turbine Efficiency

Figure 6.7 shows the expansion of a gas through a turbine on a *T-S* diagram. Gas is supplied to the turbine at a high pressure ($P_1$), expands and leaves at a lower pressure ($P_2$). Both lines of constant pressure are marked on the figure. Since entropy increases with gas temperature when pressure is held constant [see Equation (6.40)] the slope of the isobars – lines of constant pressure – is positive. If we assume that the expansion is adiabatic and reversible the final state is marked by point 2s on the figure. In reality irreversibilities due to gas expansion will generate entropy in the gas. If heat losses from the turbine casing to the surroundings are negligible this entropy will still be present in the gas as it leaves the turbine and the specific entropy will be greater than the isentropic case (point 2s) and is indicated by point 2 in Figure 6.7. The work done in reality will be less than that in the idealised isentropic case since some of the internal energy that would have been transferred as shaft work is instead used to heat the fluid leaving the turbine.

If we neglect changes in kinetic and potential energy and assume that the expansion is adiabatic, the work output of a turbine is

$$\dot{W}_t = \dot{m}\left(h_2 - h_1\right).$$

The maximum possible work output from the turbine per unit mass of fluid, for the case of isentropic expansion is

$$w_{t,s} = \left(\frac{\dot{W}_t}{\dot{m}}\right)_s = h_{2s} - h_1. \tag{6.80}$$

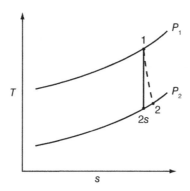

**Figure 6.7**   Gas expands in a turbine from high pressure ($P_1$) to low pressure ($P_2$). In a real process the entropy increases due to irreversibilities and the final state is point 2. Assuming isentropic expansion the final state is 2s.

In the real process, which is irreversible,

$$w_t = \frac{\dot{W}_t}{\dot{m}} = h_2 - h_1.$$  (6.81)

To quantify the magnitude of irreversibilities we define an *isentropic turbine efficiency* ($\eta_t$):

$$\eta_t = \frac{w_t}{w_{t,s}} = \frac{h_2 - h_1}{h_{2s} - h_1}.$$  (6.82)

It is quite easy to determine the isentropic efficiency of a particular turbine experimentally by measuring the actual work output and comparing it to the ideal work output calculated for an isentropic process. Turbine manufacturers routinely supply customers with data for isentropic efficiency, which vary typically from 70 to 90%. The real work output of a turbine can be estimated by multiplying its ideal work output, calculated assuming isentropic flow, by the isentropic efficiency.

## 6.13.2   Isentropic Nozzle Efficiency

Fluid accelerates in a nozzle as it goes from a high pressure to a low pressure, increasing its kinetic energy. Since there are negligible heat losses in a well-insulated nozzle, if we neglect irreversibilities we can assume that the flow is isentropic (as shown by the line 1-2s in Figure 6.7). Frictional losses in the nozzle decrease the fluid kinetic energy and raise the temperature of the fluid, increasing its entropy (line 1-2 in Figure 6.7). The *isentropic nozzle efficiency* ($\eta_n$) is defined as the ratio of the actual kinetic energy at the nozzle exit to the ideal kinetic energy assuming isentropic flow.

$$\eta_n = \frac{\mathbf{V}_2^2/2}{\mathbf{V}_{2s}^2/2} = \frac{\mathbf{V}_2^2}{\mathbf{V}_{2s}^2}.$$  (6.83)

Typical nozzles have efficiencies over 95%, so losses due to irreversibilities are very small.

### Example 6.13
**Problem:** Argon enters an adiabatic nozzle at 4 bar and 850 °C and exits at 1 bar. If the isentropic efficiency of the nozzle is 90%, find the exit velocity and temperature of the gas.
**Find:** Exit velocity $\mathbf{V}_2$ and temperature $T_2$ of argon.
**Known:** Inlet temperature $T_1 = 850\ °C = 1123.15\ K$, inlet pressure $P_1 = 4$ bar, exit pressure $P_2 = 1$ bar, isentropic efficiency $\eta_n = 0.9$.
**Assumptions:** Argon is an ideal gas with constant specific heats.
**Governing Equations:**

Isentropic nozzle efficiency

$$\eta_{nozzle} = \frac{\mathbf{V}_2^2}{\mathbf{V}_{2s}^2}$$

Nozzle exit velocity

$$\mathbf{V}_2 = \sqrt{2(h_2 - h_1)} = \sqrt{2c_p(T_2 - T_1)}$$

Isentropic process (ideal gas, constant specific heats)

$$\frac{T_2}{T_1} = \left(\frac{P_2}{P_1}\right)^{(\gamma-1)/\gamma}$$

**Properties:** Argon at 25 °C (approximation) has specific heat $c_p = 0.520 \, \text{kJ/kgK}$ (Appendix 1), specific heat ratio of argon at 25 °C (approximation) $\gamma = 1.667$ (Appendix 1).

**Solution:**
Assuming isentropic flow, the exit temperature is

$$T_{2s} = T_1 \left( \frac{P_2}{P_1} \right)^{(\gamma-1)/\gamma} = 1123 \, \text{K} \left( \frac{1 \, \text{bar}}{4 \, \text{bar}} \right)^{\frac{1.667-1}{1.667}} = 644.973 \, \text{K}.$$

The exit velocity assuming isentropic flow is

$$\mathbf{V}_{2s} = \sqrt{2c_p \left( T_1 - T_2 \right)} = \sqrt{2 \times 0.520 \times 10^3 \, \text{kJ/kg} \left( 1123.15 \, \text{K} - 644.973 \, \text{K} \right)} = 705.198 \, \text{m/s}.$$

The real velocity is then

$$\mathbf{V}_2 = \sqrt{\eta_{\text{nozzle}} \mathbf{V}_{2s}^2} = \sqrt{0.9 \left( 705.198 \, \text{m/s} \right)^2} = 669.009 \, \text{m/s}.$$

Then the real outlet temperature can be calculated:

$$\mathbf{V}_2 = \sqrt{2c_p \left( T_1 - T_2 \right)}$$

$$669.009 \, \text{m/s} = \sqrt{2 \times 0.520 \times 10^3 \, \text{kJ/kg} \left( 1123.15 \, \text{K} - T_2 \right)}$$

$$T_2 = 692.791 \, \text{K}$$

**Answer:** The exit velocity is 669.0 m/s and the exit temperature is 693 K.    ∎

## 6.13.3   Isentropic Pump and Compressor Efficiency

Pumps and compressors increase the pressure of a fluid (Figure 6.8). Neglecting heat transfer to the surroundings and changes in kinetic and potential energy the power required to compress fluid is

$$\dot{W}_c = \dot{m} \left( h_2 - h_1 \right).$$

The work per unit mass of fluid, assuming isentropic flow, is

$$w_{c,s} = \left( \frac{\dot{W}_c}{\dot{m}} \right)_s = h_{2s} - h_1, \tag{6.84}$$

whereas for the real process it is

$$w_c = \frac{\dot{W}_c}{\dot{m}} = h_2 - h_1. \tag{6.85}$$

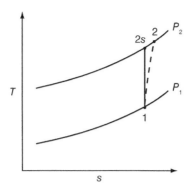

**Figure 6.8** Gas is taken in a compressor from low pressure ($P_1$) to high pressure ($P_2$). In a real process the entropy increases due to irreversibilities and the final state is point 2. Assuming isentropic expansion the final state is 2s.

The *isentropic compressor efficiency* ($\eta_c$) is

$$\eta_c = \frac{w_{c,s}}{w_c} = \frac{h_{2s} - h_1}{h_2 - h_1}.$$

(6.86)

Since the actual compression work is always greater than that calculated for an isentropic process, $\eta_c < 1$. Typical values of $\eta_c$ vary from 75 to 85%. An isentropic efficiency for pumps is defined the same way.

### Example 6.14

**Problem:** An adiabatic compressor with an isentropic efficiency of 80% takes air flowing at a rate of 0.5 kg / s from 90 kPa and 22 °C to 800 kPa. Use the air table to find the exit temperature of the air and the power input required to drive the compressor.

**Find:** Exit temperature $T_2$ of air and power required $\dot{W}$.

**Known:** Mass flow rate $\dot{m} = 0.5$ kg / s, inlet pressure $P_1 = 90$ kPa, inlet temperature $T_1 = 22$ °C $= 295.15$ K, outlet pressure $P_2 = 800$ kPa, isentropic efficiency $\eta_c = 80\%$.

**Governing equations:**

Power of compressor

$$\dot{W}_c = \dot{m}\left(h_i - h_e\right)$$

Isentropic compressor efficiency

$$\eta_c = \frac{h_{2s} - h_1}{h_2 - h_1}$$

Pressure ratio (isentropic)

$$P_{r2} = P_{r1}\left(\frac{P_2}{P_1}\right)$$

**Properties:** Air at 295 K (approximation) has specific enthalpy $h_1 = 295.17$ kJ / kgK (Appendix 7), relative pressure $P_{r1} = 1.3068$ (Appendix 7).

**Solution:**

Find the final relative pressure of the process,

$$P_{r2} = P_{r1}\left(\frac{P_2}{P_1}\right) = 1.3068\left(\frac{800\,\text{kPa}}{90\,\text{kPa}}\right) = 11.6160.$$

Interpolating from the air tables (Appendix 7), the corresponding enthalpy value is $h_{2s} = 552.083$ kJ / kg.

The actual exit enthalpy is

$$h_2 = \frac{h_{2s} - h_1}{\eta_c} + h_1 = \frac{552.083\,\text{kJ}/\text{kg} - 295.17\,\text{kJ}/\text{kg}}{0.8} + 295.17\,\text{kJ}/\text{kg} = 616.311\,\text{kJ}/\text{kg}.$$

Interpolating again from the air tables, the corresponding exit temperature is $T_2 = 608.840$ K.

The work required is

$$\dot{W}_c = \dot{m}(h_2 - h_1) = 0.5\,\text{kg}/\text{s}(616.311\,\text{kJ}/\text{kg} - 295.17\,\text{kJ}/\text{kg}) = 160.571\,\text{kW}.$$

**Answer:** The exit temperature is 609 K and the power required is 160.6 kW.                              ∎

## 6.14  Exergy

A system consists of gas contained in a cylinder that is initially at pressure $P$ and temperature $T$ while the surroundings are at pressure $P_o$ and temperature $T_o$ (Figure 6.9). The piston compressing the gas is released and the system exchanges heat with the surroundings until it reaches equilibrium when the pressure and temperature of the gas are the same as that outside. The system is now in a "dead state", where it cannot do any useful work.

If the pressure and temperature of the gas were initially higher than that of the surroundings it could, in principle, have been used to do useful work such as lifting a weight. Instead all the energy of the system was dissipated as heat or transferred as work to push back the surrounding air. Once the system reaches equilibrium it cannot do any more work. According to the first law, energy cannot be destroyed. However, *something* has been lost in this process, which is the ability of the system to do useful work. To measure the magnitude of this loss we define a new property: *Exergy is the maximum amount of useful work a system can do before it reaches equilibrium with its surroundings.*

How much work could the system have done if we had used its expansion in a more efficient way? This depends not only on the properties of the system, but also those of the surroundings. The greater the difference between the initial pressures on the two faces of the piston, the farther it could have moved.

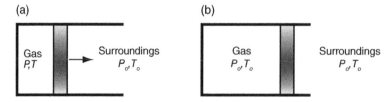

(a)

Gas
$P,T$

Surroundings
$P_o,T_o$

(b)

Gas
$P_o,T_o$

Surroundings
$P_o,T_o$

**Figure 6.9**  Gas in a cylinder (a) initially at pressure $P$ and temperature $T$, then (b) comes to equilibrium with the surroundings at pressure $P_o$ and temperature $T_o$.

## 6.14.1   Exergy of a Control Mass

To determine the exergy of a control mass let us calculate how much useful work it can do to lift a weight. As the piston in Figure 6.10 moves through an infinitesimal distance the gas in the cylinder exchanges heat ($\delta Q$) with the surroundings at temperature $T_o$. When the gas expands by $dV$, it does boundary work $\delta W_b = -P_o dV$ in pushing back the surrounding atmosphere at pressure $P_o$. It also does useful work ($\delta W_u$) in lifting the weight. An energy balance gives us

$$\delta Q + \delta W_b + \delta W_u = dE = dU + dKE + dPE. \tag{6.87}$$

Entropy is transferred to the system due to heat transfer from the surroundings and also generated within it due to internal irreversibilities. The change in entropy of the system is

$$dS = \frac{\delta Q}{T_o} + dS_{gen}. \tag{6.88}$$

Rearranging terms gives

$$\delta Q = T_o \left( dS - dS_{gen} \right). \tag{6.89}$$

Substituting Equation (6.89) into Equation (6.87),

$$\delta W_u = dE + P_o dV - T_o \left( dS - dS_{gen} \right). \tag{6.90}$$

The total useful work done during expansion is given by integrating Equation (6.90) from the initial to the final state:

$$W_u = \int_{E_1}^{E_2} dE + P_o \int_{V_1}^{V_2} dV - T_o \int_{S_1}^{S_2} dS + T_o \int_{0}^{S_{gen}} dS_{gen}, \tag{6.91}$$

resulting in

$$W_u = \left( U_2 - U_1 \right) + P_o \left( V_2 - V_1 \right) - T_o \left( S_2 - S_1 \right) + T_o S_{gen} + \left( KE_2 - KE_1 \right) + \left( PE_2 - PE_1 \right). \tag{6.92}$$

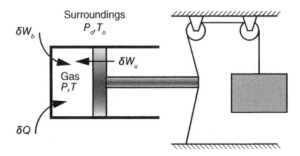

**Figure 6.10**   Gas in a cylinder initially at pressure $P$ and temperature $T$ lifts a weight as it expands and comes to equilibrium with the surroundings at pressure $P_o$ and temperature $T_o$.

The most that the system can expand is until it reaches the dead state, when it is at equilibrium with its surroundings. The exergy of the system is defined as

$$\Phi = \left(U - U_o\right) + P_o\left(V - V_o\right) - T_o\left(S - S_o\right) + \frac{m\mathbf{V}^2}{2} + mgz, \tag{6.93}$$

where the subscript $o$ indicates a property of the surroundings. Any kinetic or potential energy the system possesses can be converted to useful work, so these terms are also included in the exergy, with the velocity ($\mathbf{V}$) and height ($z$) of the system measured relative to the surroundings. The exergy of a system in equilibrium with its surroundings is zero. The useful work done by a system in going from state 1 to state 2 is

$$W_u = \Phi_2 - \Phi_1 + T_o S_{gen}. \tag{6.94}$$

When the system does net work on the surroundings $W_u < 0$, for which we require $\Phi_2 < \Phi_1$. Any entropy generated due to irreversibilities in the system will decrease the amount of useful work done by it. Exergy is therefore destroyed when entropy is created and

$$\Phi_{destroyed} = T_o S_{gen}. \tag{6.95}$$

For any reversible process $\Phi_{destroyed} = 0$, whereas for an irreversible process $\Phi_{destroyed} > 0$. If the system is reversible so that $S_{gen} = 0$ and state 2 is the dead state of the system where it is at equilibrium with the surroundings so that $\Phi_2 = 0$, the maximum useful work that is obtained from the system is $W_u = -\Phi_1$.

The exergy per unit mass of the system is

$$\phi = \left(u - u_o\right) + P_o\left(v - v_o\right) - T_o\left(s - s_o\right) + \frac{\mathbf{V}^2}{2} + gz, \tag{6.96}$$

and a change in exergy per unit mass is

$$\Delta\phi = \phi_2 - \phi_1 = \left(u_2 - u_1\right) + P_o\left(v_2 - v_1\right) - T_o\left(s_2 - s_1\right) + \frac{\left(\mathbf{V}_2^2 - \mathbf{V}_1^2\right)}{2} + g\left(z_2 - z_1\right). \tag{6.97}$$

**Example 6.15**

**Problem:** A compressed air tank contains 500 kg of air at 800 kPa and 400 K. How much work can be obtained from this if the atmosphere is at 100 kPa and 300 K?

**Find:** Maximum work $W_u$ that can be obtained from the compressed air.

**Known:** Mass of air $m = 500$ kg, air pressure $P = 800$ kPa, air temperature $T = 400$ K, atmospheric pressure $P_o = 100$ kPa, atmospheric temperature $T_o = 300$ K.

**Assumptions:** Air is an ideal gas with constant specific heats.

**Properties:** The average temperature of the air is $T_{avg} = (T_1 + T_2)/2 = (400\text{ K} + 300\text{ K})/2 = 350$ K, air has a gas constant of $R = 0.2870$ kJ / kgK (Appendix 1), and air at 350 K has specific heat at constant pressure $c_p$ = 1.008 kJ / kgK (Appendix 4), specific heat of air at constant volume at 350 K $c_v$ = 0.721 kJ / kgK (Appendix 4).

**Governing equations:**

Maximum useful work        $W_u = -m\phi = -m\left[\left(u - u_o\right) + P_o\left(v - v_o\right) - T_o\left(s - s_o\right) + \dfrac{V^2}{2} + gz\right]$

Ideal gas equation        $Pv = RT$

Entropy change (ideal gas, constant specific heats)        $\Delta s = c_p \ln\dfrac{T_2}{T_1} - R \ln\dfrac{P_2}{P_1}$

**Solution:**

Specific internal energy contribution:

$u - u_o = c_v\left(T - T_o\right) = 0.721\,\text{kJ} / \text{kgK} \times \left(400\,\text{K} - 300\,\text{K}\right) = 72.100\,\text{kJ} / \text{kg}.$

Boundary work per unit mass contribution:

$P_o\left(v - v_o\right) = P_o R\left(\dfrac{T}{P} - \dfrac{T_o}{P_o}\right) = 100\,\text{kPa} \times 0.2870\,\text{kJ} / \text{kgK}\left(\dfrac{400\,\text{K}}{800\,\text{kPa}} - \dfrac{300\,\text{K}}{100\,\text{kPa}}\right) = -71.750\,\text{kJ} / \text{kg}$

Specific entropy contribution: $T_o\left(s - s_o\right) = T_o\left(c_p \ln\dfrac{T}{T_o} - R \ln\dfrac{P}{P_o}\right)$

$= 300\,\text{K} \times \left(1.008\,\text{kJ} / \text{kgK} \times \ln\dfrac{400\,\text{K}}{300\,\text{K}} - 0.2870\,\text{kJ} / \text{kgK} \times \ln\dfrac{800\,\text{kPa}}{100\,\text{kPa}}\right) = -92.045\ \text{kJ} / \text{kg}.$

Total useful work:

$W_u = -m\phi = -500\,\text{kg}\left[72.100\,\text{kJ} / \text{kg} - 71.750\,\text{kJ} / \text{kg} + 92.045\,\text{kJ} / \text{kg}\right] = -46198\,\text{kJ}.$

**Answer:** The maximum amount of useful work that can be obtained from the compressed air is 46.2 MJ.  ∎

## 6.14.2   Exergy of a Control Volume

An energy balance on a control volume at steady state (see Figure 6.11) gives that

$$\dot{Q} + \dot{W} + \dot{m}\left(h_1 + \dfrac{V_1^2}{2} + gz_1\right) = \dot{m}\left(h_2 + \dfrac{V_2^2}{2} + gz_2\right). \tag{6.98}$$

An entropy balance on the same control volume gives

$$\dot{m}s_1 + \dfrac{\dot{Q}}{T_0} + \dot{S}_{gen} = \dot{m}s_2, \tag{6.99}$$

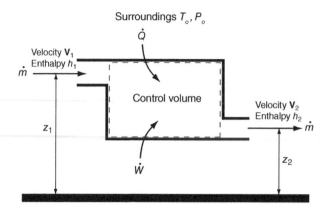

**Figure 6.11**   A fluid flows through a control volume at steady state.

where $T_o$ is the temperature of the surroundings and $\dot{S}_{gen}$ the rate of entropy generation due to irreversibilities in the control volume.

Substituting Equation (6.99) into Equation (6.98) and solving for the work done,

$$\dot{W} = \dot{m}\left[\left(h_2 - h_1\right) - T_0\left(s_2 - s_1\right) + \frac{V_2^2 - V_1^2}{2} + g\left(z_2 - z_1\right)\right] + T_0\dot{S}_{gen}.$$

$$(6.100)$$

The maximum amount of work that can be extracted from the control volume is when the fluid exits where it is at thermal and mechanical equilibrium with the surroundings. If we define the flow exergy per unit mass of the fluid as

$$\psi = \left(h - h_o\right) - T_o\left(s - s_o\right) + \frac{V^2}{2} + gz,$$

$$(6.101)$$

the work output of the control volume is

$$\dot{W} = \dot{m}\left(\psi_2 - \psi_1\right) + T_0\dot{S}_{gen}.$$

$$(6.102)$$

The maximum work output from the control volume is when flow is reversible so that $\dot{S}_{gen} = 0$. The reversible work is then

$$\dot{W}_s = \dot{m}\left(\psi_2 - \psi_1\right).$$

$$(6.103)$$

The rate at which exergy is destroyed is

$$\dot{W} - \dot{W}_s = T_0\dot{S}_{gen}.$$

$$(6.104)$$

**Example 6.16**
**Problem:** Air enters a gas turbine at 1100 K and 800 kPa with a velocity of 80 m / s. Determine the exergy of the air at the inlet to the turbine. Assume the atmosphere is at 300 K and 100 kPa.
**Find:** Flow exergy $\psi$ of the air at the turbine inlet.

**Known:** Air pressure $P = 800$ kPa, air temperature $T = 1100$ K, air velocity $\mathbf{V} = 80$ m / s, atmospheric pressure $P_o = 100$ kPa, atmospheric temperature $T_o = 300$ K.

**Assumptions:** Air is an ideal gas, negligible contribution from potential energy.

**Properties:** Air has gas constant $R = 0.2870$ kJ / kgK (Appendix 1), specific enthalpy at 1100 K $h(1100$ K$) = 1161.07$ kJ / kg (Appendix 7), specific enthalpy $h(300$ K$) = 300.19$ kJ / kg (Appendix 7), specific entropy $s^o(1100$ K$) = 3.07732$ kJ / kgK (Appendix 7), specific entropy $s^o(300$ K$) = 1.70203$ kJ / kgK (Appendix 7).

**Governing equations:**

Flow exergy per unit mass
$$\psi = (h - h_o) - T_o(s - s_o) + \frac{\mathbf{V}^2}{2} + \underbrace{gz}_{=0}$$

Entropy change (ideal gas)
$$\Delta s = s^o(T_2) - s^o(T_1) - R \ln \frac{P_s}{P_1}$$

**Solution:**

Specific enthalpy contribution:

$$h - h_o = h(1100 \text{K}) - h(300 \text{K}) = 1161.07 \text{kJ / kg} - 300.19 \text{kJ / kg} = 860.880 \text{kJ / kg}.$$

Specific entropy contribution: $s - s_o = s^o(1100 \text{K}) - s^o(300 \text{K}) - R \ln \frac{P}{P_o}$

$$= 3.07732 \text{kJ / kgK} - 1.70203 \text{kJ / kgK} - 0.2870 \text{kJ / kgK} \times \ln \frac{800 \text{kPa}}{100 \text{kPa}} = 0.778490 \text{ kJ / kgK}$$

Total flow exergy:

$$\psi = 860.880 \text{kJ / kg} - 300 \text{K} \times 0.778490 \text{kJ / kgK} + \frac{(80 \text{m / s})^2}{2 \times 1000 \text{J / kJ}} = 630.533 \text{kJ / kg}.$$

**Answer:** The flow exergy of the fluid at the turbine inlet is 630.5 kJ / kg.     ■

**Example 6.17**

**Problem:** Atmospheric air at 100 kPa and 300 K is compressed in a reversible, adiabatic process to 800 kPa. Determine the exergy of the compressed air. The air at 800 kPa is allowed to cool to 300 K. Determine the exergy destroyed during the cooling process.

**Find:** Exergy of the air at the compressor outlet $\psi_2$ and the exergy destroyed during cooling $\psi_d$.

**Known:** Outlet pressure $P_2 = 800$ kPa, atmospheric pressure $P_o = 100$ kPa, atmospheric temperature $T_o = 300$ K. final pressure $P_3 = 800$ kPa, final temperature $T_3 = 300$ K.

**Assumptions:** Air is an ideal gas with constant specific heat.

**Governing equations:**

Flow exergy per unit mass
$$\psi = (h - h_o) - T_o(s - s_o) + \frac{\mathbf{V}^2}{2} + gz$$

Enthalpy change (ideal gas, constant specific heat)
$$\Delta h = c_p(T_2 - T_1)$$

Entropy change (ideal gas, constant specific heats)     $\Delta s = c_p \ln \dfrac{T_2}{T_1} - R \ln \dfrac{P_2}{P_1}$

Isentropic process (ideal gas, constant specific heats)     $T_2 = T_1 \left( \dfrac{P_2}{P_1} \right)^{(\gamma-1)/\gamma}$

**Properties:** Air has gas constant $R = 0.2870$ kJ / kgK (Appendix 1), air at 300 K has specific heat $c_p = 1.005$ kJ / kgK (Appendix 4), specific heat ratio of air at 300 K $\gamma = 1.400$ (Appendix 4).

**Solution:**
The exit temperature is

$$T_2 = T_1 \left( \frac{P_2}{P_1} \right)^{(\gamma-1)/\gamma} = 300 \ \text{K} \left( \frac{800 \ \text{kPa}}{100 \ \text{kPa}} \right)^{(1.4-1)/1.4} = 543.434 \ \text{K}.$$

For an isentropic compression entropy does not change, and neglecting kinetic and potential energy then the flow exergy per unit mass at the outlet reduces to

$$\psi_2 = \left( h_2 - h_o \right) = c_p \left( T_2 - T_o \right) = 1.005 \ \text{kJ / kgK} \times \left( 543.434 \ \text{K} - 300 \ \text{K} \right) = 244.651 \ \text{kJ / kg}.$$

Since $T_3 = T_o$, then $h_3 = h_o$; and neglecting kinetic and potential energy changes again then the flow exergy per unit mass in the final state reduces to

$$\psi_3 = -T_o \left( s_3 - s_o \right).$$

Final specific entropy contribution:

$$s_3 - s_o = c_p \underbrace{\ln \frac{T_3}{T_o}}_{=0} - R \ln \frac{P_3}{P_o} = -0.2870 \ \text{kJ / kgK} \times \ln \frac{800 \ \text{kPa}}{100 \ \text{kPa}} = 0.596 \ 800 \ \text{kJ / kgK}.$$

So the final flow exergy per unit mass is

$$\psi_3 = -T_o \left( s_3 - s_o \right) = -300 \ \text{K} \times \left( -0.596 \ 800 \ \text{kJ / kgK} \right) = 179.040 \ \text{kJ / kg},$$

And the exergy destroyed during the process is

$$\psi_d = \psi_2 - \psi_3 = 244.651 \ \text{kJ / kg} - 179.040 \ \text{kJ / kg} = 65.611 \ \text{kJ / kg}.$$

**Answer:** The flow exergy of the fluid at the compressor outlet is 244.7kJ / kg, and the exergy destroyed during cooling is 65.6 kJ / kg.     ∎

## 6.15   Bernoulli's Equation

An incompressible fluid flows through a duct whose cross-sectional area and elevation vary (Figure 6.12). How does the pressure of the fluid change as it passes through this duct? We can answer this question using the steady flow equation. To simplify the problem, we assume that

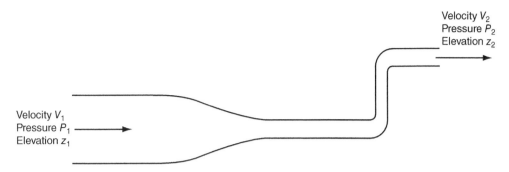

**Figure 6.12**    Steady flow of a fluid through a duct with variable cross-section.

the fluid is *inviscid*, that is, it has no viscosity. If a fluid is inviscid there is no friction between it and the walls of the duct, and we can assume that the flow is isentropic.

For an incompressible fluid undergoing an isentropic process Equation (6.28) gives

$$\Delta s = c_{avg} \ln \frac{T_2}{T_1} = 0.$$

(6.105)

Therefore $T_2 = T_1$ and the temperature of the fluid remains constant. The enthalpy change for an incompressible fluid undergoing an isothermal process is, from Equation (4.88)

$$h_2 - h_1 = v(P_2 - P_1).$$

(6.106)

As no work is done on the fluid $(\dot{W} = 0)$ and there is no heat transfer $(\dot{Q} = 0)$ the energy conservation equation for a control volume (Equation 4.96) becomes

$$(h_2 - h_1) + \frac{\mathbf{V}_2^2 - \mathbf{V}_1^2}{2} + g(z_2 - z_1) = 0.$$

(6.107)

Substituting Equation (6.106) and the density $\rho = 1/v$ into Equation (6.107) gives

$$\frac{P_1}{\rho} + \frac{\mathbf{V}_1^2}{2} + gz_1 = \frac{P_2}{\rho} + \frac{\mathbf{V}_2^2}{2} + gz_2.$$

(6.108)

Or, since we can apply the same energy balance between the inlet and any point in the duct,

$$\frac{P}{\rho} + \frac{\mathbf{V}^2}{2} + gz = \text{constant}.$$

(6.109)

Equation (6.109) was first derived by Daniel Bernoulli as early as 1738 and formed the foundation of the science of fluid mechanics. It is now known as *Bernoulli's equation* and is widely used to give a relation between velocity and pressure in incompressible fluids where viscous effects are negligible.

**Example 6.18**

**Problem:** A wind tunnel takes air from the atmosphere and uses a fan to accelerate it through the test section. A pressure transducer shows a pressure difference of 500 Pa between the inside of the wind tunnel and the outside atmosphere. Assuming that the density of air is 1.23 kg / m³, what is the velocity of air in the wind tunnel?

**Find:** Velocity of air in the wind tunnel $V_2$.

**Known:** Pressure difference between inside and outside of wind tunnel $\Delta P = 500$ Pa.

**Assumptions:** Air is an ideal gas with constant specific heat, density of air $\rho = 1.23$ kg / m³.

**Governing equation:**

Bernoulli Equation (inviscid, incompressible)
$$\frac{P_1}{\rho} + \frac{V_1^2}{2} + gz_1 = \frac{P_2}{\rho} + \frac{V_2^2}{2} + gz_2$$

**Solution:**

The air velocity in the atmosphere is $V_1 = 0$. Assuming that elevation differences are negligible, the Bernoulli equation reduce to

$$\frac{V_2^2}{2} = \frac{P_1 - P_2}{\rho} = \frac{\Delta P}{\rho},$$

$$V_2 = \sqrt{\frac{2\Delta P}{\rho}} = \sqrt{\frac{2 \times 500\,\text{Pa}}{1.23\,\text{kg} / \text{m}^3}} = 28.513\,\text{m} / \text{s}.$$

**Answer:** The air velocity in the wind tunnel is 28.5 m / s.    ∎

## Further Reading

1. W. C. Reynolds, H. C. Perkins (**1977**) *Engineering Thermodynamics*, McGraw Hill, New York.
2. H. B. Callen (**1985**) *Thermodynamics and an Introduction to Thermostatistics*, John Wiley & Sons, Ltd, London.
3. A. Bejan (**2006**) *Advanced Engineering Thermodynamics*, John Wiley & Sons, Ltd, London.

## Summary

Thermodynamic temperature ($T$) is defined as

$$T \equiv \left(\frac{\partial U}{\partial S}\right)_{m,V}.$$

Thermodynamic pressure ($P$) is defined as

$$\frac{P}{T} \equiv \left(\frac{\partial S}{\partial V}\right)_{m,U}.$$

The Gibbs equation is

$$ds = \frac{1}{T}du + \frac{P}{T}dv.$$

The change in entropy of incompressible liquids and solids is

$$\Delta s = s_2 - s_1 = c_{avg} \ln \frac{T_2}{T_1}.$$

The change in entropy of an ideal gas with constant specific heats:

$$\Delta s = s_2 - s_1 = c_v \ln \frac{T_2}{T_1} + R \ln \frac{v_2}{v_1},$$

$$\Delta s = s_2 - s_1 = c_v \ln \frac{P_2}{P_1} + c_p \ln \frac{v_2}{v_1},$$

$$\Delta s = s_2 - s_1 = c_p \ln \frac{T_2}{T_1} - R \ln \frac{P_2}{P_1}.$$

The change in entropy using air tables

$$\Delta s = s^o(T_2) - s^o(T_1) - R \ln \frac{P_2}{P_1}.$$

For an ideal gas undergoing an isentropic process:

$$\frac{T_2}{T_1} = \left(\frac{v_1}{v_2}\right)^{(\gamma-1)},$$

$$\frac{T_2}{T_1} = \left(\frac{P_2}{P_1}\right)^{(\gamma-1)/\gamma},$$

$$\frac{P_2}{P_1} = \left(\frac{v_1}{v_2}\right)^{\gamma},$$

$$Pv^{\gamma} = \text{constant.}$$

Using air tables, for an isentropic process:

$$\frac{P_2}{P_1} = \frac{P_{r2}}{P_{r1}},$$

$$\frac{v_2}{v_1} = \frac{v_{r2}}{v_{r1}}.$$

The change in entropy of a control mass undergoing an internally reversible process is

$$dS = \left(\frac{\delta Q}{T}\right)_{\text{int rev}},$$

which integrates to give

$$\Delta S = S_2 - S_1 = \int_1^2 \left(\frac{\delta Q}{T}\right)_{\text{int rev}}.$$

For a control mass at constant temperature ($T$),

$$\Delta S = \frac{Q_{12}}{T}.$$

The area under a $T$-$s$ diagram gives the heat transfer during the process,

$$Q_{12} = \int_1^2 T dS.$$

Entropy balance for a control mass is

$$\Delta S = S_{\text{in}} - S_{\text{out}} + S_{\text{gen}},$$

and written on a rate basis,

$$\frac{dS}{dt} = \dot{S}_{\text{in}} - \dot{S}_{\text{out}} + \dot{S}_{\text{gen}}.$$

In a control mass entropy enters and leaves the system with heat transfer, so

$$\frac{dS}{dt} = \sum_j \frac{\dot{Q}_j}{T_j} + \dot{S}_{\text{gen}},$$

where $T_j$ is the instantaneous temperature of the boundary across which heat transfer $\dot{Q}_j$ occurs. The entropy balance for a control volume is

$$\frac{dS}{dT} = \sum_j \frac{\dot{Q}_j}{T_j} + \sum_i \dot{m}_i s_i - \sum_e \dot{m}_e s_e + \dot{S}_{\text{gen}}.$$

For a reversible, adiabatic system at steady state,

$$s_2 = s_1.$$

The isentropic efficiency of a turbine is

$$\eta_t = \frac{w_t}{w_{t,s}} = \frac{h_2 - h_1}{h_{2s} - h_1}.$$

Isentropic nozzle efficiency,

$$\eta_{\text{nozzle}} = \frac{\mathbf{V}_2^2}{\mathbf{V}_{2s}^2}.$$

Isentropic compressor or pump efficiency,

$$\eta_c = \frac{w_{c,s}}{w_c} = \frac{h_{2s} - h_1}{h_2 - h_1}.$$

The exergy of a control mass is the maximum amount of useful work a system can do before it reaches equilibrium with its surroundings, and can be defined as

$$\Phi = \left(U - U_o\right) + P_o\left(V - V_o\right) - T_o\left(S - S_o\right) + \frac{m\mathbf{V}^2}{2} + mgz.$$

The exergy per unit mass of a control mass is

$$\phi = \left(u - u_o\right) + P_o\left(v - v_o\right) - T_o\left(s - s_o\right) + \frac{\mathbf{V}^2}{2} + gz.$$

The useful work done by a system in going from state 1 to state 2 is

$$W_u = \Phi_2 - \Phi_1 + T_o S_{\text{gen}}.$$

Exergy is destroyed when entropy is created:

$$\Phi_{\text{destroyed}} = T_o S_{\text{gen}}.$$

The flow exergy per unit mass of fluid flowing through a control volume is

$$\psi = \left(h - h_o\right) - T_o\left(s - s_o\right) + \frac{\mathbf{V}^2}{2} + gz.$$

The work output of a control volume is

$$\dot{W} = \dot{m}\left(\psi_2 - \psi_1\right) + T_o \dot{S}_{\text{gen}}.$$

The rate at which exergy is destroyed is

$$\dot{W} - \dot{W}_s = T_0 \dot{S}_{\text{gen}}.$$

Bernoulli's equation for isentropic flow in an inviscid fluid is

$$\frac{P_1}{\rho} + \frac{\mathbf{V}_1^2}{2} + gz_1 = \frac{P_2}{\rho} + \frac{\mathbf{V}_2^2}{2} + gz_2.$$

## Problems

### Control Mass Analysis

6.1 An electric motor operates on a voltage of 110 V and draws a current of 5 A. The surface temperature of the motor casing is measured to be 35 °C. What is the rate of entropy production in the motor?

6.2 An electric cable at a temperature of 70 °C loses heat to the surrounding air at 20 °C at a rate of 300 W. What is the rate of entropy generation at steady state?

6.3 An electric immersion heater placed inside a well-insulated tank is used to heat 500 l of water from 25 °C to 50 °C. The surface temperature of the heater is constant at 90 °C. Find the entropy generated during this process.

6.4 A kitchen blender operating at 110 V draws a current of 3 A when operating at steady state. The surface of the blender is at 30 °C while the surrounding air is at 24 °C. Find (a) the rate of entropy production in the blender and (b) the rate of entropy increase of the surrounding air.

6.5 A 20 mA current is passed through an insulated, 50 $\Omega$ resistor with a mass of 0.2 kg and a specific heat of 0.7 kJ / kgK. Find the entropy generated after 10 h. The initial temperature of the resistor is 20 °C.

6.6 A pitcher containing 1 l of water at 20 °C is placed inside a well-insulated refrigerator at 4 °C and left to come to equilibrium. Find the entropy generated.

6.7 Five kilograms of water at 20 °C are poured into an insulated bucket that already contains 10 kg of water at 80 °C. What is the entropy generated by the hot and cold water mixing?

6.8 A steel container with a mass of 2 kg is filled with 1 l of water, both initially at a temperature of 80 °C. The container and water cool down to the temperature of the surrounding atmosphere at 17 °C. Calculate the entropy change of the steel, the water and the surroundings.

6.9 A 2 kg steel ingot at 400 °C is heat treated by plunging it into a 10 l tank filled with water initially at 10 °C. The tank is well insulated on all sides. Find the entropy generated during this process. Assume constant specific heats for steel ($c_{steel}$ = 0.500 kJ / kg K) and water ($c_{water}$ = 4.186 kJ / kg K).

6.10 A rigid, well-insulated tank containing 60 kg of water at 10 °C has a 1 kW immersion heater placed in it. How much entropy has been generated after the heater has been switched on for 4 h?

6.11 A 1 kg block of ice at 0 °C is dropped into a lake at 20 °C. What is the entropy generated? Assume that the energy released by ice as it melts equals 333 kJ / kg.

6.12 Two identical solid blocks, each with mass $m$ and specific heat $c$, are placed together in an insulated box and allowed to reach thermal equilibrium. If their initial temperatures were $T_A$ and $T_B$ respectively what is the entropy generated?

6.13  Air at 100 kPa and 425 K is compressed to 300 kPa and 375 K. Can this process be adiabatic?

6.14  An ideal gas contained in a cylinder is compressed by a piston from $V_1 = 2$ m$^3$ and $T_1 = 400$ K to $T_2 = 300$ K. If the specific entropy of the gas decreases by 0.34 kJ / kgK, calculate the final volume of the gas. Assume $c_v = 0.25$ kJ / kgK and $\gamma = 1.6$.

6.15  A cylinder fitted with a piston contains 2 kg of air at 1.0 MPa and 500 K. The air expands in an isothermal process until its volume doubles. Calculate the change in entropy of the air.

6.16  Ammonia at 200 kPa and 100 °C is heated at constant pressure to 300 °C. Find the heat added and the change in entropy per kilogram of gas.

6.17  Argon at 290 K is cooled while being compressed until it reaches a final temperature of 250 K. If the specific entropy change is –0.3 kJ / kgK find the ratio of the final to the initial volume of gas.

6.18  Three kilograms of air at 5 MPa and 350 K expands to 1.5 MPa in a reversible, isothermal process. Find the heat transfer and the work done during this process.

6.19  A rigid, insulated container contains 0.2 kg of oxygen at 200 kPa and 25 °C. A paddle wheel inside the container is rotated until the pressure rises to 300 kPa. What is the change in entropy of the gas?

6.20  Find the specific entropy generated when air at 25 °C and 100 kPa in a rigid container is heated to a temperature of 100 °C by rotating a paddle wheel inside it. Find the specific entropy generated if the air is heated to the same final temperature by bringing it in contact with a thermal reservoir at 100 °C.

6.21  It takes 40 kJ of work to compress 0.5 kg of carbon dioxide from $P_1 = 200$ kPa and $T_1 = 280$ K to $P_2 = 350$ kPa and $T_2 = 320$ K. Calculate the heat transferred during this process and the entropy change of the gas.

6.22  Tank A contains 0.5 kg of air at 150 °C and 100 kPa while tank B contains 1 kg of air at 60 °C and 200 kPa. A valve connecting the two tanks is opened and their contents mixed. Assuming that both tanks are well insulated calculate the entropy generated. Assume that the specific heat of air is constant at $c_p = 1.01$ kJ / kgK.

6.23  A rigid gas tank contains 0.5 m$^3$ of nitrogen at 300 kPa and 50 °C. Another identical tank contains 0.5 m$^3$ of argon at 200 kPa and 200 °C. The two tanks are brought into close thermal contact until they reach equilibrium. Find the final temperature of the gases and the entropy generated during this process.

6.24  Air at 800 kPa and 600 °C flows through a gas turbine and emerges at 100 kPa and 300 °C. Changes in kinetic and potential energy are negligible. Is the expansion reversible?

6.25  A container filled with gas is heated reversibly in a process where the heat added depends on the temperature of the gas as

$$Q_{\text{int rev}} = aT + bT^2,$$

where $T$ is the temperature in K, $a = 100$ kJ / K and $b = 2$ kJ / K². Calculate the entropy change as the gas is heated from 300 K to 500 K.

6.26   Nitrogen at 100 kPa and 30 °C is contained in a cylinder with a freely moving piston. It is heated to a temperature of 100 °C and the volume increases by 50%. Calculate the change in internal energy, enthalpy and entropy, all per kilogram of gas.

6.27   One kilogram of a monoatomic ideal gas in a cylinder expands isentropically in a piston–cylinder from its initial pressure of 400 kPa, temperature 200 K and volume 1 m³ so that its volume triples. What is the change in internal energy?

6.28   Air at 25 °C is compressed isentropically so that its final specific volume is 20% of its initial value. What is the final temperature?

6.29   Oxygen in a cylinder is compressed by a piston in an isentropic process from an initial state of $T_1 = 30$ °C and $P_1 = 120$ kPa to a final pressure $P_2 = 320$ kPa. Find the final temperature.

6.30   Air at an initial state of $T_1 = 300$ K and $P_1 = 100$ kPa is compressed in a polytropic process during which $PV^{1.3} = $ constant to state 2 where $P_2 = 300$ kPa. It is then expanded in an isentropic expansion to state 3 where $P_3 = 100$ kPa. Determine the temperature $T_2$ and the net work done per kilogram of air in going from state 1 to 3.

6.31   Methane is compressed isentropically from 100 kPa and 25 °C to 400 kPa. Find the final temperature and the work done per kilogram of gas.

6.32   Hydrogen at 400 kPa and 80 °C is expanded to a pressure of 120 kPa in a polytropic process for which $PV^{1.2} = $ constant. Find the change in entropy per kilogram of gas.

6.33   Methane at 100 kPa and 20 °C that initially occupies a volume of 0.4 m³ is compressed in a polytropic process for which $PV^n = $ constant to a final state of 100 °C and 425 kPa. Find the work done and heat transferred during the compression and the entropy change of the gas.

6.34   Two kilograms of hydrogen are contained in one half of a rigid, insulated container that is divided into two equal parts by a partition. The other half of the container is evacuated. The partition breaks and the gas expands to fill the entire container. How much entropy is generated?

6.35   Air in a piston–cylinder system is expanded irreversibly from a pressure of 300 kPa and temperature of 400 K until its pressure is halved. During this process the air does 50 kJ / kg of work on the surroundings and 20 kJ / kg of heat is removed from it. Find the change in specific entropy of the air.

6.36   Half a kilogram of oxygen in a rigid container, initially at 130 kPa and 20 °C, is heated until its pressure reaches 500 kPa. Find the entropy change of the gas.

6.37  Air at 500 kPa and 600 K enters a copper cooling-coil immersed in a water bath at 290 K. The air flows through the coil with negligible pressure drop and emerges at a temperature of 350 K. Using the air tables, find the entropy generation per kilogram of airflow.

6.38  Ammonia at an initial pressure of 500 kPa and temperature of 200 °C is contained in a cylinder and kept compressed by a spring-loaded piston. The force acting on the piston varies as $V^2$, where $V$ is the volume of the gas, initially 0.1 m³. The system cools as it loses heat to the surrounding air at 17 °C. What is the entropy generated in this process when equilibrium is reached?

6.39  A gas storage tank with volume 5 m³ contains compressed hydrogen at a pressure of 600 kPa and a temperature of 20 °C. The tank develops a leak so that gas escapes very rapidly until the pressure drops to that of the atmosphere, 100 kPa. Assuming that the gas underwent an isentropic expansion, what is the final temperature of the gas? How much gas escaped?

## Control Volume Analysis

6.40  Air at 300 kPa and 350 K enters a well-insulated duct, 25 cm² in cross-sectional area with a mass flow rate of 0.4 kg / s, and leaves at a pressure of 100 kPa with a velocity of 200 m / s. Find the rate of entropy generation in the duct.

6.41  Two streams of air enter an insulated chamber, mix and leave with negligible pressure drop. One stream has a flow rate of 0.5 kg / s at 900 K and the other has a flow rate of 2 kg / s at 300 K. Find the exit air temperature and the rate of entropy generation.

6.42  Air at 500 kPa and 450 K enters a 20 mm diameter insulated pipe with a velocity of 180 m / s and leaves at 100 kPa and 400 K. Find the rate of entropy production in the pipe.

6.43  Hot air enters a pipe at 500 kPa and 600 K with a mass flow rate of 0.5 kg / s and exits at 450 kPa and 500 K. Find the rate of heat loss from the pipe and the rate of entropy generation in it. The ambient air temperature is 300 K.

6.44  Air at 500 K and 300 kPa enters a control volume with a mass flow rate of 2 kg / s and leaves at a pressure of 100 kPa. The control volume is adiabatic but can do work on the surroundings. Changes in kinetic and potential energy between the inlet and outlet are negligible. What are the maximum and minimum air temperatures possible at the exit of the control volume? Calculate the entropy change of the air in each case.

6.45  Air flows through an isentropic turbine operating at steady state, entering at 1 MPa and 900 K and leaving at 600 K. Determine the work done per kilogram of air and the pressure at the turbine exit.

6.46   Air enters a gas turbine at 600 kPa and 800 °C and leaves at 120 kPa. Determine the work output per kilogram of air assuming the process is isentropic.

6.47   A gas turbine receives heated gas from a combustion chamber at 1100 K and exhausts it to the atmosphere. If the turbine isentropic efficiency is 85% and it is required to supply 600 kJ of work for every kilogram of gas flowing through it, what is the exhaust temperature? What would the exhaust temperature be if the turbine were isentropic? Assume that the properties of the gas are those of air.

6.48   Measurements at the inlet and outlet of an insulated turbine show air entering at 700 kPa and 900 K and leaving at 150 kPa and 600 K. Is this process possible? If it is, how much work is done per kilogram of air flowing through the turbine?

6.49   Air enters a turbine at 500 kPa and 1000 K and exits at 100 kPa and 650 K. The exterior casing of the turbine, which is at 340 K, loses heat to the surrounding atmosphere at a rate of 50 kJ per kilogram of air flowing through it. Find the work output of the turbine and the entropy generated in it per kilogram of airflow.

6.50   Air enters an adiabatic turbine at 750 kPa and 840 K and leaves at 150 kPa. The mass flow rate of air is 5 kg / s and the work output of the turbine is 1.5 MW. Find the isentropic efficiency of the turbine and the exit temperature of the air.

6.51   Air at 100 kPa and 20 °C is compressed in a continuous process in an adiabatic compressor to a pressure of 400 kPa. Determine the exit temperature and the work required per kilogram of air.

6.52   A compressor is claimed to take argon at 100 kPa and 200 °C and compress it to 600 kPa while doing 200 kJ / kg of work. Is this possible?

6.53   Air enters an adiabatic compressor at 80 kPa and 27 °C and leaves at 480 kPa. What is the minimum power input required for an air mass-flow rate of 2 kg / s? If the isentropic efficiency of the compressor is 85% what is the temperature of the air at the exit?

6.54   Propane at 120 kPa and 25 °C is compressed to 500 kPa and 120 °C in an insulated compressor. Find the entropy generated per kilogram of gas. Assume constant specific heats.

6.55   A machine in a laboratory requires 0.2 kg / s of compressed air at 4 bar and a maximum temperature of 40 °C. This is to be provided by a compressor with 85% isentropic efficiency that takes in atmospheric air at 1 bar and 20 °C. Air from the compressor is cooled in a heat exchanger before being used. Calculate the power required to drive the compressor and the rate of heat removal in the heat exchanger.

6.56   Argon enters an adiabatic nozzle with negligible velocity at a pressure of 400 kPa and 300 °C and leaves at a pressure of 150 kPa. Assuming the expansion is reversible, what is the exit velocity?

6.57   Air enters an insulated nozzle at 800 kPa and 1000 K and exits at 200 kPa and 740 K. Find the efficiency of the nozzle.

6.58   A pipe has methane flowing through it at 400 kPa and 300 K with a velocity of 20 m / s and a mass flow rate of 0.3 kg / s. Determine the exit area of a nozzle that will accelerate the gas to 250 m / s. What is the pressure at the exit of the nozzle?

6.59   Water at a pressure of 300 kPa is pumped into a nozzle through which it jets out into the atmosphere at 100 kPa. If the inlet velocity is negligible and the nozzle efficiency is 90%, what is the exit velocity?

6.60   An aircraft cruising at a speed of 800 km / h takes air into its engine diffuser at –40°C and 70 kPa. If the air leaves the diffuser with negligible velocity what is the exit temperature and pressure?

6.61   Air enters an adiabatic, reversible diffuser at 250 °C and 100 kPa with a velocity of 250 m / s and leaves at a pressure of 150 kPa. What is the exit velocity and temperature?

6.62   A 5 kW pump takes water from a lake and raises it to a tank located at a height of 5 m. If the pressure in the tank is 500 kPa and atmospheric pressure is 100 kPa, what is the maximum water flow rate through the pump?

6.63   Water flowing at a rate of 15 kg / s enters a pump at 5 °C and 100 kPa and leaves at 3 MPa. If the isentropic efficiency of the pump is 82%, what is the power required to drive the pump?

6.64   A pump driven by a 0.5 kW motor takes in 0.2 kg / s of water at 10 °C and 120 kPa. What is the exit water pressure if the pump efficiency is 80%?

6.65   A pump for a building fire extinguishing system takes water from a storage tank and delivers it at a rate of 80 kg / s to a sprinkler located 20 m above the tank where it comes out with a velocity of 30 m / s. If the isentropic efficiency of the pump is 70%, how much power is required to run it?

### Exergy Analysis

6.66   A tank contains helium at 30 °C and 350 kPa. Determine the specific exergy of the gas if the atmosphere is at 20 °C and 100 kPa.

6.67   A 20 kg block of iron at 300 °C is dropped into an insulated tank that contains 50 l of water at 25 °C. The surrounding air is at 20 °C. Find the exergy destroyed.

6.68   Exhaust gas leaves a combustor at 300 kPa and 640 K. Calculate the specific exergy of the gas, assuming that it has the same properties as air. The atmosphere is at 300 K and 100 kPa.

6.69   A cylinder with a freely sliding piston contains 0.5 kg of air at 100 kPa and 20 °C. The air is heated for 10 min by a 100 W heater placed inside the cylinder while the pressure remains constant. If the surroundings are at 100 kPa and 20 °C, what is the exergy destroyed during the heating? Assume constant specific heats for air.

6.70   Nitrogen gas with a density of 0.5 kg / m³ at 17 °C is compressed in a cylinder by a piston to a density of 10 kg / m³ and temperature of 230 °C. What is the change in exergy per unit mass of gas? Assume that the atmosphere is at 100 kPa and 20 °C.

6.71   Air at 5 MPa and 400 °C flows through a well-insulated valve and emerges at a pressure of 3 MPa. Determine the exergy destroyed in this process per unit mass of air. The surroundings are at 20 °C.

6.72   An air compressor takes in air from the surrounding atmosphere at 100 kPa and 300 K and delivers it at 300 kPa and 500 K at a mass flow rate of 0.05 kg / s. Calculate the rate at which exergy is destroyed in this process.

6.73   Air enters an adiabatic nozzle at 500 kPa and 300 K with negligible velocity and leaves at 120 kPa with a velocity of 250 m / s. Assuming that specific heats are constant and the surrounding atmosphere is at 300 K calculate the exergy destroyed per kilogram of air.

6.74   Air enters an insulated turbine at 800 kPa and 650 K and 120 m / s with a mass flow rate of 10 kg / s. It leaves at 80 kPa, 350 K and 10 m / s. Determine the power output of the turbine and the rate at which exergy is destroyed. The atmosphere is at a temperature of 275 K.

6.75   Cold air enters a heat exchanger at 300 K and 800 kPa and leaves at 620 K and 750 kPa. Hot air enters it at 1100 K and 120 kPa and leaves at 100 kPa. If the mass flow rates of both hot and cold air streams equal 10 kg / s determine the exit temperature of the hot air. What is the rate at which exergy is destroyed in the heat exchanger?

### Bernoulli's Equation

6.76   What is the gauge pressure of air measured on the front of a car driving at 100 km / h? Assume that the density of air is constant and equal to 1.2 kg / m³.

6.77   High winds during tornadoes frequently tear the roofs off buildings. Estimate the air velocity (in km / h) necessary to lift a roof with a mass of 50 kg / m². Assume that the density of air is constant and equal to 1.2 kg / m³.

6.78   Water enters a nozzle at a pressure of 1 MPa and a velocity of 5.2 m / s. The inlet of the nozzle has a diameter of 70 mm and the outlet has a diameter of 25 mm. Find the velocity and pressure of the water at the outlet.

6.79   Water is pumped through a 50 mm diameter pipe that leads to a 10 mm diameter nozzle located 5 m above the level of the pump. If the velocity of water coming out of the nozzle is 25 m / s, what is the pressure at the pump outlet?

6.80   A duct with cross sectional area $A_1$ narrows to a constriction with cross sectional area $A_2$. A tube connected to the constriction leads to a water reservoir (Figure P6.80) that is open to the atmosphere. What mass flow rate of air $\dot{m}$ is required in the duct to lift the water in the tube by height $H$? Give the answer in terms of $A_1$, $A_2$, $H$, the densities of air $(\rho_a)$ and water $(\rho_w)$ and the acceleration due to gravity $g$.

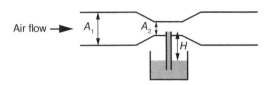

**Figure P6.80**   A duct that narrows to a constriction.

# 7

# Phase Equilibrium

---

**In this chapter you will:**

- Study phase equilibrium.
- Define the chemical potential.
- Derive a relation between pressure and temperature at the interface between two phases.
- Show phase change on pressure–volume–temperature diagrams.
- Use tables to find the properties of saturated liquid–vapour mixtures.
- Learn about other equations of state and generalised compressibility charts.

---

## 7.1 Liquid Vapour Mixtures

Matter, we have assumed, can exist in one of three states: liquid, solid or gas. Liquids and solids can reasonably be modelled as incompressible substances whose density is always constant. Gases, for the range of conditions that are typically of interest, can be assumed to follow ideal gas behaviour. But what if our system contains both liquid and vapour? How do we determine the properties of such a mixture?

Why is this important? Why do we need to deal with two-phase mixtures? An engine or turbine designer wants to minimise the mass of working fluid used since that allows the size of the device to be reduced. At the same time we want to operate at the lowest temperature possible because high working temperatures require the use of materials that are expensive and difficult to machine. A given mass of steam contains much more energy than an equal amount of liquid water. It takes 420 kJ / kg to heat water from 0 °C to 100 °C, but an additional 2257 kJ / kg is required to transform the liquid to vapour. Steam at 100 °C stores far more energy than liquid water at the same temperature.

---

*Energy, Entropy and Engines: An Introduction to Thermodynamics*, First Edition. Sanjeev Chandra.
© 2016 John Wiley & Sons, Ltd. Published 2016 by John Wiley & Sons, Ltd.
Companion website: www.wiley.com/go/chandraSol16

High energy-density makes phase change a very effective way of cooling surfaces. In power plants energy produced by combustion or a nuclear reaction is carried away from the heat source by a fluid. Boiling a liquid in contact with the heated surface produces very high rates of heat transfer, decreasing the size of the heat exchange equipment required. Reactors in chemical plants frequently evaporate liquids because vapours mix much more rapidly than liquids and chemical reactions proceed much faster in the gaseous than in the liquid state.

Applications such as these make it extremely important to be able to determine the thermodynamic properties of two-phase mixtures. Superheated steam can be treated as an ideal gas, but if it contains droplets of water it no longer satisfies the ideal gas assumption that all molecules are far apart. Similarly, liquid water can be assumed to have constant density, but if vapour bubbles are present in the water it is no longer incompressible. Neither the assumption of an ideal gas nor of incompressibility is valid for a liquid–vapour mixture. We will learn in this chapter how to analyse systems that include mixtures of phases.

## 7.2   Phase Change

Molecules in a liquid move in random paths, colliding with each other and constantly changing direction. The velocities of the particles vary – some slow, some very fast – but most travelling at speeds clustered around a mean value. Molecules in a liquid are packed close together, so those in the interior are buffeted equally from all sides (see Figure 7.1). Molecules at the free surface of the liquid, which face a gas on one side and a liquid on the other, experience many more collisions on the side with higher density, propelling them out of the liquid. Attractive forces between molecules in the liquid restrain them from escaping, but those with the highest velocities have enough momentum to leave the liquid surface and become part of the vapour phase. At a macroscopic level we observe a transformation of liquid into vapour that we call *evaporation*. Simultaneously molecules in the vapour phase strike the liquid surface, lose momentum, and are recaptured by intermolecular forces. The transformation from vapour to liquid is called *condensation*.

As the concentration of molecules in the vapour phase becomes higher the frequency with which they strike the liquid surface increases, enhancing the condensation rate. Eventually steady state is reached in which the rate of molecules leaving the liquid equals that returning to it and there is no net change in the liquid level. The liquid and vapour are then in *phase*

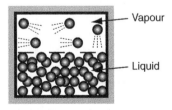

**Figure 7.1**   A rigid, sealed container is partially filled with liquid. The higher energy molecules at the free liquid surface overcome intermolecular forces and become vapour.

*equilibrium*. The liquid, in equilibrium with its own vapour, is known as a *saturated liquid* and the vapour is known as a *saturated vapour* (see Figure 7.2). The pressure and temperature everywhere in the system are constant and uniform and are known as the *saturation pressure* ($P_{sat}$) and *saturation temperature* ($T_{sat}$) respectively. The term "saturated" is a legacy of caloric theory, in which evaporation was believed to be a consequence of saturating the space between molecules with caloric, forcing them to fly apart and turn into vapour. We no longer use caloric theory but its terminology lives on.

When the temperature of the liquid is raised the average velocity of molecules increases and more break free from the bulk liquid to become part of the vapour. As the number of vapour molecules increases, more will condense back onto the liquid surface until a new equilibrium is established. If the liquid and vapour are in a closed, rigid container the pressure in the system rises as the mass of vapour increases. The saturation pressure and temperature are related: if the system is heated, more of the liquid will evaporate and the pressure inside the vessel will increase; if it is cooled vapour will condense and the pressure will decrease.

Since molecules escape from the surface of the liquid, the rate of evaporation is proportional to the free-surface area. Water poured into a flat dish evaporates much faster than if the same volume was placed in a bottle with a small mouth. The evaporation rate also increases if the air above the liquid is moving since the gas carries away vapour molecules and prevents them condensing back onto the liquid surface. Wet clothes hung out on a line dry much faster if there is a breeze blowing.

The molecules leaving the liquid during evaporation are those with the highest energy. Removing them decreases the average energy of the liquid and reduces its temperature, a phenomena that is known as *evaporative cooling*. If we want to keep the temperature of the liquid constant during evaporation we must continuously replenish the energy carried away by evaporating molecules. The energy carried away by a unit mass of evaporating liquid is known as the *latent heat of vaporisation*. The magnitude of latent heat for a particular liquid depends on the strength of intermolecular forces that have to be overcome during evaporation. Water, which has very strong intermolecular bonds, has a much higher latent heat than most hydrocarbons whose molecules are relatively weakly attracted to each other and which vaporise rapidly. Condensing molecules increase the energy of the liquid by the same amount as they decrease it when evaporating.

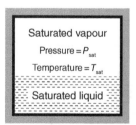

**Figure 7.2**    A rigid container is first evacuated and then partially filled with liquid. Some of the liquid will evaporate. At equilibrium the pressure in the container is known as the saturation vapour pressure ($P_{sat}$) and the temperature is the saturation vapour pressure ($T_{sat}$).

The saturation temperature of a liquid is also known as the *boiling point*. The boiling point of water at atmospheric pressure is approximately 100 °C: the temperature of water boiling in an open pan will remain at this value until all the liquid has evaporated. In a boiling liquid all the molecules have enough energy to break loose from the liquid and increasing the rate of heat transfer simply increases the rate of phase change from liquid to vapour.

When is equilibrium reached between two phases? What is the relation between the saturation temperature and the saturation pressure? For a given liquid, can we predict how the two properties will vary? These are the questions that we will try to answer in this chapter.

Though we will confine our discussion to equilibrium between solid, liquid and vapour phases, the equations we derive can be applied to far more complicated systems. Chemical compounds are different phases, so chemists use similar calculations to determine the final products of a chemical reaction. Molten metals mixed with each are also different phases, so metallurgists can predict the properties of alloys. These are only a few examples of the many applications of phase-equilibrium analysis.

## 7.3   Gibbs Energy and Chemical Potential

Figure 7.3 shows an open system held at constant temperature ($T$) and pressure ($P$) by being held in contact with constant temperature and pressure reservoirs. An infinitesimal amount of mass ($\delta m$) with specific enthalpy ($h$) and specific entropy ($s$) is added to the system in an internally reversible process, so that its mass increases by $dm$. The internal energy of the system increases by an amount $hdm$ when mass $dm$ is added, where the specific enthalpy $h$ accounts for flow work done to force the mass into the control volume (see Section 4.7).

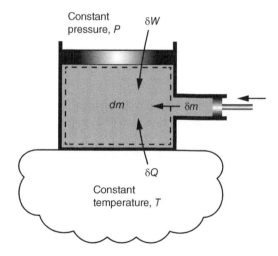

**Figure 7.3**    An open system at constant temperature ($T$) and pressure ($P$) to which infinitesimal amounts of mass and energy are added reversibly in the form of both work ($\delta W$) and heat ($\delta Q$).

Energy is exchanged in internally reversible processes between the system and surroundings as both heat ($\delta Q$) and work ($\delta W$). The net change in internal energy is

$$dU = hdm + \delta Q + \delta W. \tag{7.1}$$

The entropy of the system increases due to both mass transfer and heat transfer:

$$dS = sdm + \frac{\delta Q}{T}. \tag{7.2}$$

Rearranging Equation (7.2),

$$\delta Q = TdS - Tsdm. \tag{7.3}$$

The reversible work for a simple compressible system is $\delta W = -PdV$, where $P$ is the system pressure. Substituting for both $\delta Q$ and $\delta W$ in Equation (7.1) gives

$$dU = \underbrace{TdS}_{\text{Heat}} - \underbrace{PdV}_{\text{Work}} + \underbrace{(h - Ts)dm}_{\text{Mass Transfer}}. \tag{7.4}$$

The three terms on the right hand side of Equation (7.4) give the changes in internal energy due to heat transfer, work, and mass transfer respectively. The coefficient of the mass transfer term, $h - Ts$, is the internal energy increase per unit mass added. This is an extremely important term, and we give it a name, the *Gibbs energy* (also known as the Gibbs function), denoted by the symbol $G$ and defined as

$$G \equiv H - TS. \tag{7.5}$$

The Gibbs energy is an extensive property since it is a function of two extensive properties, enthalpy ($H$) and entropy ($S$), and has the units of energy, joules. The corresponding intensive function is the specific Gibbs energy ($g$) which represents the increase in energy of an open system per unit mass added:

$$g = \frac{G}{m} = h - Ts. \tag{7.6}$$

This is a very important property in the analysis of open systems and is also called the chemical potential. We can substitute $g$ in Equation (7.4) to give

$$dU = TdS - PdV + gdm. \tag{7.7}$$

We can rearrange Equation (7.7) as

$$dS = \frac{dU}{T} + \frac{P}{T}dV - \frac{g}{T}dm. \tag{7.8}$$

Equation (7.8) is the Gibbs equation for a system in which mass transfer can occur. It allows us to calculate changes in entropy for an open system.

## 7.4 Phase Equilibrium

A rigid, insulated container is partially filled with liquid while the remaining space is occupied by vapour of the same liquid (Figure 7.4). The system is at equilibrium so the temperature ($T$) and pressure ($P$) are uniform everywhere in the vessel. What is the condition for equilibrium in such a system?

Let us start from the state principle, which postulates that the entropy of the system is a function of its internal energy, volume and mass, so that

$$S = S(U, V, m). \tag{7.9}$$

By convention, properties of a saturated liquid are denoted by the subscript $f$ (originally from the German *flüssigkeit* for liquid) and that of saturated vapour $g$ (for gas, fortunately the same word in German). The total entropy of the system, including both liquid and vapour, is

$$S = S_f(U_f, V_f, m_f) + S_g(U_g, V_g, m_g). \tag{7.10}$$

At equilibrium the entropy is maximum so that

$$dS = dS_f(U_f, V_f, m_f) + dS_g(U_g, V_g, m_g) = 0. \tag{7.11}$$

From Equation (7.8)

$$dS_f = \frac{dU_f}{T_f} + \frac{P_f}{T_f} dV_f - \frac{g_f}{T_f} dm_f \tag{7.12}$$

and

$$dS_g = \frac{dU_g}{T_g} + \frac{P_g}{T_g} dV_g - \frac{g_g}{T_g} dm_g. \tag{7.13}$$

The combined liquid–vapour system is isolated, so that its total mass, volume and energy are fixed. A decrease of any of these properties in one phase will result in an identical increase in the other phase:

$$dU_f = -dU_g, \tag{7.14}$$

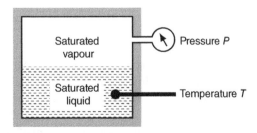

**Figure 7.4**   An insulated, rigid container filled with a mixture of liquid and vapour.

$$dV_f = -dV_g, \tag{7.15}$$

$$dm_f = -dm_g. \tag{7.16}$$

Substituting Equations (7.14) to (7.16) into Equation (7.13) gives

$$dS_g = -\frac{dU_f}{T_g} - \frac{P_g}{T_g}dV_f + \frac{g_g}{T_g}dm_f. \tag{7.17}$$

To evaluate the condition for equilibrium, we substitute Equations (7.12) and (7.17) into Equation (7.11) to get

$$dS = \left(\frac{1}{T_f} - \frac{1}{T_g}\right)dU_f + \left(\frac{P_f}{T_f} - \frac{P_g}{T_g}\right)dV_f - \left(\frac{g_f}{T_f} - \frac{g_g}{T_g}\right)dm_f = 0. \tag{7.18}$$

In Equation (7.18) the terms $dU_f$, $dV_f$ and $dm_f$ are arbitrary: since we can change mass, volume and energy independently they can have any value. To ensure that $dS = 0$, we require that each term in parentheses on the right hand side of the equation equal zero. From the first term, the condition for equilibrium is that $T_f = T_g$: the temperature in the liquid and vapour must be equal. From the second term we obtain $P_f = P_g$, or that their pressures must be equal. We have already established the conditions for thermal and mechanical equilibrium in Chapter 6, so this is not new. However, the third term gives us that

$$g_f = g_g. \tag{7.19}$$

This is the condition for phase equilibrium: *there is no exchange of mass between two phases whose chemical potentials are the same.*

Suppose we have a two-phase system that is not at equilibrium: it has uniform pressure and temperature everywhere ($P_f = P_g$ and $T_f = T_g$) but the chemical potentials of the two phases are different. For the system to reach equilibrium its entropy must increase and

$$dS = -\frac{1}{T_f}\left(g_f - g_g\right)dm_f > 0. \tag{7.20}$$

If the specific Gibbs energy of the liquid is greater than that of the vapour, so that $g_f > g_g$, we require that $dm_f < 0$ to satisfy the inequality in Equation (7.20); i.e. the mass of liquid will decrease and that of vapour will increase. *Mass transfer takes place from the phase with higher chemical potential to the phase with lower chemical potential.* In the same way that temperature determines the direction of heat transfer and pressure of work transfer, chemical potential determines the direction of mass transfer.

## 7.5 Evaluating the Chemical Potential

The chemical potential ($g$) is very important in deciding when phase equilibrium has been reached, but how do we determine its value? We have no way of measuring $g$ directly so we have to find a relation between it and other properties such as pressure and temperature. The Gibbs energy is, by definition,

$$G = H - TS = U + PV - TS. \tag{7.21}$$

Differentiating,

$$dG = dU + PdV + VdP - TdS - SdT. \tag{7.22}$$

Substituting $dU$ from Equation (7.7) into Equation (7.22) gives

$$dG = TdS - PdV + gdm + PdV + VdP - TdS - SdT. \tag{7.23}$$

After simplifying,

$$dG = gdm + VdP - SdT. \tag{7.24}$$

Equation (7.24) shows that changes in Gibbs energy of a system are due to either changes in mass ($dm$), pressure ($dP$) or temperature ($dT$). The total Gibbs energy of the system is

$$G = gm \tag{7.25}$$

and the change in it is

$$dG = mdg + gdm. \tag{7.26}$$

Combining Equation (7.26) with (7.24) gives

$$mdg = VdP - SdT. \tag{7.27}$$

Dividing through by the mass ($m$) of the system,

$$dg = vdP - sdT. \tag{7.28}$$

Equation (7.28) is known as the *Gibbs–Duhem equation* and we will use it to calculate how chemical potential varies with pressure and temperature.

## 7.6 Clausius–Clapeyron Equation

A container is partially filled with liquid while the remaining space is taken up by vapour (see Figure 7.4), with both phases at equilibrium. If the temperature of the system is raised some liquid evaporates and the pressure of the system increases until a new equilibrium is reached. Can we predict what the pressure will be at the new equilibrium state?

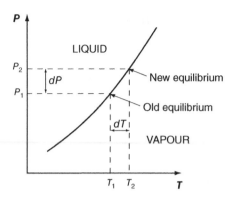

**Figure 7.5**  Phase equilibrium shown on a *P-T* diagram. The line shows combinations of pressure (*P*) and temperature (*T*) at which a two-phase system is in equilibrium. If the temperature of a system at equilibrium is increased by *dT* more vapour evaporates, increasing the pressure by *dP* until a new equilibrium state is established.

The variation of saturation pressure with temperature can be shown graphically on a diagram with *P* and *T* as axes (see Figure 7.5). The line on a *P-T* diagram shows saturated states where liquid and vapour are in equilibrium. At any given pressure if the temperature is less than the boiling point the substance is a *subcooled liquid* and, if the temperature is higher, it is a *superheated vapour*. The point $(P_1, T_1)$ in Figure 7.5 lies on the line corresponding to an equilibrium state. The temperature is raised by an infinitesimal amount *dT* while holding volume constant. Some liquid evaporates and the pressure increases by *dP* so that the system reaches equilibrium again at $(P_2, T_2)$.

If the liquid and vapour are initially at equilibrium, $g_f = g_g$. Once the temperature and pressure of the liquid and vapour are changed by *dT* and *dP*, the chemical potential of both must change by an amount *dg*. After equilibrium is re-established at the new conditions,

$$g_f + dg_f = g_g + dg_g. \tag{7.29}$$

Since $g_f = g_g$, then

$$dg_f = dg_g \tag{7.30}$$

or, from Equation (7.28),

$$v_f dP - s_f dT = v_g dP - s_g dT. \tag{7.31}$$

After rearranging Equation (7.31),

$$\frac{dP}{dT} = \frac{s_g - s_f}{v_g - v_f}. \tag{7.32}$$

Equation (7.32) gives us the slope at any point on the *P-T* curve in Figure 7.5, and in principle we could integrate it to get the entire curve of equilibrium states. However, it is not easy to evaluate this expression since we do not know how to measure $s_g$ and $s_f$. To evaluate entropy values we use the Gibbs equation, Equation (6.37),

$$Tds = dh - vdP.$$

Since pressure ($P$) and temperature ($T$) are constant for a system in equilibrium, we set $dP = 0$ and integrate while keeping $T$ constant

$$\int_{s_f}^{s_g} ds = \frac{1}{T} \int_{h_f}^{h_g} dh, \tag{7.33}$$

giving

$$s_g - s_f = \frac{h_g - h_f}{T}. \tag{7.34}$$

The difference $h_g - h_f$ is the heat required to convert saturated liquid to saturated vapour at constant pressure. This is defined as the *enthalpy of vaporisation* or the *latent heat of vaporisation*:

$$h_{fg} \equiv h_g - h_f. \tag{7.35}$$

A saturated liquid–vapour mixture behaves as a thermal reservoir since its temperature does not change when heat is added or removed. The increase in entropy ($s_g - s_f$) when a saturated liquid is completely evaporated is given by the heat added ($h_{fg}$) divided by the saturation temperature.

Substituting Equations (7.34) and (7.35) into Equation (7.32) we get

$$\frac{dP}{dT} = \frac{h_{fg}}{T\left(v_g - v_f\right)}. \tag{7.36}$$

Equation (7.36) is the *Clapeyron equation* that gives the slope of the line in Figure 7.5, relating pressure and temperature for any two pure phases in equilibrium. For the particular case of liquid–vapour equilibrium, we can simplify it further by assuming that the liquid specific volume is much less than that of the vapour so that $v_f$ is negligible compared to $v_g$, and that the vapour is an ideal gas with $v_g = RT / P$. Making these simplifying assumptions gives us the *Clausius–Clapeyron equation*,

$$\frac{dP}{dT} = \frac{h_{fg} P}{RT^2}. \tag{7.37}$$

Rearranging Equation (7.37),

$$\int \frac{dP}{P} = \int \frac{h_{fg}}{RT^2} dT. \tag{7.38}$$

Integrating Equation (7.38), assuming that the latent heat is constant, gives

$$\ln P = -\frac{h_{fg}}{RT} + C,$$    (7.39)

where $C$ is a constant of integration. Rearranging, we get an exponential variation of the saturation vapour pressure with temperature

$$P_{sat} = C \exp\left(-\frac{h_{fg}}{RT_{sat}}\right).$$    (7.40)

$C$ can be evaluated if $P_{sat}$ is known at any given value of $T_{sat}$. This equation gives us the equilibrium line on a $P$-$T$ diagram and is useful in correlating experimental data for saturation pressure.

**Example 7.1**
**Problem:** The saturation pressure of water at 100 °C is 101.35 kPa. What is the saturation pressure at 105 °C? Assume the latent heat of vaporisation $h_{fg} = 2257.06$ kJ / kgK.
**Find:** The saturation pressure $P_{sat}$ of water at $T = 105$ °C.
**Known:** At saturation temperature $T_{sat} = 100$ °C $= 373.15$ K the saturation pressure $P_{sat} = 101.35$ kPa.
**Assumptions:** Latent heat of vaporisation for water $h_{fg} = 2257.06$ kJ / kgK.
**Governing Equation:**

Clausius–Clapeyron equation          $P_{sat} = C \exp\left(-\frac{h_{fg}}{RT_{sat}}\right)$

**Solution:**
Substituting values at 100 °C,

$$101.35\,\text{kPa} = C \exp\left(-\frac{2257.06\,\text{kJ / kgK}}{0.46152\,\text{kJ / kg} \times 373.15\,\text{K}}\right).$$

Solving,

$$C = 4.98510 \times 10^7 \text{ kPa}.$$

Now using this value of $C$ for $T = 105$ °C $= 378.15$ K,

$$P_{sat} = 4.98510 \times 10^7 \text{ kPa} \times \exp\left(-\frac{2257.06\,\text{kJ / kgK}}{0.46152\,\text{kJ / kg} \times 378.15\,\text{K}}\right),$$

$$P_{sat} = 120.527\,\text{kPa}.$$

**Answer:** At 105 °C the saturation pressure is 120.5 kPa.                                    ∎

## 7.7   Liquid–Solid and Vapour–Solid Equilibria

The Clapeyron equation, Equation (7.36), is valid for any two phases that are in equilibrium – not just liquid and vapour. We could, equally well, apply them to liquid and solid phases in equilibrium. Suppose we have a glass of water with ice cubes floating in them. How does the melting point of the solid phase vary with the pressure of the system? If we denote the solid phase with the subscript $s$, we can rewrite Equation (7.36) as

$$\frac{dP}{dT} = \frac{h_{sf}}{T\left(v_f - v_s\right)}, \tag{7.41}$$

where the *enthalpy of fusion* or the *latent heat of fusion*,

$$h_{sf} \equiv h_f - h_s, \tag{7.42}$$

is the heat required to melt a unit mass of the solid under constant pressure.

By definition $h_{sf} > 0$. If the specific volume of the liquid is greater than that of the solid $(v_f > v_s)$ the slope of the equilibrium $P$-$T$ line is positive. This is true for most materials: the solid phase is denser than the liquid and has a lower specific volume. A lump of solid will sink if dropped into its own liquid. Water is a notable exception to this rule: ice has a lower density than water, which is why ice cubes float in water. For water $v_f < v_s$, so that $dP/dT < 0$. The $P$-$T$ diagram (see Figure 7.6), showing the lines of equilibrium between different saturated phases, is different for water (Figure 7.6a) than it is most other substances (Figure 7.6b) that contract upon freezing.

The negative slope of the $P$-$T$ line for ice and water has interesting consequences because it implies that the melting point of ice decreases when it is placed under pressure. If a thin wire is pulled with sufficient force onto a block of ice the pressure under it may become high enough to decrease the melting point below the ambient temperature, thawing the ice; the wire will sink into the water, which will then re-freeze once the pressure on it is relieved. The wire can pass through the block of ice without ever cutting it into two sections.

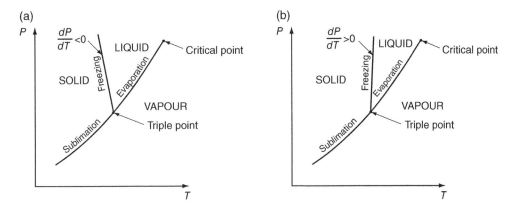

**Figure 7.6**   *P-T* diagrams, showing liquid, solid and vapour phases for (a) water, which expands when freezing and (b) a typical substance that contracts upon freezing.

If the pressure is reduced a solid will directly turn into vapour without first melting, a process known as sublimation. The slope of the solid–vapour phase equilibrium line is given by

$$\frac{dP}{dT} = \frac{h_{sg}}{T(v_g - v_s)},$$
(7.43)

where the *enthalpy of sublimation* or *latent heat of sublimation* is

$$h_{sg} \equiv h_g - h_s.$$
(7.44)

The condition where the chemical potential of all three phases – liquid, solid and vapour – is the same and all are in equilibrium is known as the *triple point*. For water the triple point occurs at 611.73 Pa and 273.16 K (0.01 °C). The triple point is useful for calibrating thermometers since we can determine when the temperature has been reached very precisely, since all three phases can coexist only at this combination of temperature and pressure.

## Example 7.2

**Problem:** The density of liquid water at atmospheric pressure and 0 °C is 1000 kg / m³. When it freezes the density decreases by approximately 8%. How much would we have to increases the pressure to reduce the melting point of ice by 1 °C? The latent heat of fusion of ice is 334 kJ / kg.

**Find:** Increase in pressure $\Delta P$ to reduce melting point by 1 °C.

**Known:** Density of water at 0 °C $\rho_f = 1000\,\text{kg} / \text{m}^3$, density of ice $\rho_s$ is 8% less than $\rho_f$, latent heat of fusion of ice $h_{sf} = 334$ kJ / kg.

**Assumptions:** Densities $\rho_s$, $\rho_f$ and latent heat $h_{sf}$ are all constant.

**Governing Equation:**

Clapeyron equation $\qquad\qquad \dfrac{dP}{dT} = \dfrac{h_{sg}}{T(v_g - v_s)}$

**Solution:**

Integrating the Clapeyron equation,

$$\int_{P_1}^{P_2} dP = \int_{T_1}^{T_2} \frac{h_{sf}}{T(v_f - v_s)}\, dT,$$

$$\Delta P = P_2 - P_1 = \frac{h_{sf}}{(v_f - v_s)} \ln\frac{T_2}{T_1}.$$

Then knowing that

$$v_f = \frac{1}{\rho_f} = \frac{1}{1000\,\text{kg} / \text{m}^3} = 0.001\,\text{m}^3 / \text{kg},$$

And the density of ice is 8% less than that of water:

$$\rho_s = (1-0.08) \times \rho_f = 0.92 \times 1000 \,\text{kg}/\text{m}^3 = 920 \,\text{kg}/\text{m}^3,$$

$$v_s = \frac{1}{\rho_s} = \frac{1}{920 \,\text{kg}/\text{m}^3} = 0.001\ 087\ 0 \,\text{m}^3/\text{kg},$$

$$\Delta P = \frac{334 \,\text{kJ}/\text{kg}}{\left(0.001 \,\text{m}^3/\text{kg} - 0.001\ 087 \,\text{m}^3/\text{kg}\right)} \ln \frac{272.15\,\text{K}}{273.15\,\text{K}} = 14.081 \,\text{MPa}.$$

**Answer:** The pressure will have to be increased by 14.1 MPa to reduce the melting point by 1 °C.                                                                                     ■

## 7.8   Phase Change on *P-v* and *T-v* Diagrams

Our discussion of phase change has focused on systems whose volumes are fixed while their pressure and temperature change. What if we hold pressure constant instead, while varying the temperature and volume? Figure 7.7 shows a liquid placed in a cylinder with a piston that moves freely as the fluid expands. As the system is heated both the volume of fluid and the temperature of the system are recorded and plotted on a graph with *T* and *V* as axes (Figure 7.8). A line of constant pressure on a *T-V* diagram is known as an *isobar*.

A liquid at a temperature below its boiling point is said to be *subcooled*. As the liquid is heated it expands slightly until it reaches its saturation temperature. Once the liquid begins to evaporate the temperature remains constant, staying at the saturation temperature corresponding to the given system pressure. There is now a *saturated mixture* of liquid and vapour in the container. The volume of the system increases significantly since the vapour typically occupies a much larger volume than liquid alone. Once all the liquid has been transformed into vapour its temperature begins to rise again, along with its volume. Further heating produces *superheated vapour*.

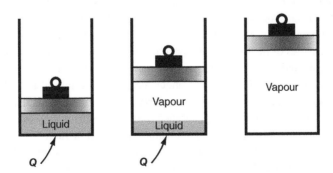

**Figure 7.7**   A liquid is heated at constant pressure. The volume increases and temperature remains constant as the liquid evaporates. Once all of the liquid has become vapour, both volume and temperature increase with heating.

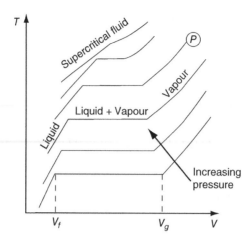

**Figure 7.8**   Lines of constant pressure (isobars) on *T-V* axes. The temperature remains constant while there is a liquid vapour mixture. As pressure increases the saturation temperature increases.

To increase the pressure in the system we place a heavier weight on the piston. Repeating the heating process at higher pressure we observe that the saturation temperature increases with pressure, as expected from our previous discussion of the Clausius–Clapeyron equation. We can depict this process as another isobar, at higher pressure, on the *T-V* diagram.

A liquid is virtually incompressible, so pressure has little effect on its volume; the higher saturation temperature, however, makes it expand slightly. The net result is that the volume of saturated liquid increases with pressure. A vapour is highly compressible: as the pressure increases its volume decreases. The horizontal portion of the isotherms in Figure 7.8, whose two ends correspond to the volume of saturated liquid ($V_f$) and saturated vapour ($V_g$), becomes smaller as pressure increases. Eventually, once we reach a sufficiently high pressure known as the *critical pressure*, there is no difference between the volume of liquid and vapour.

How do we differentiate between liquid water and steam? At atmospheric pressure the difference is obvious: steam has much lower density and occupies a far greater volume than the same mass of liquid. If we look at a transparent container containing both water and steam we can see the interface between the two phases because light passing through them is refracted by different amounts, since the index of refraction is a function of density. If pressure is increased the densities of the two phases approach each other and when the critical pressure is reached both liquid and vapour have the same density and refractive index and the meniscus between them is no longer visible. It is no longer meaningful to talk of two phases; instead, there is only a single phase known as a *supercritical fluid*. The critical pressure can be measured experimentally, since the disappearance of the meniscus is easy to see, and has been determined for a large number of fluids. Some values are listed in Table 7.1.

To represent the liquid and vapour phases graphically it is convenient to use intensive properties, temperature (*T*) and specific volume (*v*), as axes as shown in Figure 7.9a. The *T-v* diagram is representative of a given fluid, not of any particular system. We draw a line through the turning points on the isobars, corresponding to saturated vapour and saturated liquid. The locus of these points is known as the *vapour dome* that separates the single phase and two-phase

**Table 7.1**   Critical temperature ($T_c$), pressure ($P_c$) and volume ($v_c$) of common substances.

| Substance | Formula | Temperature, $T_c$ (K) | Pressure, $P_c$ (MPa) | Volume, $\bar{v}_c$ (m³/kmol) |
|---|---|---|---|---|
| Air | — | 132.5 | 3.77 | 0.0883 |
| Ammonia | $NH_3$ | 405.5 | 11.28 | 0.0724 |
| Argon | Ar | 151 | 4.86 | 0.0749 |
| Carbon dioxide | $CO_2$ | 304.2 | 7.39 | 0.0943 |
| Helium | He | 5.3 | 0.23 | 0.0578 |
| Methane | $CH_4$ | 191.1 | 4.64 | 0.0993 |
| Nitrogen | $N_2$ | 126.2 | 3.39 | 0.0899 |
| Oxygen | $O_2$ | 154.8 | 5.08 | 0.078 |
| Water | $H_2O$ | 647.3 | 22.09 | 0.0568 |

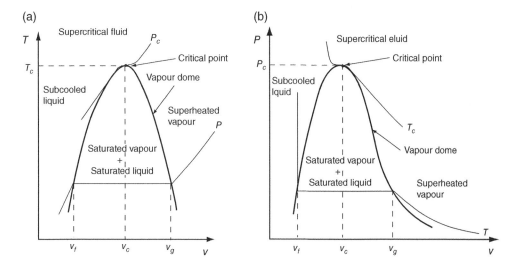

**Figure 7.9**   Liquid–vapour phase change shown on (a) $T$-$v$ and (b) $P$-$v$ diagrams.

regions. The region under the vapour dome corresponds to saturated vapour–liquid mixtures. The peak of dome is known as the *critical point* through which the critical isobar, at pressure $P_c$, passes. Above the critical pressure we have supercritical fluid. Below this isobar, to the left of the vapour dome, is subcooled liquid and to its right is superheated vapour.

It is possible to make a liquid evaporate by lowering its pressure while keeping temperature constant. Such a process is best displayed on a diagram with pressure ($P$) and specific volume ($v$) as axes, as shown in Figure 7.9b. On this we draw *isotherms*, which are lines representing equilibrium states during constant temperature processes. Starting with a subcooled liquid, as pressure is reduced volume increases very slightly, until the saturation pressure is reached. Isotherms in the subcooled liquid region are almost perfectly vertical, showing near-incompressible behaviour. The liquid begins to evaporate and the volume of the system increases as vapour is created while the pressure remains constant, so that isotherms are horizontal. Once all the liquid has been converted to vapour, the pressure of the system begins to decrease again as it expands, producing superheated vapour. Superheated vapour obeys the ideal gas equation, so pressure is inversely proportional to volume. The *critical isotherm* passes through the peak of the vapour dome corresponding to the *critical temperature* ($T_c$). Above this temperature only a single supercritical fluid phase exists. The *critical point* is the peak of the vapour dome, and the specific volume corresponding to it is known as the *critical volume* ($v_c$). Values of the critical pressure, temperature and volume are listed in Table 7.1 for some common substances and there is a more extensive list in Appendix 6.

## 7.9    Quality

When we heat a saturated liquid it begins to turn into vapour. The fraction of vapour in the two-phase mixture increases as it is heated. How do we specify the composition of such a mixture of saturated vapour and liquid? We use a new property, the *quality* of a mixture, defined by

$$x = \frac{\text{mass of vapour}}{\text{mass of mixture}} = \frac{m_g}{m}. \tag{7.45}$$

The mass of the mixture ($m$) is the sum of the masses of saturated vapour ($m_g$) and saturated liquid ($m_f$):

$$m = m_f + m_g, \tag{7.46}$$

so that

$$1 - x = \frac{\text{mass of liquid}}{\text{mass of mixture}} = \frac{m_f}{m}. \tag{7.47}$$

If the specific volume of the vapour and liquid are $v_g$ and $v_f$ respectively, the total volume of the mixture is

$$V = mv = m_g v_g + m_f v_f. \tag{7.48}$$

Dividing through by $m$, the specific volume of the mixture is

$$v = \frac{m_g}{m} v_g + \frac{m_f}{m} v_f = xv_g + (1-x)v_f, \tag{7.49}$$

or, alternatively

$$v = v_f + x(v_g - v_f). \tag{7.50}$$

Other intensive properties of mixtures can be calculated in the same way. The specific internal energy of the mixture,

$$u = xu_g + (1-x)u_f = u_f + x(u_g - u_f). \tag{7.51}$$

The specific mixture enthalpy,

$$h = xh_g + (1-x)h_f = h_f + x(h_g - h_f) = h_f + xh_{fg}.$$  (7.52)

The specific mixture entropy,

$$s = xs_g + (1-x)s_f = s_f + x(s_g - s_f).$$  (7.53)

## 7.10  Property Tables

To find the specific volume, internal energy, enthalpy and entropy of a two-phase mixture at any given saturation temperature or pressure we can use Equations (7.49) to (7.53). All we need are the properties of the saturated liquid and vapour, which are listed in tables of saturated properties, commonly known as "steam tables". The saturated property tables come in two versions. Table 7.2 has temperature as the independent property, listed in the first column in increments of 5 °C. The second column has the corresponding saturation pressure, for each value of temperature. Reading across the table gives the values of specific volume, internal energy, enthalpy and entropy, for both saturated liquid and vapour. If properties are required at temperatures lying between those listed in the tables we can use linear interpolation to calculate their values. Values of specific internal energy, enthalpy and entropy are given assuming them to be zero for liquid water at its triple point ($T = 0.01$ °C, $P = 0.6113$ kPa). The full temperature table is given in Appendix 8a, giving properties up to the critical temperature.

A second version of the saturated property table is given in Table 7.3, where the independent property is the pressure, listed in the first column. The second column gives the corresponding saturation temperature, and successive columns the specific volume, internal energy, enthalpy and entropy of both saturated vapour and liquid. The information given in both these tables is identical. We can use either one, depending on what information we have about the system, and which is more convenient to use. The full pressure table is given in Appendix 8b.

If a vapour is superheated, its pressure and temperature are independent properties and both need to be specified to fix the state. Table 7.4 gives an example of a superheated vapour table, where the properties of water vapour are given for temperatures from 100 to 1300 °C while the pressure is held constant at 0.1 MPa. Reading across each row $v$, $u$, $h$ and $s$ can be determined at any given temperature. A complete set of superheated steam tables is given in Appendix 8c.

**Example 7.3**
**Problem:** Find the specific enthalpy of a saturated water-steam mixture with 40% quality at a temperature of 50 °C.
**Find:** Specific enthalpy $h$.
**Known:** The water-steam mixture is at the saturation temperature $T_{sat} = 50$ °C, quality $x = 0.4$.
**Governing Equation:**

Mixture property ($y$)                     $y = xy_g + (1-x)y_f$

**Properties:** From Table 7.2 at $T_{sat} = 50$ °C, enthalpy of liquid $h_f = 209.33$ kJ / kg and vapour $h_g = 2592.1$ kJ / kg.

**Table 7.2** Properties of saturated water as a function of temperature.

| Temp. (°C) | Pressure (MPa) | Spec. Vol. (m³/kg) | | Int. Energy (kJ/kg) | | Enthalpy (kJ/kg) | | Entropy (kJ/kgK) | |
|---|---|---|---|---|---|---|---|---|---|
| $T_{sat}$ | $P$ | $v_f$ | $v_g$ | $u_f$ | $u_g$ | $h_f$ | $h_g$ | $s_f$ | $s_g$ |
| 0.01 | 0.000 611 3 | 0.001 000 | 206.14 | 0.00 | 2375.3 | 0.00 | 2501.4 | 0.0000 | 9.1562 |
| 5 | 0.000 872 1 | 0.001 000 | 147.12 | 20.97 | 2382.3 | 20.98 | 2510.6 | 0.0761 | 9.0257 |
| 10 | 0.001 227 6 | 0.001 000 | 106.38 | 42.00 | 2389.2 | 42.01 | 2519.8 | 0.1510 | 8.9008 |
| 15 | 0.001 705 1 | 0.001 001 | 77.93 | 62.99 | 2396.1 | 62.99 | 2528.9 | 0.2245 | 8.7814 |
| 20 | 0.002 339 | 0.001 002 | 57.79 | 83.95 | 2402.9 | 83.96 | 2538.1 | 0.2966 | 8.6672 |
| 25 | 0.003 169 | 0.001 003 | 43.36 | 104.88 | 2409.8 | 104.89 | 2547.2 | 0.3674 | 8.5580 |
| 30 | 0.004 246 | 0.001 004 | 32.89 | 125.78 | 2416.6 | 125.79 | 2556.3 | 0.4369 | 8.4533 |
| 35 | 0.005 628 | 0.001 006 | 25.22 | 146.67 | 2423.4 | 146.68 | 2565.3 | 0.5053 | 8.3531 |
| 40 | 0.007 384 | 0.001 008 | 19.52 | 167.56 | 2430.1 | 167.57 | 2574.3 | 0.5725 | 8.2570 |
| 45 | 0.009 593 | 0.001 010 | 15.26 | 188.44 | 2436.8 | 188.45 | 2583.2 | 0.6387 | 8.1648 |
| 50 | 0.012 349 | 0.001 012 | 12.03 | 209.32 | 2443.5 | 209.33 | 2592.1 | 0.7038 | 8.0763 |
| 55 | 0.015 758 | 0.001 015 | 9.568 | 230.21 | 2450.1 | 230.23 | 2600.9 | 0.7679 | 7.9913 |
| 60 | 0.019 940 | 0.001 017 | 7.671 | 251.11 | 2456.6 | 251.13 | 2609.6 | 0.8312 | 7.9096 |
| 65 | 0.025 03 | 0.001 020 | 6.197 | 272.02 | 2463.1 | 272.06 | 2618.3 | 0.8935 | 7.8310 |
| 70 | 0.031 19 | 0.001 023 | 5.042 | 292.95 | 2469.6 | 292.98 | 2626.8 | 0.9549 | 7.7553 |
| 75 | 0.038 58 | 0.001 026 | 4.131 | 313.90 | 2475.9 | 313.93 | 2643.7 | 1.0155 | 7.6824 |
| 80 | 0.047 39 | 0.001 029 | 3.407 | 334.86 | 2482.2 | 334.91 | 2635.3 | 1.0753 | 7.6122 |
| 85 | 0.057 83 | 0.001 033 | 2.828 | 355.84 | 2488.4 | 355.90 | 2651.9 | 1.1343 | 7.5445 |
| 90 | 0.070 14 | 0.001 036 | 2.361 | 376.85 | 2494.5 | 376.92 | 2660.1 | 1.1925 | 7.4791 |
| 95 | 0.084 55 | 0.001 040 | 1.982 | 397.88 | 2500.6 | 397.96 | 2668.1 | 1.2500 | 7.4159 |
| 100 | 0.101 35 | 0.001 044 | 1.6729 | 418.94 | 2506.5 | 419.04 | 2676.1 | 1.3069 | 7.3549 |

**Table 7.3** Properties of saturated water as a function of pressure.

| Pressure (MPa) | Temp. (°C) | Spec. Vol. (m³/kg) | | Int. Energy (kJ/kg) | | Enthalpy (kJ/kg) | | Entropy (kJ/kgK) | |
|---|---|---|---|---|---|---|---|---|---|
| $P$ | $T_{sat}$ | $v_f$ | $v_g$ | $u_f$ | $u_g$ | $h_f$ | $h_g$ | $s_f$ | $s_g$ |
| 0.000 611 3 | 0.01 | 0.001 000 | 206.14 | 0 | 2375.3 | 0.00 | 2501.4 | 0.0000 | 9.1562 |
| 0.0010 | 6.98 | 0.001 000 | 129.21 | 29.3 | 2385.0 | 29.30 | 2514.2 | 0.1059 | 8.9756 |
| 0.0015 | 13.03 | 0.001 001 | 87.98 | 54.71 | 2393.3 | 54.71 | 2525.3 | 0.1957 | 8.8279 |
| 0.0020 | 17.50 | 0.001 001 | 67.00 | 73.48 | 2399.5 | 73.48 | 2533.5 | 0.2607 | 8.7237 |
| 0.0025 | 21.08 | 0.001 002 | 54.25 | 88.48 | 2404.4 | 88.49 | 2540.0 | 0.3120 | 8.6432 |
| 0.0030 | 24.08 | 0.001 003 | 45.67 | 101.04 | 2408.5 | 101.05 | 2545.5 | 0.3545 | 8.5776 |
| 0.0040 | 28.96 | 0.001 004 | 34.80 | 121.45 | 2415.2 | 121.46 | 2554.4 | 0.4226 | 8.4746 |
| 0.0050 | 32.88 | 0.001 005 | 28.19 | 137.81 | 2420.5 | 137.82 | 2561.5 | 0.4764 | 8.3951 |
| 0.0075 | 40.29 | 0.001 008 | 19.24 | 168.78 | 2430.5 | 168.79 | 2574.8 | 0.5764 | 8.2515 |
| 0.010 | 45.81 | 0.001 010 | 14.67 | 191.82 | 2437.9 | 191.83 | 2584.7 | 0.6493 | 8.1502 |
| 0.015 | 53.97 | 0.001 014 | 10.02 | 225.92 | 2448.7 | 225.94 | 2599.1 | 0.7549 | 8.0085 |
| 0.020 | 60.06 | 0.001 017 | 7.649 | 251.38 | 2456.7 | 251.40 | 2609.7 | 0.8320 | 7.9085 |
| 0.025 | 64.97 | 0.001 020 | 6.204 | 271.90 | 2463.1 | 271.93 | 2618.2 | 0.8931 | 7.8314 |
| 0.030 | 69.10 | 0.001 022 | 5.229 | 289.20 | 2468.4 | 289.23 | 2625.3 | 0.9439 | 7.7686 |
| 0.040 | 75.87 | 0.001 027 | 3.993 | 317.53 | 2477.0 | 317.58 | 2636.8 | 1.0259 | 7.6700 |
| 0.050 | 81.33 | 0.001 030 | 3.240 | 340.44 | 2483.9 | 340.49 | 2645.9 | 1.0910 | 7.5939 |
| 0.075 | 91.78 | 0.001 037 | 2.217 | 384.31 | 2496.7 | 384.39 | 2663.0 | 1.2130 | 7.4564 |
| 0.100 | 99.63 | 0.001 043 | 1.694 | 417.36 | 2506.1 | 417.46 | 2675.5 | 1.3026 | 7.3594 |

**Table 7.4**   Properties of superheated water vapour at 0.1 MPa pressure.

| T (°C) | v (m³/kg) | u (kJ/kg) | h (kJ/kg) | s (kJ/kgK) |
|---|---|---|---|---|
| $T_{sat}$ | 1.694 | 2506.1 | 2675.5 | 7.3594 |
| 100 | 1.6958 | 2506.7 | 2676.2 | 7.3614 |
| 150 | 1.9364 | 2582.8 | 2776.4 | 7.6143 |
| 200 | 2.172 | 2658.1 | 2875.3 | 7.8343 |
| 250 | 2.406 | 2733.7 | 2974.3 | 8.0333 |
| 300 | 2.639 | 2810.4 | 3074.3 | 8.2158 |
| 400 | 3.103 | 2967.9 | 3278.2 | 8.5435 |
| 500 | 3.565 | 3131.6 | 3488.1 | 8.8342 |
| 600 | 4.028 | 3301.9 | 3704.4 | 9.0976 |
| 700 | 4.490 | 3479.2 | 3928.2 | 9.3398 |
| 800 | 4.952 | 3663.5 | 4158.6 | 9.5652 |
| 900 | 5.414 | 3854.8 | 4396.1 | 9.7767 |
| 1000 | 5.875 | 4052.8 | 4640.3 | 9.9764 |
| 1100 | 6.337 | 4257.3 | 4891.0 | 10.1659 |
| 1200 | 6.799 | 4467.7 | 5147.6 | 10.3463 |
| 1300 | 7.260 | 4683.5 | 5409.5 | 10.5183 |

$$P = 0.10 \text{ MPa } (T_{sat} = 99.63 \text{ °C})$$

**Solution:**

$$h = xh_g + (1-x)h_f = 0.4 \times 2592.1 \text{ kJ / kg} + (1-0.4) \times 209.33 \text{ kJ / kg} = 1162.438 \text{ kJ / kg}$$

**Answer:** The mixture enthalpy is 1162.4 kJ / kg.                                                      ∎

**Example 7.4**
**Problem:** What is the temperature and specific volume of a saturated water-steam mixture with 1% quality at a pressure of 50 kPa?
**Find:** Saturation temperature $T_{sat}$ and specific volume $v$.
**Known:** Mixture is at saturation pressure $P_{sat} = 50$ kPa, quality $x = 0.4$.
**Governing Equation:**

Mixture property (y)                              $y = xy_g + (1-x)y_f$

**Properties:** From Table 7.3 at $P_{sat} = 50$ kPa, saturation temperature $T_{sat} = 81.33$ °C, specific volume of liquid $v_f = 0.001030$ m³ / kg and vapour $v_g = 3.240$ m³ / kg.

**Solution:**

$$v = xv_g + (1-x)v_f = 0.01 \times 3.240 \text{ m}^3 \text{ / kg} + (1-0.01) \times 0.001 \ 030 \text{ m}^3 \text{ / kg} = 0.033 \ 420 \text{ m}^3 \text{ / kg}$$

**Answer:** The mixture temperature is 81 °C and its specific volume is 0.0334 m³ / kg.             ∎

**Example 7.5**
**Problem:** What is the quality of a saturated steam–water mixture at 40 °C with a specific volume of 15.2 m³ / kg?

**Find:** Quality of mixture $x$.

**Known:** Saturation temperature $T_{sat} = 40\,°C$, specific volume $v = 15.2\ m^3\,/\,kg$.

**Governing Equation:**

Mixture property ($y$)                                         $y = xy_g + (1-x)y_f$

**Properties:** From Table 7.2 at $T_{sat} = 40\,°C$, specific volume of liquid $v_f = 0.001\ 008\ m^3\,/\,kg$ and vapour $v_g = 19.52\ m^3\,/\,kg$.

**Solution:**

$$v = xv_g + (1-x)v_f = v_f + x(v_g - v_f)$$

$$x = \frac{v - v_f}{v_g - v_f} = \frac{15.2\,m^3\,/\,kg - 0.001\ 008\,m^3\,/\,kg}{19.52\,m^3\,/\,kg - 0.001\ 008\,m^3\,/\,kg} = 0.778\ 68$$

**Answer:** The quality is 77.9%.                                                                            ■

**Example 7.6**

**Problem:** The specific entropy of water at 0.01 MPa is 8.320 kJ / kgK. What state (subcooled, superheated or saturated) is it in?

**Find:** State of the water.

**Known:** Saturation pressure $P = 0.01$ MPa, specific entropy $s = 8.320$ kJ / kgK

**Properties:** From Table 7.2 at $P_{sat} = 0.01$ MPa, specific entropy of vapour $s_g = 8.1502$ kJ / kgK.

**Solution:**

Since $s > sg$, it is in the superheated steam state.

**Answer:** The water is superheated vapour.                                                              ■

**Example 7.7**

**Problem:** Superheated water vapour at 0.1 MPa and 250 °C is expanded in an isentropic process to a temperature of 35 °C. Show that the final state is a saturated mixture and find its quality.

**Find:** The quality $x_2$ of the final saturated mixture.

**Known:** Initial saturation pressure $P_{1,sat} = 0.1$ MPa, initial saturation temperature $T_{1,sat} = 250\,°C$, final saturation temperature $T_{2,sat} = 35\,°C$, isentropic process.

**Properties:** From Table 7.4 at $P_{1,sat} = 0.1$ MPa and $T_{1,sat} = 250\,°C$, specific entropy $s_1 = 8.0333$ kJ / kgK. From Table 7.2 at $T_{2,sat} = 35\,°C$, specific entropy of liquid $s_f = 0.5053$ kJ / kg and vapour $s_g = 8.3531$ kJ / kg.

**Solution:**

Since the water undergoes an isentropic process, $s_2 = s_1 = 8.0333$ kJ / kgK. Since $s_{2,f} < s_2 < s_{2,g}$, the final state is a saturated mixture.

$$s_2 = x_2 s_g + (1-x_2)s_f = s_f + x_2(s_g - s_f)$$

$$x_2 = \frac{s_2 - s_f}{s_g - s_f} = \frac{8.0333\ kJ\,/\,kgK - 0.5053\ kJ\,/\,kgK}{8.3531\ kJ\,/\,kgK - 0.5053\ kJ\,/\,kgK} = 0.95925$$

**Answer:** The quality of the final saturated mixture is 95.9%.                                ■

## Control Mass Analysis

**Example 7.8**

**Problem:** A piston-cylinder system contains 0.1 kg of water at 0.5 MPa and 400 °C. It is cooled at constant pressure so that 100 kJ of energy is removed from the system. What is the final state of the water?

**Find:** Final state of the water.

**Known:** Mass of the water $m = 0.1$ kg, initial pressure $P_1 = 0.5$ MPa, initial temperature $T_1 = 400$ °C, constant pressure so final pressure $P_2 = P_1$, heat removed $Q_{12} = -100$ kJ.

**Diagram:**

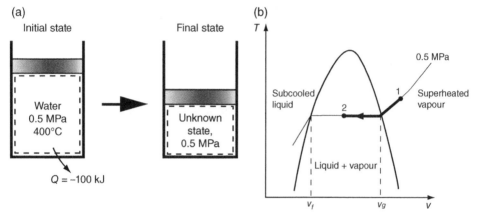

**Figure E7.8** (a) A piston–cylinder system containing water is cooled at constant pressure. (b) $T$-$V$ diagram for water being cooled at constant pressure.

**Governing Equations:**

For a constant pressure process

$$q_{12} = \frac{Q_{12}}{m} = h_2 - h_1.$$

**Solution:**

This is a first-law control mass problem of the type that we have seen in Chapter 4, the only difference being that we now also consider the possibility of a phase change occurring within the fluid. To determine whether this happens, we have to fix the initial and final state of the water in the system.

To fix the state of a simple compressible system we need two independent properties. In the initial state we know the pressure ($P_1 = 0.5$ MPa) and temperature ($T_1 = 400$ °C). From the superheated steam tables, at a pressure of 0.5 MPa, $T_{sat} = 151.86$ °C. Since $T_1 > T_{sat}$, we know that the water is in the form of superheated steam. We can therefore mark the initial state (1) on a $T$-$v$ diagram. It is cooled at constant pressure, so we can represent the process by a line that follows the isobar of $P = 0.5$ MPa on $T$-$v$ axes.

How do we determine the final state (2)? We know one property, pressure, but we need to determine one more, which we can get from an energy balance. For heat transfer at constant pressure,

$$q_{12} = \frac{Q_{12}}{m} = h_2 - h_1.$$

From the superheated steam tables (Appendix 8c) at $P_1 = 0.5$ MPa and $T_1 = 400\,°C$, specific enthalpy $h_1 = 3271.9$ kJ / kgK. Therefore,

$$h_2 = h_1 + \frac{Q_{12}}{m} = 3271.9 \,\text{kJ} / \text{kgK} + \frac{(-100\,\text{kJ})}{0.1\,\text{kg}} = 2271.9 \,\text{kJ} / \text{kg}.$$

From the saturated water tables (Appendix 8b) at $P_1 = 0.5$ MPa, specific enthalpy of liquid $h_f = 640.23$ kJ / kg and vapour $h_g = 2748.7$ kJ / kg. Since $h_f < h_2 < h_g$, the final state is a saturated mixture:

$$h_2 = x_2 h_g + (1 - x_2) h_f = h_f + x_2 (h_g - h_f),$$

$$x_2 = \frac{h_2 - h_f}{h_g - h_f} = \frac{2271.9 \,\text{kJ}/\text{kg} - 640.23 \,\text{kJ/kg}}{2748.9 \,\text{kJ}/\text{kg} - 640.23 \,\text{kJ/kg}} = 0.773 \ 79.$$

**Answer:** The final state will be a saturated mixture with a quality of 77.4%. ∎

## Example 7.9
**Problem:** An insulated piston-cylinder device contains 2 kg of saturated steam at 0.5 MPa. The steam is allowed to expand reversibly until its pressure drops to 0.1 MPa. Find the work done during this process.
**Find:** Work done during expansion $W$.
**Known:** Saturated steam, mass of steam $m = 2$ kg, initial pressure $P_1 = 0.5$ MPa, final pressure $P_2 = 0.1$ MPa, reversible and insulated (adiabatic) means isentropic.
**Diagram:**

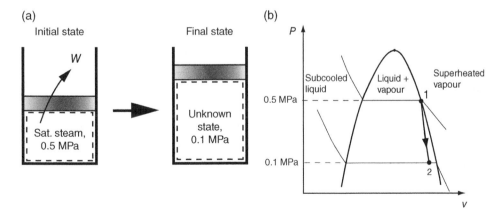

**Figure E7.9** (a) A piston–cylinder system containing saturated steam is expanded. (b) P-v diagram for steam being expanded.

**Governing Equations:**

Energy balance                               $Q + W = \Delta U + \Delta KE + \Delta PE$

**Solution:**
In this problem the initial state of the system is known, since we know the pressure ($P_1 = 0.5$ MPa) and the quality ($x = 1$). To fix the final state we know the final pressure ($P_2 = 0.1$ MPa). We can determine a second property from the fact that the expansion process is adiabatic and reversible, which means that it is isentropic ($s_1 = s_2$). Therefore, from the saturated water tables (Appendix 8b) at $P_1 = 0.5$ MPa, specific enthalpy of liquid $s_1 = s_g = 6.8213$ kJ / kgK. At $P_2 = 0.1$ MPa, specific entropy of liquid $s_f = 1.3026$ kJ / kgK and vapour $s_g = 7.3594$ kJ / kgK. Since $s_f < s_2 < s_g$, we know that the final state is a saturated vapour–liquid mixture. We can sketch the process on a $P$-$v$ diagram.
The final quality is

$$x_2 = \frac{s_2 - s_f}{s_g - s_f} = \frac{6.8213\,\text{kJ} / \text{kgK} - 1.3026\,\text{kJ} / \text{kgK}}{7.3594\,\text{kJ} / \text{kgK} - 1.3026\,\text{kJ} / \text{kgK}} = 0.911158.$$

From an energy balance,

$$W + \underset{=0}{\underbrace{Q}} = \Delta U = m\left(u_2 - u_1\right).$$

At $P_1 = 0.5$ MPa, specific internal energy of vapour $u_1 = u_g = 2561.2$ kJ / kg. At $P_2 = 0.1$ MPa, specific internal energy of liquid $u_f = 417.36$ kJ / kg and vapour $u_g = 2506.1$ kJ / kg. Using the quality of $x_2 = 0.911158$ from earlier,

$u_2 = u_f + x_2\left(u_g - u_f\right),$
$u_2 = 417.36\,\text{kJ} / \text{kg} + 0.911158 \times \left(2506.1\,\text{kJ} / \text{kg} - 417.36\,\text{kJ} / \text{kg}\right) = 2320.53\,\text{kJ} / \text{kg}.$
$W = m\left(u_2 - u_1\right) = 2\,\text{kg} \times \left(2320.2\,\text{kJ} / \text{kg} - 2561.2\,\text{kJ} / \text{kg}\right),$
$W = -481.340\,\text{kJ}.$

**Answer:** The system does 481.3 kJ of work on the surroundings while expanding.                    ■

## *Control Volume Analysis*

**Example 7.10**
**Problem:** Steam enters an adiabatic nozzle at 1 MPa and 400 °C with negligible velocity. It exits at a pressure of 0.6 MPa. Assuming that the expansion is reversible, calculate the exit temperature and velocity.
**Find:** Exit temperature $T_2$ and velocity $\mathbf{V}_2$.
**Known:** Inlet pressure $P_1 = 1$ MPa, inlet temperature $T_1 = 400$ °C, exit pressure $P_2 = 0.6$ MPa, nozzle is adiabatic.
**Diagram:**
**Assumptions:** Expansion is reversible, and since the nozzle is adiabatic, the process is isentropic.

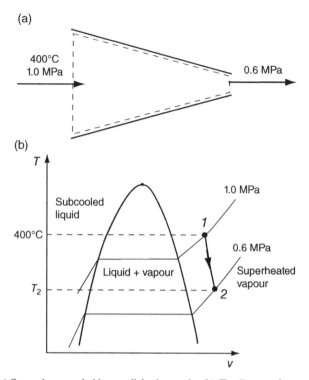

**Figure E7.10** (a) Steam is expanded in an adiabatic nozzle. (b) *T-v* diagram for steam being expanded.

**Governing Equation:**

Adiabatic nozzle
$$V_2 = \sqrt{2(h_1 - h_2)}$$

**Solution:**
At the inlet we know the pressure and temperature, which fixes the state. At the outlet we know the pressure, and we can obtain a second property by assuming that the expansion is isentropic so that $s_1 = s_2$. The inlet conditions are $P_1 = 1$ MPa and $T_1 = 400\ °C$. From the superheated steam tables (Appendix 8c) specific entropy $s_1 = 7.4651$ kJ / kgK and specific enthalpy $h_1 = 3263.9$ kJ / kg. For saturated vapour at $P_2 = 0.6$ MPa (Appendix 8b) specific entropy of vapour $s_g = 6.7600$ kJ / kg K. Since $s_2 > s_g$, the steam is superheated at the outlet. We can represent the process on a *T-v* diagram.

We know the pressure and specific entropy at the outlet. To find the temperature we must interpolate in the superheated steam table at 0.6 MPa. The value of $s_2 = 7.4651$ kJ / kgK lies between those for $s$ at 300 and 350 °C. From the superheated steam tables, at 0.6 MPa:

| $T\,(°C)$ | $h$ (kJ / kg) | $s$ (kJ / kgK) |
| --- | --- | --- |
| 300 | 3061.6 | 7.3724 |
| $T_2 = ?$ | $h_2 = ?$ | $s_2 = 7.4651$ |
| 350 | 3165.7 | 7.5464 |

Interpolating for temperature,

$$\frac{T_2 - 300 \ ^\circ C}{350 \ ^\circ C - 300 \ ^\circ C} = \frac{7.4651 \, kJ / kgK - 7.3724 \, kJ / kgK}{7.5464 \, kJ / kgK - 7.3724 \, kJ / kgK},$$

$$T_2 = 326.638 \ ^\circ C.$$

Interpolating for enthalpy,

$$\frac{h_2 - 3061.6 \, kJ / kg}{3165.7 \, kJ / kg - 3061.6 \, kJ / kg} = \frac{7.4651 \, kJ / kgK - 7.3724 \, kJ / kgK}{7.5464 \, kJ / kgK - 7.3724 \, kJ / kgK},$$

$$h_2 = 3117.06 \, kJ / kg.$$

From an energy balance for an adiabatic nozzle,

$$\mathbf{V}_2 = \sqrt{2(h_1 - h_2)} = \sqrt{2(3263.9 \, kJ / kg - 3117.1 \, kJ / kg)10^3 \, J / kJ} = 541.849 \, m / s.$$

**Answer:** The exit temperature is 327 °C and the exit velocity is 541.8 m / s.  ∎

### Example 7.11
**Problem:** Steam enters an insulated turbine at 7 MPa and 500 °C and leaves at 75 kPa with a mass flow rate of 10 kg / s. Assuming that expansion in the turbine is reversible, what is the work output?
**Find:** Work output of turbine $W$.
**Known:** Inlet pressure $P_1 = 7$ MPa, inlet temperature $T_1 = 500$ °C, exit pressure $P_2 = 75$ kPa, mass flow rate $\dot{m} = 10 \, kg / s$.
**Diagram:**

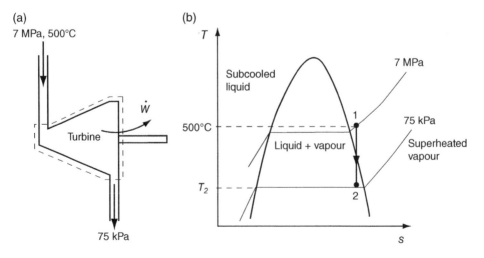

**Figure E7.11** (a) Steam is expanded in an adiabatic turbine. (b) $T$-$s$ diagram for steam being expanded isentropically.

**Assumptions:** Expansion is reversible, process is adiabatic due to insulated turbine, together make an isentropic process.
**Governing Equation:**

For a turbine $$\dot{W} = \dot{m}\left(h_2 - h_1\right)$$

**Solution:**
At the inlet pressure $P_1 = 7$ MPa the saturation temperature of steam $T_{sat} = 285.88$ °C (Appendix 8b), and since $T_1 > T_{sat}$ the steam is superheated. From the superheated steam tables (Appendix 8c) specific entropy $s_1 = 6.7975$ kJ / kgK and specific enthalpy $h_1 = 3410.3$ kJ / kg. We assume that the expansion is adiabatic and reversible, which means that it is isentropic so that $s_1 = s_2$. It is convenient to show an isentropic process on a $T$-$s$ diagram where it appears as a vertical line. The initial state (1) lies in the superheated vapour region. It is not clear as yet whether state (2), which lies on the 75 kPa isobar, lies above or below the vapour dome – whether the exhaust from the turbine is superheated steam or a saturated mixture. From the saturated water tables (Appendix 8b) at $P_2 = 75$ kPa, specific entropy of fluid $s_f = 1.2130$ kJ / kgK and gas $s_g = 7.4564$ kJ / kgK. Since $s_f < s_2 < s_g$, the final state is a saturated vapour–liquid mixture. We can therefore draw the isentropic expansion line on a $T$-$s$ diagram:
We can find the quality at state 2:

$$x_2 = \frac{s_2 - s_f}{s_g - s_f} = \frac{6.7975\,\text{kJ} / \text{kgK} - 1.2130\,\text{kJ} / \text{kgK}}{7.4564\,\text{kJ} / \text{kgK} - 1.2130\,\text{kJ} / \text{kgK}} = 0.894\ 46.$$

Then we can solve for the final specific enthalpy,

$$h_2 = h_f + x_2 h_{fg},$$

At 75 kPa specific enthalpy of fluid $h_f = 384.39$ kJ.kg and latent heat of vaporisation $h_{fg} = 2278.6$ kJ / kg from Appendix 8b, so

$$h_2 = 384.39\,\text{kJ} / \text{kg} + 0.89446 \times \left(2278.6\,\text{kJ} / \text{kg}\right) = 2422.5\,\text{kJ} / \text{kg},$$

$$\dot{W} = \dot{m}\left(h_2 - h_1\right) = 10\,\text{kg} / \text{s}\left(2422.5\,\text{kJ} / \text{kg} - 3410.3\,\text{kJ} / \text{kg}\right),$$

$$\dot{W} = -9878.0\,\text{kW} = -9.88\,\text{MW}.$$

The negative sign means that the work is done by the system on the surroundings.
**Answer:** The work output from the turbine is 9.88 MW.                                    ∎

**Example 7.12**
**Problem:** An insulated mixing chamber has two inlets: the first has saturated water entering at 100 kPa with a flow rate of 4 kg / s, while the second has 2 kg / s of superheated steam at

100 kPa and 150 °C. The mixture leaves the chamber at a pressure of 100 kPa. Find the rate of entropy generation in this process.

**Find:** Rate of entropy generation $\dot{s}$.

**Known:** Inlet pressure $P_1 = 100$ kPa, inlet temperature $T_1 = 500$ °C, mass flow rate $\dot{m}_1 = 4$ kg / s, inlet pressure $P_2 = 100$ kPa, inlet temperature $T_2 = 150$ °C, mass flow rate $\dot{m}_2 = 2$ kg / s, exit pressure $P_3 = 100$ kPa.

**Diagram:**

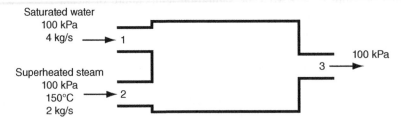

**Figure E7.12**   An insulated mixing chamber.

**Governing Equations:**

Energy rate balance
$$\dot{Q} + \dot{W} = \sum_e \dot{m}_e h_e - \sum_i \dot{m}_i h_i$$

Entropy rate balance
$$\dot{S}_{gen} + \underbrace{\dot{S}_{heat}}_{=0} = \sum_e \dot{m}_e s_e - \sum_i \dot{m}_i s_i$$

**Solution:**

From a mass balance,

$$\dot{m}_1 + \dot{m}_2 = \dot{m}_3,$$

$$\dot{m}_3 = 4 \, \text{kg} / \text{s} + 2 \, \text{kg} / \text{s} = 6 \, \text{kg} / \text{s}.$$

From an energy balance for a control volume, assuming negligible changes in kinetic and potential energy, and that the chamber is adiabatic and that there is no work done in it

$$\dot{m}_1 h_1 + \dot{m}_2 h_2 = \dot{m}_3 h_3.$$

For saturated water at 100 kPa (Appendix 8b), specific enthalpy of liquid $h_1 = h_f = 417.46$ kJ/kg. For superheated steam at 100 kPa and 150 °C, specific enthalpy $h_2 = 2776.4$ kJ / kg. So the exit enthalpy is

$$4 \, \text{kg} / \text{s} \times 417.46 \, \text{kJ} / \text{kg} + 2 \, \text{kg} / \text{s} \times 2776.4 \, \text{kJ} / \text{kg} = 6 \, \text{kg} / \text{s} \times h_3,$$

$$h_3 = 1203.8 \, \text{kJ} / \text{kg}.$$

Since $h_f < h_3 < h_g$, we know that the exit state is a saturated vapour–liquid mixture, with quality

$$x_3 = \frac{h_3 - h_f}{h_g - h_f} = \frac{1203.8\,\text{kJ}/\text{kg} - 417.46\,\text{kJ}/\text{kg}}{2675.5\,\text{kJ}/\text{kg} - 417.46\,\text{kJ}/\text{kg}} = 0.348\ 24.$$

For saturated fluid at 100 kPa, specific entropy of liquid $s_1 = s_f = 1.3026$ kJ / kgK and vapour $s_g = 7.3594$ kJ / kgK. For superheated steam at 100 kPa and 150 °C, specific entropy $s_2 = 7.6143$ kJ / kg, so entropy at the exit is

$$s_3 = s_f + x_3\left(s_g - s_f\right) = 1.3026\,\text{kJ}/\text{kgK} + 0.34824 \times \left(7.3594\,\text{kJ}/\text{kgK} - 1.3026\,\text{kJ}/\text{kgK}\right),$$

$$s_3 = 3.4118\,\text{kJ}/\text{kgK}.$$

From an entropy balance,

$$\dot{S}_{gen} = \dot{m}_3 s_3 - \left(\dot{m}_1 s_1 + \dot{m}_2 s_2\right),$$

$$\dot{S}_{gen} = 6\,\text{kg/s} \times 3.4118\,\text{kJ}/\text{kgK} - \left(4\,\text{kg/s} \times 1.3026\,\text{kJ}/\text{kgK} + 2\,\text{kg/s} \times 7.6143\,\text{kJ}/\text{kg}\right),$$

$$\dot{S}_{gen} = 0.031\ 800\,\text{kW/kgK}.$$

**Answer:** Entropy is generated at a rate of 0.0318 kW / kgK.                           ■

## 7.11   Van der Waals Equation of State

Using property tables requires effort to look up property values, interpolate where necessary, and use them in calculations. Using the ideal gas equation to calculate properties, as we did earlier, was much easier. Why cannot we use the same equation to calculate the properties of liquids or liquid–vapour mixtures?

The ideal gas equation is a very simplified model of real gas behaviour that assumes that molecules are far apart and inter-molecular forces are negligible, which is not true if the gas density becomes high. Figure 7.10a shows isotherms for an ideal gas, which are a family of curves with the shape $P = \text{constant} / v$. They resemble isotherms for a real substance, shown in Figure 7.10b, only in the case of a superheated vapour at low pressure, which has large intermolecular spacing. Once liquid starts condensing and isotherms follow constant pressure lines in the two-phase region the ideal gas model is no longer valid. Liquids are virtually incompressible ($v = \text{constant}$) and the isotherms in the subcooled region are almost vertical.

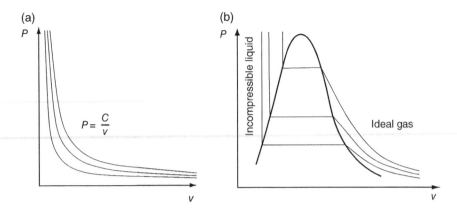

**Figure 7.10**  *P-v* diagram showing isotherms for (a) an ideal gas and (b) a real substance.

The ideal gas model assumes that gas molecules occupy negligible space themselves, and that there are no attractive forces between molecules. As a gas is compressed and the spacing between molecules becomes small, both these assumptions are no longer valid. The first attempt to derive an improved equation of state, which would work even at high gas density, was made by Johannes van der Waals in 1873. He reasoned that if the volume occupied by the molecules in a mole of gas equals $b$, the volume available for movement of molecules would not be $V$, but $V - Nb$. The ideal gas equation,

$$P = \frac{NR_uT}{V},$$

can then be modified to

$$P = \frac{NR_uT}{V - Nb}. \tag{7.54}$$

Attractive forces between molecules are short range, so their effect is most significant between molecules immediately next to each other. Molecules in the bulk of the gas are surrounded on all sides by other identical molecules and find themselves pulled equally in all directions. Molecules near the walls are pulled back into the gas since they experience attractive forces on only one side (see Figure 7.11). The net effect is to reduce the pressure on the container created by molecules hitting the walls.

The force exerted by molecular collisions on the container walls is proportional to gas molecular density, $N/V$. The force pulling back the surface molecules is proportional to the density of molecules in the gas immediately behind the surface layer. The pressure is therefore lowered by an amount proportional to $(N/V)^2$. Modifying Equation (7.54) to account for this pressure reduction gives

$$P = \frac{NR_uT}{V - Nb} - a\left(\frac{N}{V}\right)^2, \tag{7.55}$$

where $a$ is a constant of proportionality. Rewriting Equation (7.55) in terms of the specific molar volume $\bar{v} = V / N$,

$$P = \frac{R_u T}{\bar{v} - b} - \frac{a}{\bar{v}^2}, \tag{7.56}$$

or

$$\left(P + \frac{a}{\bar{v}^2}\right)(\bar{v} - b) = R_u T. \tag{7.57}$$

Equation (7.57) is known as the van der Waals equation of state. The constants $a$ and $b$ can be determined experimentally: Table 7.5 gives values for some common gases.

In the limit of low pressure, as $\bar{v} \rightarrow \infty$, the term $a / \bar{v}^2 \rightarrow 0$ in Equation (7.57) and, since $\bar{v} \gg b$, the van der Waals equation reduces to the ideal gas equation:

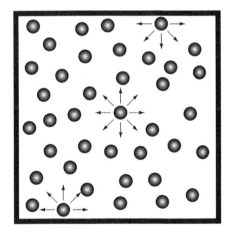

**Figure 7.11**   Molecules in the bulk of the gas experience attractive forces on all sides: those near the wall are pulled back from one direction, reducing the pressure.

**Table 7.5**   Values of van der Waals constants for common gases.

| Substance | Formula | $a$ (kPa m⁶/kmol²) | $b$ (m³/kmol) |
|---|---|---|---|
| Air | — | 136 | 0.0365 |
| Carbon dioxide | $CO_2$ | 364.7 | 0.0428 |
| Nitrogen | $N_2$ | 137 | 0.0386 |
| Hydrogen | $H_2$ | 24.5 | 0.0263 |
| Water | $H_2O$ | 554 | 0.0305 |
| Oxygen | $O_2$ | 136.9 | 0.0319 |
| Helium | He | 3.47 | 0.0238 |
| Argon | Ar | 135 | 0.0320 |
| Methane | $CH_4$ | 229.3 | 0.048 |

$$P\overline{v} = R_u T.$$

As pressure increases, in the limit $P \rightarrow \infty$, Equation (7.57) gives

$$\overline{v} - b = \frac{R_u T}{P + a / \overline{v}^2} \rightarrow 0$$

or

$$\overline{v} = b.$$

The specific volume is constant, as is the case for an incompressible substance. The van der Waals equation therefore serves as a bridge between an ideal gas and an incompressible liquid.

We can find a relationship between the van der Waals constants and the critical temperature and pressure by examining the isotherm that passes through the critical point. At the critical point the isotherm passing through it has an inflection point. Seen on a $P$-$\overline{v}$ diagram (Figure 7.12), at the inflection point,

$$\left( \frac{\partial P}{\partial \overline{v}} \right)_T = 0 \text{ and } \left( \frac{\partial^2 P}{\partial \overline{v}^2} \right)_T = 0 \text{ at } T = T_c, P = P_c \text{ and } v = \overline{v}_c.$$

At the critical point the van der Waals equation, Equation (7.56), is

$$P_C = \frac{R_u T}{\overline{v}_c - b} - \frac{a}{v_c^2} \tag{7.58}$$

and

$$\left( \frac{\partial P}{\partial \overline{v}} \right)_T = -\frac{R_u T_c}{\left( \overline{v}_c - b \right)^2} + \frac{2a}{\overline{v}_c^3} = 0, \tag{7.59}$$

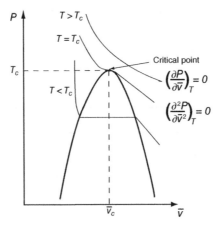

**Figure 7.12** $P - \overline{v}$ diagram showing three isotherms, one corresponding to the critical temperature ($T_c$), one above and one below. The isotherm corresponding to the critical temperature has an inflection point at the critical point.

$$\left(\frac{\partial^2 P}{\partial \bar{v}^2}\right)_T = \frac{2R_u T_c}{\left(\bar{v}_c - b\right)^3} - \frac{6a}{\bar{v}_c^4} = 0. \tag{7.60}$$

Simultaneously solving Equations (7.58) to (7.60) gives expressions for the van der Waals constants as a function of the critical properties:

$$a = \frac{27}{64} \frac{R_u T_c^2}{P_c}, \tag{7.61}$$

$$b = \frac{R_u T_c}{8P_c}, \tag{7.62}$$

and a relation for the critical volume:

$$\bar{v}_c = \frac{3}{8} \frac{R_u T_c}{P_c}. \tag{7.63}$$

## 7.12  Compressibility Factor

We define the *reduced pressure* ($P_r$), *reduced temperature* ($T_r$) and *reduced specific volume* ($\bar{v}_r$):

$$P_r = \frac{P}{P_c}, T_r = \frac{T}{T_c} \text{ and } \bar{v}_r = \frac{\bar{v}}{\bar{v}_c}. \tag{7.64}$$

Combining these definitions with Equations (7.61) to (7.63) and substituting into the van der Waals equation gives

$$P_r = \frac{8T_r}{3\bar{v}_r - 1} - \frac{3}{\bar{v}_r^2}. \tag{7.65}$$

This is a universal equation of state, valid for all gases, since it does not contain either the van der Waals constants or the critical properties. To verify this equation experimentally we define the *compressibility factor* of a gas

$$Z = \frac{Pv}{RT} = \frac{P\bar{v}}{R_u T}. \tag{7.66}$$

Figure 7.13 shows the variation of the compressibility factor as a function of $P_r$ for values of $T_r$ ranging from 1.0 to 2.0. Data for all gases lie on the same curves, confirming the *principle of corresponding states*, that the compressibility of all gases is the same at the same reduced pressure and temperature.

From the ideal gas equation, the compressibility of an ideal gas is $Z = 1$. The generalized compressibility chart (Figure 7.13) allows us to see under what conditions a gas behaves in an ideal manner. At low pressures ($P_r \ll 1$) or at temperatures well above critical ($T_r > 2$) we can reasonably assume all gases are ideal. Deviations from ideal gas behaviour are largest near the critical point ($P_r = T_r = 1$).

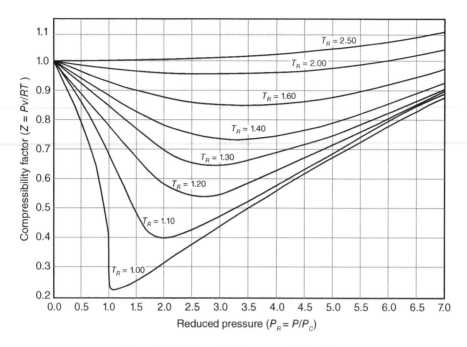

**Figure 7.13**   Generalised compressibility chart.

Compressibility charts can be used to correct calculations made using the ideal gas equation under conditions that $Z$ is not close to 1.

## 7.13   Other Equations of State

At the critical point ($P_r = T_r = 1$) the van der Waals equation predicts [by combining Equations (7.66) and (7.63)]

$$Z_c = \frac{R_u T_c}{P \bar{v}_c} = \frac{3}{8} = 0.375. \tag{7.67}$$

In reality the compressibility of most gases at the critical point lies between 0.25 and 0.35 (see Figure 7.13). The van der Waals equation is therefore not very accurate quantitatively, though it is the simplest equation that can give qualitative information about phase change.

### 7.13.1   Redlich–Kwong Equation of State

The van der Waals equation assumes that intermolecular forces are independent of temperature and pressure. It can be improved by modifying the molecular attraction term, giving us the *Redlich–Kwong* equation of state,

$$P = \frac{R_u T}{\bar{v} - b_{RK}} - \frac{a_{RK}}{\bar{v}\left(\bar{v} + b_{RK}\right)\sqrt{T}}. \tag{7.68}$$

The constants $a_{RK}$ and $b_{RK}$ can be evaluated from the critical properties:

$$a_{RK} = \frac{0.42748 R_u^2 T_C^{2.5}}{P_C} \quad \text{and} \quad b_{RK} = \frac{0.08664 R_u T_C}{P_C}.$$

## 7.13.2 Virial Equation of State

In general, the larger the number of constants in an equation of state, the more accurate it is. Of course, increasing the number of constants makes it more difficult to use since all of them have to be determined either experimentally or theoretically. The most general equation of state is the virial equation that has the form of an infinite series:

$$P = \frac{R_u T}{\bar{v}}\left(1 + \frac{B(T)}{\bar{v}} + \frac{C(T)}{\bar{v}^2} + \dots\right), \tag{7.69}$$

where $B$, $C$ and so on are known as the virial coefficients and are functions of temperature. They can be determined experimentally or, for simple substances, from quantum mechanical models. In the limit of low pressure they all go to zero, leaving the ideal gas equation. The series can be truncated when it is sufficiently accurate.

## Example 7.13

**Problem:** Calculate the pressure of nitrogen that has a specific volume $v = 0.025$ m³ / kg and a temperature of 252 K using (a) the ideal gas equation and (b) the van der Waals equation of state.

**Find:** Pressure of nitrogen $P$.

**Known:** Specific volume $v = 0.025$ m³ / kg, temperature $T = 252$ K.

**Governing Equations:**

Ideal gas equation

$$P\bar{v} = R_u T$$

van der Waals equation

$$\left(P + \frac{a}{\bar{v}^2}\right)(\bar{v} - b) = R_u T$$

**Properties:** For nitrogen, the critical temperature $T_C = 126.2$ K (Table 7.1), critical pressure $P_C = 3.39$ MPa (Table 7.1) and molar mass $M = 28.013$ kg / kmol (Appendix 1), van der Waals constants $a = 137$ kPa m⁶ / kmol² and $b = 0.0386$ m³ / kmol (Table 7.5).

## Solution:

The molar specific volume of nitrogen is

$$\bar{v} = vM = 0.025\,\text{m}^3 \text{ / kg} \times 28.013\,\text{kg / kmol} = 0.700\ 33\,\text{m}^3 \text{ / kmol}.$$

(a) Ideal gas equation,

$$P = \frac{R_u T}{\bar{v}} = \frac{8.314\,\text{kJ / kmolK} \times 252\,\text{K}}{0.700\ 33\,\text{m}^3 \text{ / kmol}} = 2991.6\,\text{kPa} = 2.99\,\text{MPa}.$$

(b) van der Waals equation,

$$P = \frac{R_u T}{(\bar{v} - b)} - \left(\frac{a}{\bar{v}^2}\right),$$

$$P = \frac{8.314 \times 10^3 \text{ J/kmolK} \times 252 \text{ K}}{\left(0.700 \ 33 \text{ m}^3 / \text{kmol} - 0.0386 \text{m}^3 / \text{kmol}\right)} - \left[\frac{137 \times 10^3 \text{ Pam}^6 / \text{kmol}^2}{\left(0.700 \ 33 \text{ m}^3 / \text{kmol}\right)^2}\right],$$

$$P = 2.8868 \text{ MPa}.$$

**Answer:** The pressure calculated using the (a) ideal gas equation is 2.99 MPa, (b) van der Waals equation is 2.89 MPa.                                                                                                    ∎

**Example 7.14**
**Problem:** Calculate the specific volume of steam at a temperature of 600 °C and pressure of 10 MPa using (a) steam tables, (b) the ideal gas equation and (c) the generalised compressibility charts.
**Find:** Specific volume of steam $v$.
**Known:** Temperature $T = 600 \text{ °C} = 873.15 \text{ K}$, pressure $P = 10 \text{ MPa}$.
**Governing Equations:**

Ideal gas equation                                                       $Pv = RT$

Compressibility factor                                              $Z = \dfrac{P\bar{v}}{R_u T}$

**Properties:** For water, gas constant $R = 0.4615 \text{ kJ/kgK}$ (Appendix 1), critical temperature $T_C = 647.3 \text{ K}$ and critical pressure $P_C = 22.09 \text{ MPa}$ (Table 7.1).

**Solution:**
(a) From the superheated steam tables (Appendix 8c) at $T = 600 \text{ °C}$ and $P = 10 \text{ MPa}$, specific volume $v = 0.038 \ 37 \text{ kg/m}^3$.
(b) From the ideal gas equation,

$$v = \frac{RT}{P} = \frac{0.4615 \times 10^3 \text{ J/kgK} \times 873.15 \text{K}}{10 \times 10^6 \text{ Pa}} = 0.040 \ 296 \text{m}^3 / \text{kg}.$$

(c) Using the generalised compressibility charts (Appendix 10), with reduced temperature

$$T_r = \frac{T}{T_C} = \frac{873.15 \text{ K}}{647.3 \text{K}} = 1.348 \ 95 = 1.35$$

and reduced pressure

$$P_r = \frac{P}{P_C} = \frac{10 \text{MPa}}{22.09 \text{MPa}} = 0.452 \ 69 = 0.45,$$

then the compressibility factor $Z = 0.95$.

$$P = Z\frac{RT}{P} = 0.95 \times \frac{0.4615 \times 10^3 \, \text{J/kgK} \times 873.15\,\text{K}}{10 \times 10^6 \, \text{Pa}} = 0.038 \, 281\,\text{m}^3/\text{kg}.$$

**Answer:** The specific volume calculated using (a) superheated steam tables is $0.0384 \, \text{kg} / \text{m}^3$, (b) ideal gas equation is $0.0403 \, \text{kg} / \text{m}^3$ and (c) generalised compressibility charts is $0.0383 \, \text{kg} / \text{m}^3$. ∎

## Further Reading

1.  Y. A. Cengel, M. A. Boles (**2015**) *Thermodynamics – An Engineering Approach*, McGraw Hill, New York.
2.  M. J. Moran, H. N. Shapiro, D. D. Boettner, M. B. Bailey (**2014**) *Fundamentals of Engineering Thermodynamics*, John Wiley & Sons, Ltd, London.
3.  C. Borgnakke, R. E. Sonntag (**2012**) *Fundamentals of Thermodynamics*, John Wiley & Sons, Ltd, London.

## Summary

*Evaporation* is the transformation of liquid into vapour while the transformation from vapour to liquid is called *condensation*. The energy carried away by a unit mass of evaporating liquid is known as the *latent heat of vaporisation*.

  *Gibbs energy* (also known as the Gibbs function), denoted by the symbol $G$ and defined as

$$G \equiv H - TS.$$

Gibbs energy per unit mass, also known as the *chemical potential* is

$$g = \frac{G}{m} = h - Ts.$$

The *Gibbs equation* for a system in which mass transfer can occur is

$$dS = \frac{dU}{T} + \frac{P}{T}dV - \frac{g}{T}dm.$$

  There is no exchange of mass between two phases whose chemical potentials are the same. The condition for phase equilibrium is that

$$g_f = g_g.$$

The *Gibbs–Duhem* equation gives the variation of chemical potential with temperature and pressure:

$$dg = vdP - sdT.$$

The *Clapeyron* equation gives the slope of the liquid–vapour equilibrium line:

$$\frac{dP}{dT} = \frac{h_{fg}}{T\left(v_g - v_f\right)}.$$

The *Clausius–Clapeyron equation* is derived assuming that the vapour is an ideal gas:

$$\frac{dP}{dT} = \frac{h_{fg}P}{RT^2}.$$

The *quality* of a liquid–vapour mixture is defined as

$$x = \frac{\text{mass of vapour}}{\text{mass of mixture}} = \frac{m_g}{m}.$$

This allows us to find the mixture properties of specific volume, specific internal energy, specific enthalpy and specific entropy:

$$v = xv_g + (1-x)v_f,$$

$$u = xu_g + (1-x)u_f,$$

$$h = xh_g + (1-x)h_f = h_f + xh_{fg},$$

$$s = xs_g + (1-x)s_f.$$

The van der Waals equation of state is

$$P = \frac{R_u T}{\overline{v} - b} - \frac{a}{\overline{v}^2}.$$

where

$$a = \frac{27}{64}\frac{R_u^2 T_c^2}{P_c} \quad \text{and} \quad b = \frac{R_u T_c}{8P_c}.$$

The *compressibility factor* of a gas is

$$Z = \frac{Pv}{RT} = \frac{P\overline{v}}{R_u T}.$$

The *Redlich–Kwong* equation of state is

$$P = \frac{R_u T}{\overline{v} - b_{RK}} - \frac{a_{RK}}{\overline{v}\left(\overline{v} + b_{RK}\right)\sqrt{T}},$$

with

$$a_{RK} = \frac{0.42748 R_u^2 T_C^{2.5}}{P_C} \quad \text{and} \quad b_{RK} = \frac{0.08664 R_u T_C}{P_C}.$$

The *virial equation* of state is

$$P = \frac{R_u T}{\overline{v}}\left(1 + \frac{B(T)}{\overline{v}} + \frac{C(T)}{\overline{v}^2} + \cdots\right).$$

## Problems

### *Clausius–Clapeyron Equation*

7.1  Given that the saturation pressure of water at 200 °C is 1.5538 MPa, use the Clausius–Clapeyron equation to determine the saturation pressure at 205 °C. Estimate the value of the enthalpy of vaporisation from the saturated water property tables. Compare the calculated saturation pressure with the value given in the tables.

7.2  Estimate the value of $dP/dT$ at 100 °C from the saturated water property tables. Use $dP/dT$ to calculate $h_g - h_f$ and $s_g - s_f$ at the same temperature and compare your answer with the values given in the tables.

7.3  From the saturation property tables for refrigerant 134a look up values for $P_{sat}$ and $h_{fg}$ at 0 °C. Use these values to determine $P_{sat}$ at 4 °C. The molar mass of refrigerant 134a is 102.03 kg / kmol.

7.4  The triple point of water is at a temperature of 273.16 K and a pressure of 0.6113 kPa. Given that the latent heat of sublimation of water at the triple point is 2834.8 kJ / kg find the saturation pressure of water vapour in equilibrium with ice at –20 °C.

7.5  Ice melts at 0 °C and atmospheric pressure. Given that the specific volume of ice $v_s$ = $1.0911 \times 10^{-3}$ m³ / kg, the specific volume of liquid water $v_s$ = $1.000 \times 10^{-3}$ m³ / kg and the latent heat of fusion $h_{sf}$ = 334 kJ / kg, what is the melting temperature of ice at a gauge pressure of 10 MPa?

7.6  The edge of an ice-skate blade is approximately 280 mm long and 3 mm wide. Assuming that a skater has a mass of 75 kg that is evenly distributed over two skates, how much will the pressure exerted on the ice under the skates reduce its melting point? Sharpening the blades decreases their area and increases the pressure. If we want to reduce the melting temperature of the ice by 2 °C, what should the surface area of the skates be? Use the data for ice given in the previous problem.

### *Calculating Properties from Tables*

7.7  An aluminium water bottle has walls that buckle if the pressure difference across them exceeds 30 kPa. The bottle is half filled with water, heated on a stove until the water is boiling, and sealed. It is then removed from the stove and allowed to cool to the ambient temperature of 25 °C. Will it collapse?

7.8  Five kilograms of water are boiling in an open pot placed over a 500 W heater. Estimate the minimum amount of time required to evaporate all the water.

7.9  A cylindrical pot, 15 cm diameter, is filled with water and covered with a well-fitting lid on which a 10 kg mass is placed. At what temperature will the water boil? Assume atmospheric pressure is 100 kPa.

7.10   Find the specific entropy of a saturated water-steam mixture at a temperature of 40 °C and 50% quality.

7.11   If water is at a temperature of 300 °C and a pressure of 300 kPa, what phase is it in and what is its specific enthalpy?

7.12   Steam at a pressure of 200 kPa has a specific volume of 0.05 m³ / kg. What are the internal energy and entropy for 1 kg of steam under these conditions?

7.13   A rigid tank contains 5 kg of water at a pressure of 500 kPa and a temperature 400 °C. Find the volume of the vessel and the enthalpy of the water.

7.14   A rigid tank with a volume of 0.3 m³ contains 5 kg of refrigerant 134a at a temperature of –4°C. Determine the internal energy of the refrigerant.

7.15   Saturated steam at a temperature of 100 °C is heated in a rigid tank until its temperature reaches 500 °C. What is the final pressure?

7.16   A rigid container is filled with 2 kg of a saturated liquid–vapour water mixture at a temperature of 200 °C. When the water is heated it passes through the critical point. What is the initial quality of the mixture?

7.17   A closed container in the shape of a vertical cylinder is 0.1 m high. At the bottom of the container is a 1 cm deep layer of saturated water while the rest is filled with saturated steam. If this system is in equilibrium at 300 kPa, what is its quality?

7.18   Superheated steam at a pressure of 0.2 MPa and temperature of 270 °C is expanded in an isentropic process to a temperature of 40 °C. If the final state is a saturated mixture, find its specific volume.

7.19   A tank is separated into two sections by a partition. One side of the tank is filled with 0.01 m³ of water that is a saturated liquid at a pressure of 1000 kPa, whereas the other side is evacuated. The partition is removed and the water then expands to fill the entire tank. The final state of the water is a temperature of 150 °C and pressure of 300 kPa. What is the volume of the tank?

*Control Mass Analysis*

7.20   An insulated, rigid tank contains 3 kg of a saturated water-steam mixture at a pressure of 125 kPa. Initially 65% of the mass of the mixture is vapour. Determine the amount of heat required to evaporate the remainder of the liquid.

7.21   A 2 kg mass of saturated water vapour at a pressure of 100 kPa is heated at constant pressure to 400 °C. What is the work done?

7.22   A 5 kg mass of saturated refrigerant 134a vapour at a pressure of 200 kPa undergoes a constant pressure process in which its volume increases by 50%. Determine the work done in expansion and the heat transfer during the process.

7.23   A rigid tank with a volume of 0.2 m³ contains a saturated steam-water mixture at a pressure of 225 kPa and a quality of 80%. The tank is heated until the steam pressure in it becomes 400 kPa. How much heat was added?

7.24   A rigid tank containing 1 kg of water at a temperature of 250 °C and a pressure of 1.4 MPa is cooled until the pressure in it is 100 kPa. Find the amount of heat removed.

7.25   Two kilograms of refrigerant 134a vapour at an intial pressure of 100 kPa and temperature of 20 °C are compressed in a process for which $Pv$ = constant to a final pressure of 320 kPa. Find the work done during this process and the final temperature.

7.26   A cylinder with a frictionless piston contains 2 l of saturated liquid water at a pressure of 100 kPa. The water is heated until its quality is 80%. Find the amount of heat added.

7.27   A cylinder contains 0.5 m³ of saturated steam at a pressure of 200 kPa, confined by a piston that is initially resting on stops and requires 400 kPa of pressure to lift (Figure P7.27). The steam is heated until its volume becomes 0.8 m³. Determine the final temperature of the steam and the heat added.

Steam
200 kPa
0.5 m³

**Figure P7.27**   A cylinder containing saturated steam confined by a piston resting on stops.

7.28   A rigid tank with a volume of 0.6 m³ contains 10 kg of water at a pressure of 2 MPa. Heat is added to the tank until the pressure in it is 3.5 MPa. Determine the final temperature of the water and the total heat transfer.

7.29   A cylinder with a freely moving piston contains 10 kg of superheated steam at a pressure of 200 kPa and volume of 12.5 m³. The steam is kept at constant pressure while 5 MJ of heat are added to it. What is the final temperature?

7.30   Two rigid, well-insulated tanks, labelled $A$ and $B$, are connected by a valve that is initially closed. Tank $A$ is filled with 4 kg of saturated liquid water at a pressure of 300 kPa while tank $B$ is evacuated. The valve is opened and the water allowed to expand into the empty tank. When equilibrium is reached the pressure in both tanks is 75 kPa. What is the volume of each tank?

7.31 A cylinder with a piston contains 2 kg of superheated steam at a pressure of 200 kPa and temperature of 150 °C. The piston just touches a linear spring that is initially uncompressed (Figure P7.31). The steam is heated until its pressure rises to 400 kPa and temperature to 600 °C. Find the work done and the heat added in this process.

Steam
2 kg
200 kPa
150°C

**Figure P7.31**   Superheated steam in a cylinder with a piston pushing against a linear spring.

7.32 A cylinder contains 0.75 m³ of refrigerant 134a at a temperature of 90 °C and a pressure of 1.2 MPa. The refrigerant is confined by a piston that requires a minimum pressure of 0.4 MPa to hold it against a set of stops (Figure P7.32). What is the work done and the heat transfer when the refrigerant is cooled until its quality is 80%?

R-134a
0.75 m³
1.2 MPa
90°C

**Figure P7.32**   A cylinder containing refrigerant 134a confined by a piston held against a set of stops.

7.33 A cylinder with a freely moving piston is fitted with a set of stops that confine the enclosed volume to a maximum of 0.2 m³ (Figure P7.33). It initially contains 2 kg of water with a volume of 0.05 m³ and pressure of 250 kPa. The water is heated until its pressure reaches 3.0 MPa. Find the final temperature and the work done in this process.

**Figure P7.33** A cylinder containing water confined by a piston whose movement its limited by a set of stops.

7.34 A spherical rubber balloon contains a saturated liquid–vapour mixture of refrigerant 134a with a mass of 0.5 kg, quality of 50% and initial pressure of 120 kPa. The balloon is heated until the pressure in it reaches 180 kPa. During expansion the pressure in the balloon is proportional to its surface area. How much work was done and heat added in this process?

7.35 A cylinder with a piston containing 0.6 kg of superheated steam at 300 kPa and 150 °C is connected by a closed valve to a tank with a volume of 0.2 m³ containing a saturated water–steam mixture at 500 kPa with a quality of 60% (Figure P7.35). Neglecting heat losses to the surroundings, find the final equilibrium state of the water after the valve is opened.

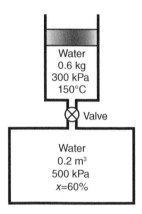

**Figure P7.35** A cylinder with a piston connected by a valve to a tank.

7.36 One kilogram of a saturated water–steam mixture at 200 °C expands against a piston in a cylinder. It stops when it reaches a pressure of 100 kPa, when the steam quality is 60%. Assuming the process is reversible and adiabatic, determine the initial pressure of the water and the work done in expansion.

7.37  Superheated refrigerant 134a at a pressure of 240 kPa and a temperature of 30 °C undergoes a process in which the final state is saturated vapour at a pressure of 100 kPa. Is it possible for this process to be adiabatic?

7.38  Water at a pressure of 2.5 MPa and a temperature of 300 °C is compressed in an iso-thermal process until it becomes saturated liquid. Calculate the heat transfer and work done during this process per kilogram of water.

7.39  An insulated cylinder with a freely moving piston contains 0.5 kg of saturated refrig-erant 134a vapour at a pressure of 140 kPa. A 1 kW electrical heater installed inside the cylinder is switched on for 54 s. Find the change in entropy of the refrigerant and its final temperature.

7.40  A closed, partially insulated cylinder is divided into two parts by a movable, insulated piston. One part is filled with air and completely insulated, while the other part is filled with a saturated water-steam mixture and not insulated. Initially, the system is at equilibrium and the volume of each section is 1 m³, with the air at a temperature of 40 °C and the steam at a temperature of 80 °C and 30% quality. Heat is added to the steam until the system pressure reaches 400 kPa. Determine the amount of heat transfer.

7.41  Superheated steam contained in a rigid tank is initially at a temperature of 300 °C and a pressure of 0.3 MPa. What is the change in entropy per kilogram of the steam if it is allowed to cool until its temperature drops to 80 °C?

7.42  A valve connects two identical tanks, one of which contains 2 kg of a saturated steam–water mixture at a pressure of 150 kPa and 80% quality while the other tank is evacu-ated. The valve is opened so that the steam expands while losing heat to the surroundings until the final equilibrium pressure in both tanks is 50 kPa. What is the change in entropy of the steam?

7.43  A rigid tank that initially contains 10 kg of refrigerant 134a vapour at a pressure of 1 MPa and a temperature of 40 °C is allowed to cool to the ambient air temperature of 20 °C. Find the change in entropy of the refrigerant, the surroundings and the entropy generated.

7.44  A cylinder with a frictionless piston containing saturated steam at 400 kPa is allowed to expand isentropically until the pressure in it drops to 100 kPa. Find the final temper-ature and the work done per kilogram of steam during expansion.

*Control Volume Analysis*

7.45  Saturated steam enters a well-insulated heat exchanger with a pressure of 200 kPa and a quality of 80% and leaves as saturated liquid at the same pressure. The steam is cooled by a stream of ethylene glycol with a mass flow rate of 15 kg / s that enters at a temperature of 15 °C and leaves at 80 °C. Find the mass flow rate of steam.

7.46  Superheated steam at 0.5 MPa and 600 °C enters an insulated chamber with a mass flow rate of 2.5 kg / s. Liquid water at 5 MPa and 20 °C is sprayed through a nozzle into the chamber to condense the steam. What should the mass flow rate of water be so that saturated liquid at 0.5 MPa leaves the chamber?

7.47   An adiabatic compressor takes 0.4 kg / s of saturated vapour refrigerant 134a at a pressure of 140 kPa and compresses it to a pressure of 600 kPa and a temperature of 50 °C. How much power is required to drive the compressor?

7.48   A condenser uses cooling air to condense 2500 kg / h of saturated steam that enters at a temperature of 80 °C and leaves as saturated liquid at 30 °C. The air temperature rises from 4 to 30 °C as it goes through the condenser. What is the mass flow rate of air required?

7.49   Saturated refrigerant 134a vapour at a pressure 200 kPa enters a heated tube with a flow rate of 0.2 kg / s. It exits at a pressure of 140 kPa and temperature of 0 °C. If the tube diameter is 50 mm determine the vapour velocity at the inlet and exit and the heat supplied to the refrigerant.

7.50   A saturated steam-water mixture flows in a pipe at a pressure of 2.25 MPa. The temperature of a small amount of the steam that is vented through a valve and released to the atmosphere is measured to be 124 °C. What is the quality of steam in the pipe? Assume atmospheric pressure is 100 kPa.

7.51   Steam at a pressure of 4 MPa and a temperature of 500 °C enters an adiabatic nozzle with a velocity of 10 m / s and leaves at a pressure of 500 kPa and velocity of 250 m / s. Determine the exit temperature.

7.52   Saturated steam enters a diffuser at a pressure of 300 kPa with a velocity of 150 m / s and leaves at a pressure of 1.2 MPa and a temperature of 350 °C. Determine the rate of heat transfer to the steam if the mass flow rate is 0.2 kg / s and the exit area of the diffuser is twice that of the inlet.

7.53   Steam at a pressure of 2.5 MPa and a temperature of 400 °C enters a turbine with a velocity of 80 m / s and mass flow rate of 10 kg / s. It leaves the turbine as saturated vapour at 75 kPa with a velocity of 30 m / s. If the turbine generates 5 MW of power, what is the rate of heat loss from its casing to the surroundings?

7.54   A two-stage turbine receives steam from two boilers, one of which supplies 3 kg / s of steam at a pressure of 1.8 MPa and a temperature of 700 °C while the other provides 8 kg / s at a pressure of 600 kPa and a temperature of 500 °C. The turbine exhausts saturated vapour at a pressure of 50 kPa. What is the power output of the turbine?

7.55   A pump with an isentropic efficiency of 80% takes 0.1 kg / s of water at a temperature of 10 °C and a pressure of 100 kPa, and raises its pressure to 3 MPa. What is the power supplied to the pump?

7.56   Refrigerant 134a enters a compressor as saturated vapour at a temperature of –12 °C and exits at a pressure of 800 kPa. What is the work required to drive the compressor per kilogram of refrigerant if its isentropic efficiency is 90%?

7.57   An insulated mixing tank is used to mix two water streams. Stream 1 enters at a pressure of 500 kPa and a temperature of 900 °C and stream 2 is saturated liquid water at a pressure of 500 kPa. At steady state water flows out of the tank with a flow rate of 40 kg/, a pressure of 500 kPa and temperature of 300 °C. Calculate the rate of entropy change for this process.

7.58   Steam at a pressure of 850 kPa and a temperature of 300 °C enters a nozzle with a mass flow rate of 0.8 kg / s and negligible velocity and is discharged at a pressure of 500 kPa. Assuming isentropic expansion, determine the exit velocity.

7.59   Superheated steam at a temperature of 200 °C and a pressure of 1 MPa expands through a nozzle to a pressure of 200 kPa. Assuming the process is reversible and adiabatic, determine the exit enthalpy of the steam.

7.60   Steam enters a nozzle with a velocity of 40 m / s at a pressure of 1.6 MPa and a temperature of 300 °C and leaves at a pressure of 0.3 MPa. Determine the exit velocity, assuming a reversible and adiabatic process.

7.61   Steam enters an adiabatic turbine at a pressure of 800 kPa and a temperature of 600 °C and leaves at a pressure of 50 kPa. What is the maximum amount of work that can be obtained from the turbine per kilogram of steam that flows through it?

7.62   Steam enters a turbine at a pressure of 1 MPa and a temperature of 400 °C with a velocity of 50 m / s and leaves at a pressure of 100 kPa with a velocity of 200 m / s. Determine the work output of the turbine per kg of steam passing through it, assuming the process is reversible and adiabatic.

7.63   Steam enters a turbine at a pressure of 1 MPa and a temperature of 500 °C. It expands in a reversible, adiabatic process and exits at a pressure of 50 kPa. The power output of the turbine is 800 kW. Determine the mass flow rate of steam, neglecting changes in kinetic and potential energy.

7.64   An adiabatic steam turbine produces 3.25 MW of power when steam enters it at a pressure of 2.4 MPa and a temperature of 500 °C and leaves as saturated vapour at a pressure of 50 kPa. Find the mass flow rate of steam through the turbine and the turbine isentropic efficiency.

7.65   Steam enters an adiabatic turbine at a pressure of 8 MPa and a temperature of 450 °C and leaves at a pressure of 50 kPa. If the isentropic efficiency of the turbine is 85%, find the work done by the turbine per kilogram of steam flowing through it.

7.66   Steam enters an adiabatic turbine at a pressure of 3.5 MPa and a temperature of 500 °C and leaves at a pressure of 100 kPa. If the isentropic efficiency of the turbine is 95% find the outlet quality of the steam.

### Exergy Analysis

7.67   What is the specific exergy of water kept at a pressure of 200 kPa and a temperature of 200 °C? Assume that the surrounding temperature and pressure are 20 °C and 100 kPa, respectively.

7.68   A rigid tank filled with refrigerant 134a at a pressure of 140 kPa and a temperature of 30 °C is cooled until its temperature is –24 °C. Find the change in exergy per unit mass assuming that the surrounding temperature and pressure are 20 °C and 100 kPa, respectively.

7.69   Superheated steam at a pressure of 200 kPa and a temperature of 250 °C, contained in a cylinder fitted with a freely moving piston, is allowed to cool until it becomes saturated vapour. Calculate the change in exergy per kilogram of steam assuming that the surrounding temperature and pressure are 20 °C and 100 kPa, respectively.

7.70   Steam contained in a tank at a temperature of 350 °C and a pressure of 1.6 MPa leaks out at a rate of 1.5 kg / s through a small hole and is released into a larger containment vessel that is at a pressure of 300 kPa. Calculate the rate of exergy destruction assuming a reference temperature $T_0 = 20$ °C.

7.71   Steam at 4.5 MPa and 500 °C enters an insulated turbine and leaves as saturated vapour at 75 kPa. Determine the work done by the turbine and the exergy destroyed per kilogram of steam. Assume that the surrounding atmosphere is at 20 °C.

7.72   Refrigerant 134a enters a compressor as saturated vapour at –12 °C and leaves at a pressure of 320 kPa and a temperature of 80 °C. Determine the power required to drive the compressor and the rate of exergy destruction if the mass flow rate of refrigerant is 0.1 kg / s.

7.73   An insulated mixing chamber has two inlets: the first lets in 3 kg/s of saturated liquid water at a pressure of 100 kPa while the second supplies superheated steam at a pressure of 100 kPa and temperature of 300 °C. A saturated liquid–vapour mixture with 95% quality at a pressure of 100 kPa leaves the chamber. Find the mass flow rate of steam into the chamber and the rate of exergy destruction.

## Compressibility Charts

7.74   A 0.75 m³ volume cylinder contains nitrogen at a pressure of 5 MPa and a temperature of 125 K. What is the mass of gas in the cylinder?

7.75   Calculate the specific volume of steam at a pressure of 20 MPa and a temperature of 600 °C using (a) the ideal gas equation, (b) compressibility charts and (c) superheated steam tables.

7.76   Find the compressibility factor for steam at a pressure of 30 MPa and a temperature of 500 °C from generalised compressibility charts. Compare the value with that calculated from superheated steam tables.

7.77   When the ideal gas equation is used to calculate the specific volume of hydrogen at a temperature of 35 K and a pressure of 1.5 MPa, what is the error in the answer? State the error as a percentage of the actual specific volume.

7.78   Find the pressure of 2 kg of oxygen occupying a volume of 0.1 m³ at a temperature of 250 K using both the ideal gas equation and van der Waals equation.

7.79   What is the density of carbon dioxide at a pressure of 6.5 MPa and a temperature of
       30 °C? Compare the values obtained using the ideal gas equation and the compress-
       ibility charts.

7.80   Eight kilograms of propane, initially at a temperature of 400 K and a pressure of 4 MPa,
       are expanded in a reversible polytropic process for which $PV^n$ = constant to a final
       pressure of 2 MPa and a temperature of 370 K. Using the generalized compressibility
       charts find (a) the initial and final volume of the gas, (b) the value of the polytropic
       exponent $n$ and (c) the work done during expansion.

# 8

# Ideal Heat Engines and Refrigerators

---

**In this chapter you will:**

- Give a definition of a heat engine.
- Learn about perpetual motion machines – heat engines that cannot exist since they violate the laws of thermodynamics.
- Analyse Carnot engines – the most efficient heat engines possible in theory.
- Show how Carnot engines operated in reverse act as refrigerators or heat pumps.
- State the Carnot principles, that reversible engines are the most efficient engines possible and that all reversible engines have the same efficiency.

---

## 8.1 Heat Engines

Much of our discussion so far has been about the fundamental laws of thermodynamics, but we have now learned enough that we can address the question that initially drove the development of these laws: what is the most efficient engine that can be built?

To begin, let us identify what we are discussing when we talk about an engine. *A heat engine is any device that operates in a cycle and does work ($W_{net}$) on the surroundings as long as heat ($Q_{in}$) is added to it.* The engine must do a net amount of work (the work output less any work used internally by the engine) on its surroundings. The requirement that the engine work in a cycle is essential for it to operate continuously, so that as long as we supply heat to the device we can extract work from it. Energy storage devices, such as batteries, spinning flywheels or compressed springs are not considered engines since they will cease to work once we have extracted all available energy.

---

*Energy, Entropy and Engines: An Introduction to Thermodynamics*, First Edition. Sanjeev Chandra.
© 2016 John Wiley & Sons, Ltd. Published 2016 by John Wiley & Sons, Ltd.
Companion website: www.wiley.com/go/chandraSol16

How much work can we obtain from an engine for a given amount of heat? Our goal is to design an engine with maximum *thermal efficiency*, which we define as

$$\eta_{th} = \frac{\text{net work output}}{\text{heat input}} = \frac{W_{net}}{Q_{in}}. \tag{8.1}$$

It is frequently convenient to define the thermal efficiency of an engine on a rate basis as

$$\eta_{th} = \frac{\dot{W}_{net}}{\dot{Q}_{in}}. \tag{8.2}$$

We will disregard our sign convention for work and heat when discussing engine efficiencies, because we want to compare only their magnitudes: considering positive and negative values would only be confusing. In this chapter $W$ and $Q$ will always be positive, and arrows will indicate their direction in system diagrams.

## 8.2   Perpetual Motion Machines

What is the maximum value of thermal efficiency ($\eta_{th}$) that an engine can have in theory? Let us start by eliminating the types of engines that we cannot build, even in principle. Any engine that violates one of the laws of thermodynamics cannot exist and is known as a *perpetual motion machine*.

A *perpetual motion machine of the first kind* is a device that does work but does not interact with the surroundings in any other way (Figure 8.1). Such an engine would, in theory, have infinite efficiency but we can quickly dismiss the feasibility of building this machine since it violates the first law of thermodynamics. We cannot extract energy endlessly from a closed system to which there is no energy input. Eventually all available energy in the control mass would be removed and the engine would cease to function.

A *perpetual motion machine of the second kind* is an engine that violates the second law of thermodynamics, and understanding why it cannot be built requires a little more reflection. To develop an engine with 100% efficiency an inventor proposes to build the device shown in Figure 8.2. According to this design, heat ($Q_{in}$) from a high temperature source is supplied to the engine and all of it is converted to work ($W_{net}$). This engine does not violate the principle of energy conservation since $W_{net} = Q_{in}$. However, for continuous operation we require that the engine operate in a cycle in which all properties are restored to their initial value once the

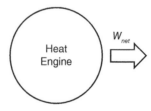

**Figure 8.1**   A perpetual motion machine of the first kind, which does work without any heat input.

cycle is completed. This means all the energy and entropy transferred into the engine must also be removed in the same cycle. Removing energy presents no difficulty: all the energy added as heat can be transferred out in the form of work done by the engine. Restoring the entropy of the engine to its initial value, however, is impossible. When heat is added to the engine shown in Figure 8.2 its entropy increases. There is no way of reducing entropy since the only output from the engine is work: reversible work transfers have no effect on entropy and irreversible work would generate even more entropy. To eliminate the entropy added to the engine we must transfer heat from it to the surroundings. We conclude: *It is impossible for any device operating in a cycle to receive heat from a high temperature source and produce a net amount of work without rejecting heat to a low temperature sink.* This is known as the Kelvin–Planck statement of the second law of thermodynamics that can be shown to be equivalent to our second law postulate.

## 8.3   Carnot Engine

We have ruled out the possibility of building an engine with either infinite or 100% efficiency. What, then, is the highest efficiency that a heat engine can have while still satisfying the laws of thermodynamics?

A real engine must look like the one shown schematically in Figure 8.3, that takes heat ($Q_H$) from a high temperature ($T_H$) source and rejects heat ($Q_C$) to a sink at a lower temperature ($T_C$). The discarded heat ($Q_C$) represents a loss of energy, but is required to carry away entropy and keep the engine functioning. We must maximise the amount of entropy transported out with waste heat while minimising the entropy increase during heat addition. Irreversibilities also create entropy so we will postulate that all processes in our engine are reversible: the engine is frictionless and heat and work transfers are all quasi-equilibrium.

Over a cycle the change in entropy of the engine due to reversible heat addition must equal zero:

$$\Delta S_{\text{cycle}} = \oint \frac{\delta Q_{\text{int rev}}}{T} = 0. \tag{8.3}$$

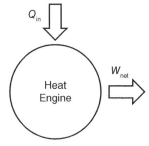

**Figure 8.2**   A perpetual motion machine of the second kind, which does work while receiving heat, but does not reject any heat.

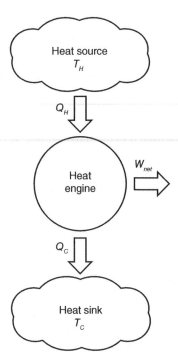

**Figure 8.3**  A heat engine takes heat ($Q_H$) from a high temperature ($T_H$) heat source and rejects some ($Q_C$) to a low temperature ($T_C$) heat sink while doing work ($W_{net}$).

We can divide the cycle into two parts, head addition and rejection, where

$$\Delta S_{\text{heat addition}} = \int_{\substack{\text{heat} \\ \text{addition}}} \frac{\delta Q_{\text{int rev}}}{T} \tag{8.4}$$

and

$$\Delta S_{\text{heat rejection}} = \int_{\substack{\text{heat} \\ \text{rejection}}} \frac{\delta Q_{\text{int rev}}}{T}. \tag{8.5}$$

To minimise the entropy increase of the engine during heat addition it should be done at the highest temperature possible, which is that of the heat source ($T_H$). If all heat transfer is done isothermally at constant temperature $T_H$ we can take it out of the integral in Equation (8.4) so that

$$\Delta S_{\text{heat addition}} = \frac{1}{T_H} \int_{\substack{\text{heat} \\ \text{addition}}} \delta Q_{\text{int rev}} = \frac{Q_H}{T_H}, \tag{8.6}$$

where $Q_H$ is the total heat added over a cycle. Similarly, to maximize the entropy decrease during heat rejection by the engine, heat transfer should be done at the lowest temperature possible.

If all heat rejection is done in an isothermal process at $T_C$,

$$\Delta S_{\text{heat rejection}} = \frac{1}{T_c} \int\limits_{\substack{\text{heat} \\ \text{rejection}}} \delta Q_{\text{int rev}} = \frac{Q_C}{T_C}. \tag{8.7}$$

Heat addition and rejection in the engine are both isothermal processes. To complete the cycle we need intermediate steps that take the engine from the high temperature to the low temperature and back again. To avoid adding any more entropy to the engine both these processes must be reversible and adiabatic, which will keep entropy constant. To reduce its temperature from $T_H$ to $T_C$ the engine goes through an isentropic expansion while an isentropic compression raises the temperature from $T_C$ to $T_H$. Figure 8.4 shows this thermodynamic cycle, named after Sadi Carnot who first proposed it, on a $T$-$S$ diagram. Any engine that works on this cycle is known as a *Carnot engine*.

Changes in both energy and entropy are zero over a cycle, so that

$$\Delta E_{\text{cycle}} = Q_H - Q_C - W_{\text{net}} = 0 \tag{8.8}$$

and

$$\Delta S_{\text{cycle}} = \Delta S_{\text{heat addition}} - \Delta S_{\text{heat rejection}} = \frac{Q_H}{T_H} - \frac{Q_C}{T_C} = 0. \tag{8.9}$$

Rearranging Equation (8.8) gives

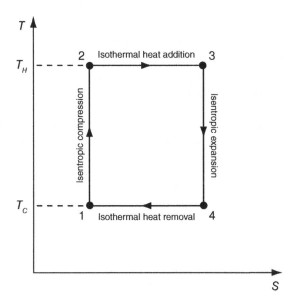

**Figure 8.4** A Carnot cycle shown on a $T$-$S$ diagram.

$$W_{net} = Q_H - Q_C,$$ (8.10)

and from Equation (8.9)

$$\frac{Q_C}{Q_H} = \frac{T_C}{T_H}.$$ (8.11)

The efficiency of an engine is defined as

$$\eta_{th} = \frac{W_{net}}{Q_{in}} = \frac{Q_H - Q_C}{Q_H} = 1 - \frac{Q_C}{Q_H}.$$ (8.12)

Substituting Equation (8.11) into Equation (8.12) gives the thermal efficiency of a Carnot engine:

$$\eta_{th,\text{Carnot}} = 1 - \frac{T_C}{T_H}.$$ (8.13)

The thermal efficiency of a Carnot engine depends only on the temperatures it operates between, and is independent of the working fluid in the engine or details of its construction. Typically, the maximum temperature an engine can withstand depends on the material it is made of. In a steam power plant the maximum temperatures are usually restricted to 1000 K, while gas turbines made of more exotic materials can operate at temperatures up to 1400 K. The heat sink for most heat engines is the atmosphere at a temperature of approximately 300 K. If we set $T_H = 1000$ K and $T_C = 300$ K, the thermal efficiency

$$\eta_{th,\text{Carnot}} = 1 - \frac{T_C}{T_H} = 1 - \frac{300}{1000} = 0.7.$$

Even under the very highly idealised conditions of a Carnot engine, the maximum efficiency we would expect in a steam power plant would be about 70%. In reality plant thermal efficiencies are about half of this value. The additional losses are due to irreversibilities that are always present in real engines, caused by frictional losses and rapid compression, expansion and heat transfer processes.

The $T$-$S$ diagram (Figure 8.4) gives us information about heat transfer to and from the engine. The area under line 2-3 represents the heat added and the area under line 4-1 the heat rejected:

$$Q_H = T_H \left( S_3 - S_2 \right),$$ (8.14)

$$Q_C = T_C \left( S_4 - S_1 \right) = T_C \left( S_3 - S_2 \right).$$ (8.15)

So, the net work output is

$$W_{net} = Q_H - Q_C = \left( T_H - T_C \right) \left( S_3 - S_2 \right).$$ (8.16)

In the $T$-$S$ diagram (Figure 8.4) the area enclosed by the rectangle 1-2-3-4 represents $W_{net}$.

**Example 8.1**

**Problem:** A thermal power plant generates 300 MW of electricity with an efficiency of 35%. What is the rate of heat rejection?

**Find:** Rate of heat rejected by power plant $\dot{Q}_C$.

**Known:** Thermal efficiency $\eta_{th} = 0.35$, power output $\dot{W} = 300\,W$.

**Governing Equation:**

Thermal efficiency

$$\eta_{th} = \frac{\dot{W}_{net}}{\dot{Q}_{in}}$$

**Solution:**

Rearranging the thermal efficiency equation with heat input from the high temperature source,

$$\dot{Q}_H = \frac{\dot{W}}{\eta_{th}} = \frac{300\,MW}{0.35} = 857.14\,MW.$$

Then using an energy balance over the power generation cycle,

$$\dot{Q}_C = \dot{Q}_H - \dot{W} = 857.14 - 300 = 557.14\,MW.$$

**Answer:** Heat is rejected at a rate of 557 MW. ∎

## 8.3.1 Two-Phase Carnot Engine

How do we build a Carnot engine? Well, we cannot in reality. We can, however, think of several different idealised systems that execute the Carnot cycle. Such ideal engines cannot be built, but they serve as a very useful benchmark against which we can compare real engines to gauge their performance.

We can carry out the isothermal heat addition and removal required for a Carnot engine by making use of the fact that the temperature of a two-phase mixture remains constant while it is being heated or cooled. Heat transfer changes the quality of the mixture but its temperature remains constant, equal to the saturation temperature. We can achieve isentropic compression and expansion by using a reversible, adiabatic pump and turbine. Figure 8.5 shows an example of a Carnot engine with a working fluid that undergoes phase change during the cycle. The system has four components: evaporator, turbine, condenser and pump.

The Carnot cycle is shown graphically on a *T-s* diagram in Figure 8.6 in which the vapour dome and isobars representing the pressure of the evaporator and condenser are drawn. The numbers on the diagram marking the start and end of processes refer to the same stages in the Carnot cycle that are labelled in Figure 8.5. The four processes that make up the cycle are listed below.

*1→2* A pump compresses a liquid–vapour mixture in a reversible, adiabatic process. The compression is isentropic and fluid emerges as saturated liquid at the pressure of the evaporator.

*2→3* Saturated liquid is heated reversibly in an evaporator in contact with the higher temperature heat source ($T_H$). The fluid is assumed to be at temperature $T_H - \Delta T$ where $\Delta T \rightarrow 0$, so heat transfer is reversible. In practice this would imply infinitesimally slow heat transfer, so we can never achieve it in a real engine. The liquid emerges from the evaporator as saturated vapour.

3→4    A reversible, adiabatic turbine takes the high-pressure vapour and expands it isentropically. The turbine supplies enough work to run the compressor pump and the remainder ($W_{net}$) is the net work output of the engine.

4→1    The two-phase mixture leaving the turbine is cooled reversibly in a condenser where it loses heat to the low temperature heat sink ($T_c$). The fluid temperature is $T_c + \Delta T$ where $\Delta T \to 0$, so heat transfer takes place reversibly.

In Figure 8.6 the constant temperature heating and cooling process are seen to lie on the saturation temperatures corresponding to the pressure of the evaporator and the condenser. To maximise the efficiency of the cycle we have to (i) increase the temperature at which heat is added by increasing the evaporator pressure as much as possible and (ii) lower the condenser pressure. The maximum temperature and pressure are usually limited by the strength of the materials used to construct the turbine. In real power plants the condenser may be operated at a partial vacuum, so that its pressure is lower than that of the surrounding atmosphere.

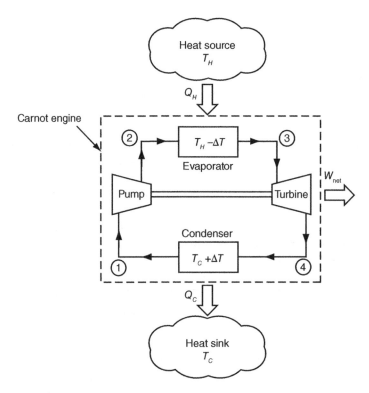

**Figure 8.5**    A two-phase Carnot engine consisting of a fluid flowing through a turbine, pump, evaporator and condenser.

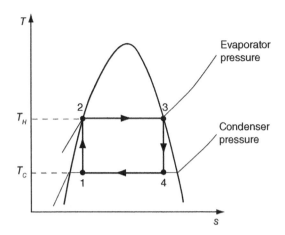

**Figure 8.6**   A two-phase Carnot engine cycle shown on a $T$-$s$ diagram.

## Example 8.2

**Problem:** A Carnot engine using 10 kg / s of water as the working fluid operates between an evaporator temperature of 300 °C and a condenser temperature of 80 °C. Find the rate of heat addition in the boiler and the steam quality at the turbine exhaust.

**Find:** Rate of heat addition $\dot{Q}_H$ and steam quality $x_4$ at the turbine exhaust.

**Known:** Evaporator temperature $T_H = 300$ °C, condenser temperature $T_C = 80$ °C, mass flow rate of water $\dot{m} = 10$ kg / s.

**Diagram:** The Carnot cycle on a $T$-$s$ diagram (Figure E8.2).

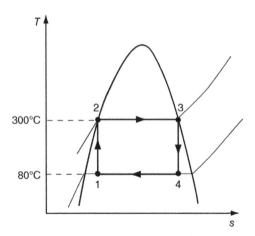

**Figure E8.2**   $T$-$s$ diagram for a Carnot cycle using water as the working fluid.

## Solution:

We can find the rate of heat addition from an energy balance at the high temperature heat source,

$$\dot{Q}_H = \dot{m}\left(h_3 - h_2\right) = \dot{m}\left(h_g - h_f\right).$$

For saturated water at $T_H = 300\,°\text{C}$ (Appendix 8a), the specific enthalpy of liquid $h_f = 1344.0$ kJ / kg and vapour $h_g = 2749.0$ kJ / kg. So,

$$\dot{Q}_H = 10\,\text{kg/s} \times \left(2749.0\,\text{kJ/kg} - 1344.0\,\text{kJ/kg}\right) = 14\ 050\,\text{kJ/s} = 14.05\,\text{MW}.$$

Also, specific enthalpy $s_3 = s_g = 5.7045$ kJ / kgK. Since expansion in the turbine is isentropic, $s_4 = s_3$. At $80\,°\text{C}$, the specific entropy of liquid $s_f = 1.0753$ kJ / kgK and vapour $s_g = 7.6122$ kJ / kgK, so the quality of the exhaust steam is

$$x_4 = \frac{s_4 - s_f}{s_g - s_f} = \frac{5.7045\,\text{kJ/kgK} - 1.0753\,\text{kJ/kgK}}{7.6122\,\text{kJ/kgK} - 1.0753\,\text{kJ/kgK}} = 0.707\ 84.$$

**Answer:** The rate of heat addition is 14.1 MW. The steam quality at the turbine exit is 70.8%.                                                                                          ■

## 8.3.2   Single Phase Carnot Engine

It is not necessary that the working fluid in a Carnot engine undergo a phase change since it is possible to execute a Carnot cycle using an ideal gas contained in a cylinder fitted with a piston. Figure 8.7 shows how such a cycle would operate, in which the cylinder is brought alternately into contact with the heat source and heat sink. The Carnot cycle is shown on a $P$-$v$ diagram in Figure 8.8, which complements the $T$-$s$ diagram (Figure 8.4) and graphically shows the work done during the cycle.

$1 \rightarrow 2$   The piston compresses the gas in a reversible, adiabatic process (for which $s$ = constant) until it reaches the temperature of the heat source $(T_H)$.

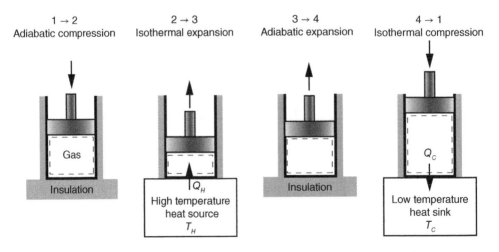

**Figure 8.7**   A Carnot cycle carried out by a gas in a piston with a sliding piston.

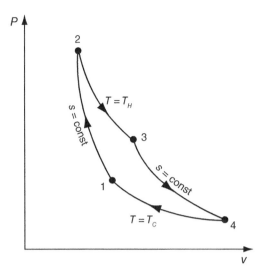

**Figure 8.8**   A Carnot cycle shown on a *P-v* diagram.

2→3   The gas is brought into contact with the heat source at temperature $T_H$ and allowed to expand in an isothermal process. Heat $(Q_H)$ is transferred to the gas that is at temperature $T_H - \Delta T$ where $\Delta T \to 0$ during this process.

3→4   The cylinder is insulated and the gas allowed to expand adiabatically until its temperature drops to that of the low temperature heat sink $(T_C)$.

4→1   The gas is cooled isothermally. It loses heat $(Q_C)$ to the low temperature sink at $T_C$. The gas temperature is $T_C + \Delta T$ where $\Delta T \to 0$.

On the *P-v* graph (Figure 8.8) the area under the curve 4-1-2 shows the work done on the gas, per unit mass, as it is compressed, first isothermally (4-1) and then adiabatically (1-2). The area under the curve 2-3-4 represents the work done by the gas as it expands, first isothermally (2-3) and then adiabatically (3-4). The difference between these two areas, shown by the area enclosed by the cycle 1-2-3-4, is the net work output from the engine, per unit mass of gas.

**Example 8.3**
**Problem:** A Carnot engine using air as the working fluid works between temperatures of 573 K and 293 K. The pressure at the start of isothermal expansion is 100 kPa and at the end is 50 kPa. Find the work output and heat added per kilogram of air.
**Find:** Work output *w* and heat added *q* both per kilogram of air in the Carnot engine.
**Known:** Temperature of heat source $T_H = 573$ K, temperature of heat sink $T_C = 293$ K, pressure at the start of isothermal expansion $P_2 = 100$ kPa, Pressure at the end of isothermal expansion $P_3 = 50$ kPa.
**Diagram:** The Carnot cycle on a *T-s* diagram (Figure E8.3).
**Assumptions:** Air is an ideal gas with constant specific heats.

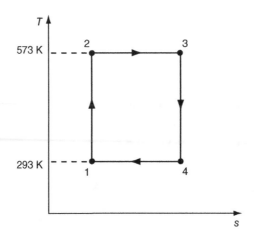

**Figure E8.3**   *T-s* diagram for a Carnot cycle using air as the working fluid.

**Governing Equations:**

Carnot cycle heat input

Carnot cycle net work

$$q_H = T_H \Delta s_{23}$$
$$w_{net} = (T_H - T_C)(\Delta s_{23})$$

**Properties:** Gas constant of air $R = 0.287$ kJ / kgK (Appendix 1).

**Solution:**

The entropy increase during isothermal expansion is, since $T_2 = T_3$,

$$\Delta s_{23} = \underbrace{c_p \ln \frac{T_3}{T_2}}_{=0} - R \ln \frac{P_3}{P_2},$$

$$\Delta s_{23} = -0.287 \, \text{kJ / kgK} \times \ln \frac{50 \, \text{kPa}}{100 \, \text{kPa}} = 0.198 \ 933 \, \text{kJ / kgK}.$$

The heat added is

$$q_H = T_H \Delta s_{23} = 573 \, \text{K} \times 0.198 \ 933 \, \text{kJ / kgK} = 113.989 \, \text{kJ / kg}.$$

The work output is

$$w_{net} = (T_H - T_C)(\Delta s_{23}) = (573 \, \text{K} - 293 \, \text{K}) \times 0.198 \ 933 \, \text{kJ / kgK} = 55.7012 \, \text{kJ / kg}.$$

**Answer:** The heat added is 114.0 kJ / kg and the work output is 55.7 kJ / kg.   ∎

## 8.4   Refrigerators and Heat Pumps

A Carnot engine operates in a reversible cycle, so what happens if we reverse it? An engine operating in reverse becomes a *refrigerator*, defined as a *device that takes heat from a low temperature region and transfers it to a high temperature region while being supplied with work* (Figure 8.9).

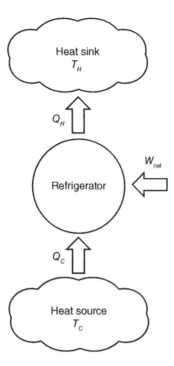

**Figure 8.9**  Schematic diagram of a refrigerator.

Refrigerators are familiar devices, transferring heat from their cold interior compartments to the warmer surrounding air. An air conditioner is another example of a refrigerator, cooling the interior of a building and rejecting heat to the outside atmosphere. But how does heat travel from a low to a high temperature?

A refrigerator makes use of two phenomena that occur when liquid–vapour phase change occurs. First, a liquid absorbs latent heat during evaporation and yields it while condensing. Second, the temperature at which the phase change occurs increases with pressure. The working fluid in a refrigerator is typically a liquid that boils at a temperature much below 0 °C at atmospheric pressure. As the low-pressure refrigerant passes through a heat exchanger coil inside the refrigerator cabinet it removes heat as it changes from liquid to vapour. This vapour is compressed to a high pressure and temperature, where its saturation temperature is well above the temperature of the ambient air. Then, as the refrigerant vapour passes through another heat exchanger coil on the exterior of the refrigerator, it loses heat to the surroundings and condenses. Repeating this cycle allows heat to be extracted continuously from a low temperature region.

## 8.4.1   Carnot Refrigerator

We can convert a Carnot engine to a Carnot refrigerator by reversing the flow direction of the fluid in the system, as shown in Figure 8.10. The turbine operated in reverse becomes a compressor, while the compressor running backwards acts as a turbine. The turbine does not provide all the work required to run the compressor, so we must supply it with additional work ($W_{net}$).

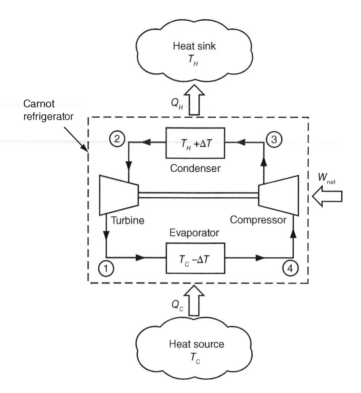

**Figure 8.10**  A Carnot refrigerator made by reversing the flow in a Carnot engine and changing the temperature in the evaporator and condenser by infinitesimal amounts.

The temperature at the exit of the compressor (point 3 in Figure 8.10) is increased by an infinitesimal amount to $T_H + \Delta T$ so that it is greater than that of the high temperature region ($T_H$), while the temperature at the exit of the turbine (point 1) is decreased to $T_C - \Delta T$, making it fall below the low temperature $T_C$. The roles of the heat sink and source are now reversed: the cycle is taking heat from the low temperature source and rejecting it at a higher temperature. Instead of obtaining work from the machine, we are supplying work to it.

Figure 8.11 shows a two-phase reverse Carnot cycle for a refrigerator on a *T-s* diagram. The fluid enters the condenser as saturated vapour (state 3), condenses isothermally and leaves as saturated liquid (state 2). It expands isentropically in the turbine and leaves as a low-pressure two-phase mixture (state 1). In the evaporator it takes heat isothermally from the low-temperature region and leaves as a higher quality mixture than it entered (state 4). The compressor increases its pressure and temperature in an isentropic process and returns it to the condenser, completing the cycle.

To gauge the performance of a refrigerator we compare the heat it removes ($Q_C$) to the work it requires for operation ($W_{net}$) and define the *coefficient of performance* (*COP*) of a refrigerator (*COP*$_R$):

$$COP_R = \frac{Q_C}{W_{net}} = \frac{Q_{out}}{W_{net}}. \tag{8.17}$$

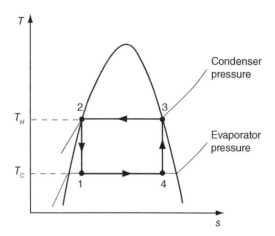

**Figure 8.11**    A two-phase reverse Carnot cycle for a refrigerator shown on a *T-s* diagram.

Energy conservation tells us that

$$W_{net} = Q_H - Q_C \tag{8.18}$$

or, substituting into Equation (8.17),

$$COP_R = \frac{Q_C}{Q_H - Q_C} = \frac{1}{Q_H / Q_C - 1}. \tag{8.19}$$

If the net entropy of the refrigerator does not change over a cycle we require that

$$\Delta S_{cycle} = \Delta S_{heat\ rejection} - \Delta S_{heat\ addition} = \frac{Q_H}{T_H} - \frac{Q_C}{T_C} = 0 \tag{8.20}$$

or

$$\frac{Q_H}{Q_C} = \frac{T_H}{T_C}. \tag{8.21}$$

Therefore, from Equation (8.19),

$$COP_R = \frac{1}{T_H / T_C - 1}. \tag{8.22}$$

Since $T_H > T_C$, the coefficient of performance of a refrigerator will always be positive, and is usually greater than 1. That is why we use the term coefficient of performance for refrigerators, rather than efficiency, since efficiencies are by convention less than unity. As the temperature of the cold and hot spaces become closer (so that $T_H \rightarrow T_C$), $COP_R$ becomes larger, approaching infinity for $T_H = T_C$. The lower the temperature maintained in a refrigerator, the poorer its performance. The $COP_R$ of a typical domestic refrigerator varies between 2 to 3, while that of a freezer lies between 1 and 1.5.

In the limit that $T_C \to 0$, $COP_R$ also goes to zero, implying that the work required to drive a refrigerator becomes infinitely large. Cooling a volume to absolute zero using a refrigerator requires an infinite amount of work, and is therefore not possible.

Since $Q_C$ is always less than $Q_H$, Equation (8.18) gives us that $W_{net} > 0$. Work always has to be supplied to a refrigerator for it to operate. This conclusion leads to yet another statement of the second law, known as the *Clausius statement*: *It is impossible to build a device that operates in a cycle and transfers heat from a low temperature region to a high temperature region without work being done on the device by the surroundings.*

This statement is intuitively obvious – we do not expect heat to spontaneously be conducted from a cold body to a hot body. Heat can be transported in a direction opposite to the temperature gradient only by a refrigerator that consumes energy.

## Example 8.4

**Problem:** A Carnot refrigerator removes 500 kW from a space at –20 °C and transfers it to the atmosphere at 25 °C. What is the power required and the coefficient of performance?

**Find:** The power required $\dot{W}_{net}$ by the refrigerator and its coefficient of performance $COP_R$.

**Known:** Heat removed from cold space $\dot{Q}_C = 500$ kW, temperature of heat source $T_C = -20\ °C = 253.15$ K, temperature of heat sink $T_H = 25\ °C = 298.15$ K.

**Governing Equation:**

Property of Carnot refrigerator (entropy balance)
$$\frac{Q_H}{Q_C} = \frac{T_H}{T_C}$$

**Solution:**

For a Carnot refrigerator,

$$\frac{\dot{Q}_H}{\dot{Q}_C} = \frac{T_H}{T_C},$$

$$\dot{Q}_H = \dot{Q}_C \frac{T_H}{T_C} = 500\,\text{kW} \times \left( \frac{298.15\,\text{K}}{253.15\,\text{K}} \right) = 588.88\,\text{kW}.$$

Then the power input is

$$\dot{W}_{net} = \dot{Q}_H - \dot{Q}_C = 588.88\,\text{kW} - 500\,\text{kW} = 88.880\,\text{kW}.$$

$$COP_R = \frac{\dot{Q}_C}{\dot{W}} = \frac{500\,\text{kW}}{88.880\,\text{kW}} = 5.6256$$

**Answer:** The refrigerator requires 88.9 kW of power, and operates with a coefficient of performance of 5.63. ∎

## Example 8.5

**Problem:** A Carnot refrigerator using 0.005 kg / s of refrigerant 134a has a condenser pressure of 700 kPa and requires 500 W to operate. What is the rate at which it cools the low temperature region?

**Find:** Cooling rate of Carnot refrigerator $\dot{Q}_C$.

**Known:** Mass flow rate of Refrigerant 134a $\dot{m} = 0.005\,\text{kg}/\text{s}$, condenser pressure $P_{32} = 700$ kPa, work input $\dot{W}_{net} = 0.5\,\text{kW}$.

**Diagram:**

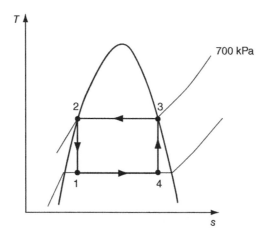

**Figure E8.5** *T-s* diagram for a Carnot refrigerator using refrigerant 134a as the working fluid.

**Solution:**

From an energy balance for the condenser,

$$Q_H = \dot{m}\left(h_2 - h_3\right).$$

For a Carnot refrigerator, $h_2 = h_f$ and $h_3 = h_g$. Using the saturation tables for R-134a (Appendix 9b) at $P_{23} = 700$ kPa, specific enthalpy of liquid $h_f = 86.78$ kJ / kg and gas $h_g = 261.85$ kJ / kg, so

$$Q_H = \dot{m}\left(h_2 - h_3\right) = 0.005\,\text{kg}/\text{s} \times \left(281.85\,\text{kJ}/\text{kg} - 86.78\,\text{kJ}/\text{kg}\right) = 0.875\ 35\,\text{kW}.$$

Energy balance for the refrigerator,

$$\dot{Q}_C = \dot{Q}_H - \dot{W}_{net} = 0.875\ 35\,\text{kW} - 0.5\,\text{kW} = 0.375\ 35\,\text{kW}.$$

**Answer:** The refrigerator removes 0.375 kW of heat from the low temperature region. ■

### 8.4.2 Carnot Heat Pump

A refrigerator is not only useful for cooling, but, contrary to intuition, it can be quite an effective means of heating a space. The simplest way of using electrical energy ($W_{net}$) to generate heat ($Q_H$) is to pass the current through a resistance heater, in which case the heat provided equals the electric work ($Q_H = W_{net}$). Instead, suppose that in winter we set up an air conditioner backwards, where it takes heat ($Q_C$) from the cold atmosphere and transfers it to the interior of a house. The energy available for heating the house equals $Q_H = W_{net} + Q_C$, which is more than could be obtained from a resistance heater. Such a device is known as a *heat pump*, which takes heat from the cold exterior of a building and rejects it to the warmer interior. Another advantage of this approach is that it is possible to make a single appliance that can switch from operating as an air conditioner in summer to acting as a heat pump in winter by reversing the direction of refrigerant flow.

The purpose of a heat pump is to add heat to the high temperature sink, so we compare the heat rejected ($Q_H$) to the work input ($W_{net}$). The coefficient of performance of a heat pump is

$$COP_{HP} = \frac{Q_H}{W_{net}},\qquad(8.23)$$

or, substituting $W_{net} = Q_H - Q_C$,

$$COP_{HP} = \frac{Q_H}{Q_H - Q_C} = \frac{1}{1 - Q_C / Q_H} = \frac{1}{1 - T_C / T_H}.\qquad(8.24)$$

Since $T_C / T_H < 1$, the coefficient of performance of a heat pump $COP_{HP} > 1$. The $COP_{HP}$ for typical heat pumps lies between 2 and 3, which explains why heat pumps are useful, since the heat they supply is significantly greater than the electrical work input to them. A heat pump provides two to three times the heat that would be obtained by simply dissipating the electricity in a resistance heater.

## Example 8.6

**Problem:** A house is maintained at a temperature of 23 °C by a heat pump when the outside air temperature is 5 °C. If the house loses 25 kW of heat to the atmosphere through its exterior surfaces, what is the power required by the heat pump?

**Find:** Power $\dot{W}_{net}$ required by the heat pump.

**Known:** Heat supplied to the house $Q_H = 25$ kW, high temperature $T_H = 23$ °C $= 296.15$ K, low temperature $T_C = 5$ °C $= 278.15$ K.

**Assumptions:** The heat pump is a Carnot refrigerator.

**Governing Equation:**

Coefficient of performance for a heat pump $\qquad\qquad COP_{HP} = \dfrac{Q_H}{W_{net}} = \dfrac{1}{1 - T_C / T_H}$

**Solution:**

$$COP_{HP} = \frac{1}{1 - T_C / T_H} = \frac{1}{1 - 278.15\,\text{K} / 296.15\,\text{K}} = 16.453$$

$$\dot{W}_{net} = \frac{\dot{Q}_H}{COP_{HP}} = \frac{25\,\text{kW}}{16.453} = 1.5195\,\text{kW}$$

**Answer:** The heat pump requires 1.52 kW to operate.     ■

## 8.5 Carnot Principles

We have described a Carnot engine, which is one example of a reversible engine. But is this the most efficient engine possible? We can think of many other engine cycles, both reversible and irreversible. Are any of them more efficient? Going through every possible cycle and comparing their efficiencies would be a tremendously difficult task. Fortunately, we can prove two statements known as the *Carnot Principles* that make this unnecessary. The two principles are:

1. The efficiency of a reversible heat engine is always greater than that of an irreversible engine operating between the same two temperatures.
2. The efficiencies of all reversible heat engines operating between the same two temperatures are the same.

We can prove the first statement by assuming the reverse to be true and showing that this assumption leads to a provably false conclusion. Figure 8.12 shows two engines, one reversible and the other irreversible, operating between the same high ($T_H$) and low ($T_C$) temperatures. Both take the same amount of heat ($Q_H$) from the high temperature heat source. Let us assume that the irreversible engine is the more efficient of the two, so that its work output ($W_I$) is greater than that of the reversible engine ($W_R$). The heat rejected by the irreversible engine ($Q_{C,I}$) will therefore be less than that lost by the reversible engine ($Q_{C,R}$).

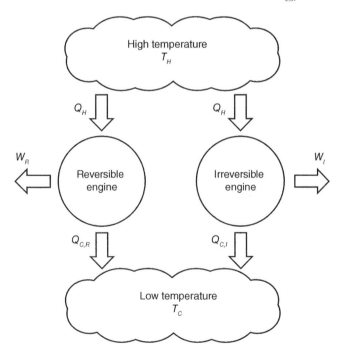

**Figure 8.12** Two engines, one reversible and the other irreversible, operating between the same high ($T_H$) and low ($T_C$) temperatures. The work output of the irreversible engine is assumed to be greater than that of the reversible engine ($W_I > W_R$). The heat lost by the irreversible engine is therefore less than that lost by the reversible engine ($Q_{C,I} < Q_{C,R}$).

The reversible engine can be operated in reverse as a refrigerator (Figure 8.13), so that it takes heat ($Q_{C,R}$) from the low temperature region and transfers it to the high temperature region. The work required by the refrigerator ($W_R$) can be supplied by the irreversible engine, still leaving net work ($W_I - W_R$) to be transferred to the surroundings.

Let us put a system boundary around both the reversible refrigerator and irreversible engine and treat them as components of one machine (Figure 8.14). While the refrigerator adds heat ($Q_H$) to the high temperature reservoir, the irreversible engine takes the same amount from it. We can connect the refrigerator to the engine and eliminate the high temperature reservoir entirely, without any difference to the operation of our machine (Figure 8.14). Also, the heat rejected by the engine ($Q_{C,I}$) can be transferred directly to the refrigerator, and only the amount $Q_{C,R} - Q_{C,I}$ taken from the low temperature reservoir. Our composite machine therefore takes a net amount of heat $Q_{C,R} - Q_{C,I}$ from a single heat source and does net work $W_I - W_R$ on the surroundings. However, this is a perpetual motion machine of the second kind, which violates the second law of thermodynamics and cannot exist. We must therefore reject the assumption – that the irreversible engine is more efficient than the reversible engine – that led us to this erroneous conclusion. An irreversible engine can never be more efficient than a reversible engine, as stated in the first Carnot principle.

We can prove the second Carnot principle in much the same way that we proved the first. Assume that there are two reversible engines operating between the same two temperatures, and that one engine is more efficient than the other. We can then reverse the less efficient engine

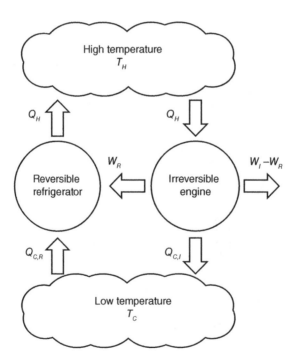

**Figure 8.13**   The reversible engine is operated as a reversible refrigerator with the work ($W_R$) for it supplied by the irreversible engine, leaving net work ($W_I - W_R$) to be transferred to the surroundings.

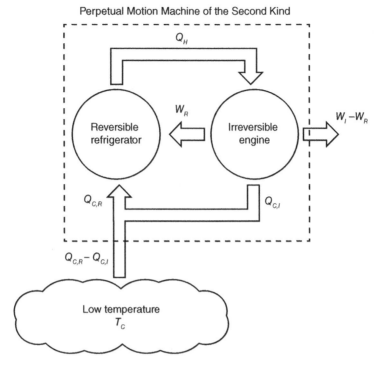

**Figure 8.14** The irreversible engine and reversible refrigerator together make a perpetual motion machine of the second kind.

and run it as a refrigerator. Following exactly the same argument as above, we can show that the engine and refrigerator combined constitute a perpetual motion machine of the second kind. The only way we can avoid violating the second law is to conclude that all reversible engines have the same thermal efficiency and no particular one can be more efficient than another.

Following our proofs of the Carnot principles we can conclude that the efficiency of a Carnot engine is greater than that of any irreversible engine and equal to that of any other reversible engine. The Carnot efficiency is therefore the highest attainable by any heat engine.

**Example 8.7**
**Problem:** An inventor claims to have developed a heat engine that receives heat from a source at 900 K and rejects 200 kW of heat to a heat sink at 400 K, while providing 300 kW of work. Can this claim be valid?
**Find:** Check if the inventor's claim is valid. See if the efficiency of the proposed engine exceeds the Carnot efficiency.
**Known:** Heat source temperature $T_H = 900$ K, heat sink temperature $T_C = 400$ K, heat rejected $\dot{Q}_C = 200\,\text{kW}$, power output $\dot{W} = 300\,\text{kW}$.
**Governing Equation:**

Carnot thermal efficiency $$\eta_{\text{Carnot}} = 1 - \frac{T_C}{T_H}$$

**Solution:**

From the First Law,

$$\dot{Q}_H = \dot{Q}_C + \dot{W} = 300\,\text{kW} + 200\,\text{kW} = 500\,\text{kW}.$$

Then the thermal efficiency of the proposed cycle would be

$$\eta_{th} = \frac{300\,\text{kW}}{500\,\text{kW}} = 60\%.$$

However, the Carnot efficiency is only

$$\eta_{\text{Carnot}} = 1 - \frac{T_C}{T_H} = 1 - \frac{400\,\text{K}}{900\,\text{K}} = 55.556\%,$$

so the proposed engine is not valid.

**Answer:** Since $\eta_{th} > \eta_{\text{Carnot}}$ the engine is not possible.     ∎

## Further Reading

1.  W. C. Reynolds, H. C. Perkins (**1977**) *Engineering Thermodynamics*, McGraw Hill, New York.
2.  Y. A. Cengel, M. A. Boles (**2015**) *Thermodynamics – An Engineering Approach*, McGraw Hill, New York.
3.  M. J. Moran, H. N. Shapiro, D. D. Boettner, M. B. Bailey (**2014**) *Fundamentals of Engineering Thermodynamics*, John Wiley & Sons, London.
4.  C. Borgnakke, R. E. Sonntag (**2012**) *Fundamentals of Thermodynamics*, John Wiley & Sons, London.

## Summary

A heat engine is any device that operates in a cycle and does work ($W_{net}$) on the surroundings as long as heat ($Q_{in}$) is added to it. The *thermal efficiency* of a heat engine

$$\eta_{th} = \frac{\text{net work output}}{\text{heat input}} = \frac{W_{net}}{Q_{in}}.$$

On a rate basis the thermal efficiency is

$$\eta_{th} = \frac{\dot{W}_{net}}{\dot{Q}_{in}}.$$

A *perpetual motion machine* is an engine that violates one of the laws of thermodynamics. A *perpetual motion machine of the first kind* is a device that does work but does not interact with the surroundings in any other way. A *perpetual motion machine of the second kind* takes heat from a single thermal reservoir and does work on the surroundings.

The *Kelvin–Planck statement* states that it is impossible for any device operating in a cycle to receive heat from a high temperature source and produce a net amount of work without rejecting heat to a low temperature sink.

The thermal efficiency of a Carnot engine is

$$\eta_{th,\text{Carnot}} = 1 - \frac{Q_C}{Q_H} = 1 - \frac{T_C}{T_H}.$$

A refrigerator is a device that takes heat from a low temperature region and transfers it to a high temperature region while being supplied with work. The *coefficient of performance* of a refrigerator is

$$COP_R = \frac{Q_C}{W_{net}} = \frac{1}{T_H/T_C - 1}.$$

The coefficient of performance of a heat pump is

$$COP_{HP} = \frac{Q_H}{W_{net}} = \frac{1}{1 - T_C/T_H}.$$

The *Clausius statement* states that it is impossible to build a device that operates in a cycle and transfers heat from a low temperature region to a high temperature region without work being done on the device by the surroundings.

The *Carnot Principles*:

1. The efficiency of a reversible heat engine is always greater than that of an irreversible engine operating between the same two temperatures.
2. The efficiencies of all reversible heat engines operating between the same two temperatures are the same.

## Problems

### Heat Engines

8.1 What is the maximum thermal efficiency of an engine that operates between a heat source at 850 °C and a heat sink at 35 °C?

8.2 Ocean Thermal Energy Conversion (OTEC) plants use the temperature difference between cold water near the ocean floor and warmer surface water to run a heat engine. What is the maximum efficiency of such a plant operating between deep water at 5 °C and surface water at 25 °C?

8.3 An engine receives 400 MJ of heat from a fuel burner. It loses 150 MJ to the atmosphere. Find the work done and the efficiency of the engine.

8.4 An engine with thermal efficiency 30% receives 120 kJ of heat from a high temperature source at 800 K. Find the temperature of the heat sink and the heat rejected.

8.5 An engine rejects heat to the atmosphere at 5 °C. If the work done is twice the heat rejected, find the cycle efficiency and the temperature of the heat source.

8.6 An engine is supposed to operate between two thermal reservoirs at 800 °C and 30 °C, producing 55 kW of power while using 75 kW of heat. Is this feasible?

8.7    A power plant with an efficiency of 30% generates 90 kW while rejecting heat to a lake at 8 °C. Environmental regulations restrict the amount of heat dissipated in the lake to 6000 MJ per day. How long can the power plant operate each day before it reaches this limit? How much does the entropy of the lake increase?

8.8    A car engine has a power output of 95 kW and a thermal efficiency of 25%. If gasoline gives a heat output of 47 MJ / kg when burned, find the rate of fuel consumption in kg / h.

8.9    A power plant that uses coal as a fuel produces 110 MW. It consumes 1500 t of coal per day. If the heat of combustion of coal is 20 MJ / kg, find the efficiency of the plant.

8.10   A Carnot engine receives 200 kW of heat from a heat source at 500 °C and rejects heat to a heat sink at 60 °C. Determine the power developed and the heat rejected.

8.11   A heat engine produces 8500 kW of power when operating between thermal reservoirs at 650 K and 250 K. Determine the rates at which heat is taken from the hot reservoir and rejected to the cold reservoir if the engine is (a) a Carnot engine and (b) a real engine with a thermal efficiency of 40%.

8.12   A Carnot engine with a power output of 20 kW takes 60 kW of heat from a high-temperature reservoir at 800 K. What is the temperature of the heat sink?

8.13   A nuclear power plant generates 80 MW when taking heat from a reactor at 615 K. Heat is discarded to a river at 298 K. If the thermal efficiency of the plant is 73% of the maximum possible value, how much heat is discarded to the river?

8.14   A 2 MW power plant has an overall efficiency of 45%. How much fuel oil does the plant burn a day, if the heat of combustion of the fuel is 46 000 kJ / kg?

8.15   A 600 MW steam power plant is built on the banks of a river with a flow rate of $10^5$ kg / s of water. The boiler of the power plant operates at 600 °C and the condenser at 40 °C. The condenser is cooled with water drawn from the river, which is then returned to the main flow. Find the rise in temperature of the river.

8.16   An enclosure at 35 °C is to be heated at a rate of 5500 kJ / h with energy from seawater at 4 °C. What power is required to provide this heat using a Carnot heat pump?

8.17   Two Carnot engines work in series between a high temperature reservoir at 900 K and a low temperature reservoir at 300 K, with the second engine taking all the heat rejected by the first. If the efficiency of the first engine is twice that of the second, find the efficiency of both engines.

8.18   A Carnot engine operates between a heat source at $T_H$ and a heat sink at $T_{C,1}$. If we want to double the efficiency of the engine, what should the temperature of the heat sink $T_{C,2}$ be, keeping $T_H$ constant?

8.19   Show that decreasing $T_C$ by an amount $\Delta T$ while keeping $T_H$ constant is a more effective way to increase the thermal efficiency of a Carnot engine than increasing $T_H$ by $\Delta T$ while keeping $T_C$ constant.

8.20 A Carnot heat engine receives heat from a reservoir at $T_0$ through a heat exchanger in which the heat transferred is proportional to the temperature difference: $\dot{Q}_H = K(T_0 - T_H)$ where $K$ is a constant and $T_H$ the temperature at which the engine receives heat. Show that the work output of the engine is maximum if $T_H = \sqrt{T_C T_0}$ where TC is the temperature at which it rejects heat.

## Refrigerators and Heat Pumps

8.21 A test sample in a laboratory is cooled to a temperature of $10^{-6}$ K. How much work is required to remove 1 J more energy from the sample if the energy is rejected to the surrounding atmosphere at 17 °C?

8.22 A refrigerator is set to maintain a temperature of –5 °C while rejecting heat to the surrounding air at 25 °C. If the setting is changed to –8 °C, find the percentage change in the work required by the refrigerator for the same amount of heat removed.

8.23 A heat pump and refrigerator are installed between the same two thermal reservoirs. If both devices work on a reverse Carnot cycle and transfer the same amount of heat show that $COP_{HP} = COP_R + 1$.

8.24 A Carnot engine operates between 500 K and 200 K and drives a Carnot refrigerator, which provides cooling at 150 K and rejects heat at 200 K. Find the ratio of heat extracted by the refrigerator to heat delivered to the engine.

8.25 A Carnot refrigerator using carbon dioxide as a working fluid transfers heat from 0 to 100 °C. During isothermal compression the density of the fluid doubles. Determine the coefficient of performance of the refrigerator and the amount of heat transferred from the cold region per unit mass of refrigerant.

8.26 A Carnot refrigerator is to provide 100 kW of refrigeration at –80 °C and reject heat to a room at 25 °C. This refrigerator is to be powered by a Carnot engine, which extracts heat from a thermal reservoir at 100 °C at a rate of 150 kW. The Carnot engine also rejects heat at 25 °C. Is this plan feasible? Explain your answer.

8.27 A tray containing 0.4 kg of water at 25 °C is placed in a freezer with $COP_R = 2.0$ and a power consumption of 500 W. What is the minimum amount of time for the water to freeze? The latent heat of fusion of water is 334 kJ / kg.

8.28 A refrigerator cools an enclosure at –40 °C and rejects heat to the surrounding atmosphere at 25 °C. To provide power for the refrigerator, a heat engine works between a 300 °C reservoir and the atmosphere. Assuming all processes are reversible, determine the ratio of the heat supplied from the 300 °C reservoir and the heat extracted from the –40 °C reservoir.

8.29 A furnace delivers heat $Q_{H,1}$ at $T_H$ to heat a room. Instead of heating the room directly, the following plan is proposed: use the heat generated by the furnace to drive a Carnot heat engine rejecting heat to the atmosphere outside the room; then use the power generated by this engine to drive a Carnot heat pump extracting heat from the atmosphere outside to deliver heat $Q_{H,2}$ to the room. The room temperature is $T_R$ and the outside temperature is $T_O$. Determine the ration of $Q_{H,1}/Q_{H,2}$ and judge if this plan is better than heating room directly from the furnace, assuming $T_H > T_R$.

*Carnot Cycle Analysis*

8.30   A Carnot engine with 40% thermal efficiency has an evaporator that operates as a temperature of 240 °C. What is the power output of the engine if it uses 0.4 kg / s of water as a working fluid?

8.31   A Carnot engine draws heat from a tank containing 100 kg of hot air, initially at 800 K. It rejects heat to the atmosphere at 300 K. The air in the tank cools as it loses heat, until eventually the engine stops working when the air temperature drops to 350 K. How much work was done?

8.32   A Carnot engine uses a 10 kg steel block, initially at 800 K, as a heat source, and another 10 kg steel block, initially at 300 K, as a heat sink. As the engine operates the high temperature block cools while the other heats up. Finally, when both blocks are at the same temperature, the engine stops working. What is the final temperature of the two blocks? How much work was done? Assume that the specific heat of steel is 0.5 kJ / kgK.

8.33   A Carnot engine using 1.5 kg / s of water as a working fluid has its evaporator at 250 °C. If the engine efficiency is 40% find the power output.

8.34   A Carnot engine using air as a working fluid works between temperatures of 400 °C and 25 °C. The ratio of the specific volume at the start of isothermal expansion to that at the end of isothermal expansion is 1:3. Assuming air behaves as an ideal gas, determine the work output of the engine per kilogram of air.

8.35   In a Carnot engine saturated water vapour at 300 °C enters the turbine and expands isentropically to 35 °C. Heat is rejected to a reservoir at 35 °C in an isothermal process. Determine the heat added and rejected and the work output by the engine per kilogram of water.

8.36   Steam at 800 °C with a volume of 10 m$^3$ is compressed in a cylinder by a piston on which a constant pressure of 3.5 MPa is applied until it becomes saturated liquid. The heat released during this process is used to power a cyclic heat engine, which rejects heat to the ambient temperature of 25 °C. If the overall process is isentropic, determine the work output of the heat engine.

8.37   A Carnot engine with air as the working fluid operates between a high temperature reservoir at 1200 K and a low temperature reservoir at 300 K. The pressure at the start and end of isothermal heat rejection are 100 kPa and 600 kPa respectively. Assuming that air has constant specific heat, find the heat added and the work done by the engine per kilogram of air.

8.38   A Carnot refrigerator using refrigerant 134a as a working fluid takes heat at –4 °C and rejects it a 24 °C, while using 380 W of power. What is the mass flow rate of the refrigerant?

8.39  A Carnot heat pump is used to maintain a room at 24 °C. Refrigerant 134a enters the heat pump evaporator at 8 °C with a mass flow rate of 0.02 kg / s and quality of 20%, and leaves with a quality of 75%. If the compressor is supplied 1.5 kW of power what is the COP of the heat pump and at what rate is it transferring heat to the room?

8.40  A Carnot cycle is executed by 0.2 kg of air operating between a heat source at 30 °C and a heat sink at 0 °C. The heat source transfers 5 kJ of heat to the air in each cycle and the volume at the end of the isothermal compression is 0.05 m³. Find the pressure and volume at the start and end of each process in the cycle, the work done in each process, and the net work done in a cycle. Assume constant specific heat for air with $c_v = 0.717$ kJ / kgK and $\gamma = 1.4$.

# 9

# Vapour Power and Refrigeration Cycles

**In this chapter you will:**

- Study the Rankine cycle used by vapour–power heat engines.
- Analyse different methods to improve the performance of the Rankine cycle, such as superheat, reheat and regeneration.
- Learn how to build a refrigerator or heat pump using a reverse Rankine cycle.

## 9.1 Rankine Cycle

The Carnot cycle, we have demonstrated, produces the most efficient heat engine possible; it seems reasonable to expect that steam power plants all around the world would be operating using this cycle. Granted that real turbines are not perfectly reversible and adiabatic, but it should still be possible to reasonably approximate the four processes that make up the Carnot cycle. However, this is not the case since the Carnot cycle, though theoretically ideal, proves impractical to implement in reality.

One major difficulty in building a two-phase Carnot engine is that it is difficult to compress a liquid–vapour mixture. In Figure 8.6, which showed a Carnot cycle on a $T$-$s$ diagram, a two-phase mixture entered a pump at state 1, was compressed to a higher pressure and temperature, and left as saturated liquid at state 2. In practice it is difficult to pump a two-phase mixture, for two reasons. First, recall from Chapter 4 that the work done by a pump is

$$W_p = \int_{P_1}^{P_2} v dP.$$

*Energy, Entropy and Engines: An Introduction to Thermodynamics*, First Edition. Sanjeev Chandra.
© 2016 John Wiley & Sons, Ltd. Published 2016 by John Wiley & Sons, Ltd.
Companion website: www.wiley.com/go/chandraSol16

The work required for driving the pump increases with the specific volume of the fluid. If we introduce vapour into a liquid the specific volume of the mixture increases significantly, and with it the work required by the pump. If there are vapour bubbles in the liquid they have to be compressed by the pump, which requires additional work. The second problem is that a two-phase mixture creates a fluctuating load in a pump or compressor as vapour bubbles and liquid slugs impinge alternately on the impeller blades, eroding surfaces and damaging the machinery.

We can avoid all of these problems by letting the liquid-vapour mixture entering the condenser cool down sufficiently until it is a saturated liquid. Then, only liquid enters the pump, reducing the work required to raise its pressure and avoiding any damage. This modified cycle, shown in Figure 9.1, is known as the *Rankine cycle* and consists of the following four processes.

1→2    Saturated liquid at the condenser pressure enters the pump at state 1 and is compressed in an isentropic process. Subcooled liquid at the boiler pressure exits the pump (state 2).

2→3    Subcooled liquid enters the boiler at state 2 and is heated at constant pressure until it becomes saturated vapour (state 3).

3→4    Saturated vapour expands isentropically in a turbine until it becomes a liquid–vapour mixture at the pressure of the condenser (state 4).

4→1    The two-phase mixture leaving the turbine is cooled at constant pressure in a condenser until it becomes a saturated liquid.

Figure 9.2 shows a schematic diagram of a vapour power system using the Rankine cycle.

We can perform an energy balance for the working fluid as it passes through each stage of the cycle, neglecting kinetic and potential energy changes and any heat losses to the surroundings. The work and heat terms are positive in the pump and boiler, and negative in the turbine and condenser. In calculating cycle efficiencies we are only interested in the magnitudes of the

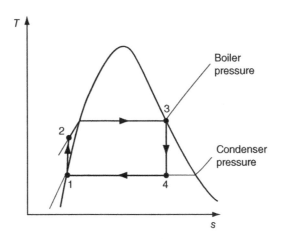

**Figure 9.1**    Rankine cycle on a *T-s* diagram.

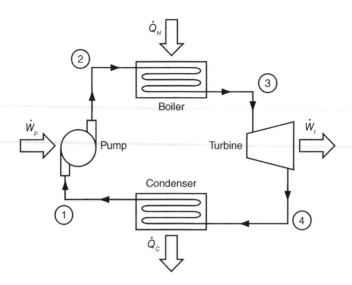

**Figure 9.2**   Components of a Rankine cycle.

energy transfers, not their direction. Therefore the energy balances for the turbine and condenser energy transfers will be written so as to give positive values. The four components are:

*Pump*: The pump compresses the saturated liquid leaving the condenser to the pressure of the
boiler. If $\dot{W}_p$ is the power supplied to the pump and $\dot{m}$ the mass flow rate of working fluid, the
work input per unit mass of fluid

$$w_p = \frac{\dot{W}_p}{\dot{m}} = h_2 - h_1.$$   (9.1)

Assuming that flow is isentropic and that the liquid is incompressible with constant specific
volume ($v$),

$$w_p = \int_{P_1}^{P_2} v dP = v_1 \left( P_2 - P_1 \right).$$   (9.2)

*Boiler*: The boiler takes in subcooled liquid and heats it until it is saturated vapour. If $\dot{Q}_H$ is the
rate that heat is supplied to the boiler, the heat input per unit mass of fluid is

$$q_H = \frac{\dot{Q}_H}{\dot{m}} = h_3 - h_2.$$   (9.3)

*Turbine*: Saturated vapour enters the turbine and is expanded isentropically. If $\dot{W}_t$ is the power
output of the turbine, the work done per unit mass of fluid is

$$w_t = \frac{\dot{W}_t}{\dot{m}} = h_3 - h_4.$$   (9.4)

Note that the energy balance is written so that $w_t$ will have a positive value.

*Condenser*: The exhaust from the turbine is cooled down until it becomes saturated liquid. If $\dot{Q}_C$ is the rate at which heat is transferred from the condenser the heat removed per unit mass of fluid is

$$q_C = \frac{\dot{Q}_C}{\dot{m}} = h_4 - h_1. \tag{9.5}$$

This equation makes $q_C$ positive.

The work required by the pump is usually much smaller than the work output of the turbine. We define the *back work ratio*:

$$bwr = \frac{w_p}{w_t} = \frac{(h_2 - h_1)}{(h_3 - h_4)}. \tag{9.6}$$

The back work ratio for a typical Rankine cycle is of the order of 1–3% so that the pump work can often be neglected in calculations.

The thermal efficiency of the Rankine cycle is given by

$$\eta_{th,Rankine} = \frac{w_{net}}{q_H} = \frac{w_t - w_p}{q_H}. \tag{9.7}$$

In terms of fluid enthalpies, the thermal efficiency is

$$\eta_{th,Rankine} = \frac{(h_3 - h_4) - (h_2 - h_1)}{(h_3 - h_2)} = 1 - \frac{(h_4 - h_1)}{(h_3 - h_2)}. \tag{9.8}$$

Alternatively,

$$\eta_{th,Rankine} = \frac{q_H - q_C}{q_H} = \frac{(h_3 - h_2) - (h_4 - h_1)}{(h_3 - h_2)} = 1 - \frac{(h_4 - h_1)}{(h_3 - h_2)}. \tag{9.9}$$

**Example 9.1**

**Problem:** Saturated steam at a pressure of 6 MPa enters the turbine of a Rankine cycle and leaves at a condenser pressure of 30 kPa. What is the back work ratio? Find the thermal efficiency of the cycle and compare it to that of a Carnot cycle operating between the same temperatures.

**Find:** Back work ratio $bwr$ of the Rankine cycle, thermal efficiency of the Rankine cycle $\eta_{th,Rankine}$, thermal efficiency of a Carnot cycle $\eta_{th,Carnot}$ operating between the same temperatures.

**Known:** Turbine inlet pressure $P_2 = P_3 = 6$ MPa, condenser outlet pressure $P_1 = P_4 = 30$ kPa.

**Process Diagram:** The Rankine cycle on a *T-s* diagram (Figure E9.1).

**Assumptions:** Expansion through the turbine is isentropic so $\Delta S_{34} = 0$.

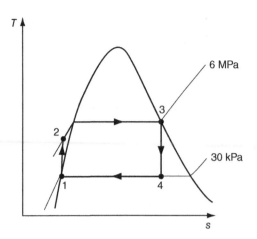

**Figure E9.1**   *T-s* diagram for an ideal Rankine cycle.

**Governing Equations:**

Back work ratio of Rankine cycle

$$bwr = \frac{(h_2 - h_1)}{(h_3 - h_4)}$$

Thermal efficiency of Rankine cycle

$$\eta_{th,\text{Rankine}} = 1 - \frac{(h_4 - h_1)}{(h_3 - h_2)}$$

**Solution:**

We need to find the enthalpy at each of the four states marked in the process diagram.

For saturated steam at $P_3$ = 6 MPa (Appendix 8b), specific enthalpy of vapour $h_3 = h_g$ = 2784.3 kJ / kg and specific entropy of vapour $s_3 = s_g$ = 5.8892 kJ / kgK.

For saturated water at $P_4$ = 30 kPa (Appendix 8b), specific enthalpy of liquid $h_{4,f}$ = 289.23 kJ / kg and vapour $h_{4,g}$ = 2625.3 kJ / kg, and specific entropy of liquid $s_{4,f}$ = 0.9439 kJ / kgK and vapour $s_{4,g}$ = 7.7686 kJ / kgK. Expansion through the turbine is isentropic, so that $s_4 = s_3$ = 5.8892 kJ / kgK; so the quality of the mixture in state 4 can be found:

$$x_4 = \frac{s_4 - s_{4,f}}{s_{4,g} - s_{4,f}} = \frac{5.8892\,\text{kJ} / \text{kgK} - 0.9439\,\text{kJ} / \text{kgK}}{7.7686\,\text{kJ} / \text{kgK} - 0.9439\,\text{kJ} / \text{kgK}} = 0.724\ 62.$$

The enthalpy at the turbine exit is then

$$h_4 = h_{4,f} + x_4 \left( h_{4,g} - h_{4,f} \right),$$

$$h_4 = 289.23\,\text{kJ} / \text{kg} + 0.7246 \times \left( 2625.3\,\text{kJ} / \text{kg} - 289.23\ \text{kJ} / \text{kg} \right) = 1982.0\,\text{kJ} / \text{kg}.$$

The enthalpy at the condenser exit is $h_1 = h_{4,f}$ = 289.23 kJ / kg.
The enthalpy at the pump exit is

$$h_2 = h_1 + w_p = h_1 + v_1 \left( P_2 - P_1 \right).$$

For saturated water at $P_1 = 30$ kPa (Appendix 8b), specific volume $v1 = 0.001022$ m$^3$ / kg, then

$$h_2 = 289.23 \, kJ / kgK + 0.001 \ 022 \ m^3 / kg \times (6000 \, kPa - 30 \, kPa) = 295.33 \, kJ / kg.$$

The back work ratio is

$$bwr = \frac{(h_2 - h_1)}{(h_3 - h_4)} = \frac{295.33 \, kJ / kg - 289.23 \, kJ / kg}{2784.3 \, kJ / kg - 1982.0 \, kJ / kg} = 0.007 \ 603 \ 1.$$

The thermal efficiency of a Rankine cycle is

$$\eta_{th,Rankine} = 1 - \frac{(h_4 - h_1)}{(h_3 - h_2)} = 1 - \frac{1982.0 \, kJ / kg - 289.23 \, kJ / kg}{2784.3 \, kJ / kg - 295.33 \, kJ / kg} = 0.319 \ 89.$$

For a Carnot engine using the same working fluid pressures, at 6 MPa $T_H = T_{sat} = 275.64 \ °C = 548.79$ K, and 30 kPa $T_C = T_{sat} = 69.10 \ °C = 342.25$ K, so the Carnot efficiency would be

$$\eta_{th,Carnot} = 1 - \frac{T_C}{T_H} = 1 - \frac{342.25 \, K}{548.79 \, K} = 0.376 \ 36.$$

**Answer:** The back work ratio is 0.760%, and the thermal efficiency of the Rankine cycle is 32.0%, while the related thermal efficiency of the Carnot cycle operating between the same temperatures is 37.6%. ∎

## 9.2 Rankine Cycle with Superheat and Reheat

The efficiency of a Rankine cycle is less than that of a Carnot cycle operating between the same two temperatures because part of the heat added in a Rankine cycle is at a temperature less than the maximum available. In practice the Rankine cycle is preferable to the Carnot cycle because it requires only liquid to be pumped, which consumes very little power. It also avoids having a two-phase mixture in the pump, which can lead to damage. However, the Rankine cycle still has a two-phase mixture flowing through the turbine, which can also cause serious problems. Water droplets that condense inside a steam turbine impinge on turbine blades that are rotating at high speeds and very rapidly erode them. In practice it is not advisable to allow steam quality in a turbine to drop below 90%.

One way to increase the quality of steam at the turbine exit is to introduce superheated, rather than saturated, steam into the turbine. This has two advantages: the average temperature of heat addition in the boiler increases when steam is superheated in it, thereby increasing cycle efficiency; and the quality at the turbine exhaust increases, avoiding damage to the turbine blades due to impacting water droplets. Figure 9.3 shows a Rankine cycle with superheat, in which superheated steam at state 3 enters the turbine and after isentropic expansion leaves at state 4s. The steam exiting the turbine has higher quality than it did in the ordinary Rankine cycle (compare with Figure 9.1).

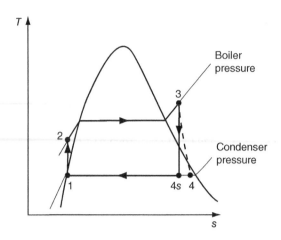

**Figure 9.3**  Rankine cycle with superheat.

In reality irreversibilities in the turbine will increase both the specific entropy of the steam leaving (to state 4) and its quality. The ratio of the actual work output ($w_{t,a}$) of the turbine to the ideal work output ($w_{t,s}$) can be calculated from the isentropic efficiency of the turbine:

$$\eta_t = \frac{w_{t,a}}{w_{t,s}} = \frac{h_3 - h_4}{h_3 - h_{4,s}}.$$  (9.10)

**Example 9.2**
**Problem:** Superheated steam at a pressure of 6 MPa and temperature of 400 °C enters the turbine of a Rankine cycle and leaves at a condenser pressure of 30 kPa. Find the thermal efficiency of the cycle if (a) the turbine is isentropic and (b) the turbine isentropic efficiency is 92%.
**Find:** Thermal efficiency of the Rankine cycle for (a) an isentropic turbine $\eta_{R,s}$ (b) a non-isentropic turbine $\eta_{R,a}$.
**Known:** Turbine inlet pressure $P_2 = P_3 = 6$ MPa, turbine inlet temperature $T_3 = 400$ °C, condenser pressure $P_1 = P_4 = 30$ kPa, turbine isentropic efficiency $\eta_t = 92\%$.
**Process Diagram:** The Rankine cycle on a $T$-$s$ diagram (Figure E9.2).
**Assumptions:** Expansion through the turbine is isentropic.
**Governing Equations:**

Thermal efficiency of Rankine cycle

$$\eta_{th,Rankine} = 1 - \frac{(h_4 - h_1)}{(h_3 - h_2)}$$

Isentropic efficiency

$$\eta_t = \frac{h_3 - h_4}{h_3 - h_{4,s}}$$

**Solution:**
We need to find the enthalpy at each of the four states marked in the process diagram.
    For superheated steam at $P_3 = 6$ MPa and $T_3 = 400$ °C (Appendix 8c), specific enthalpy $h_3 = 3177.2$ kJ / kg and specific entropy $s_3 = 6.5408$ kJ / kgK.

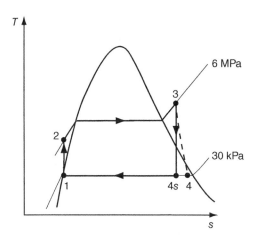

**Figure E9.2** *T-s* diagram for a Rankine cycle with superheat.

For saturated water at $P_4 = 30$ kPa (Appendix 8b), specific enthalpy of liquid $h_{4,f} = 289.23$ kJ / kg and vapour $h_{4,g} = 2625.3$ kJ / kg, and specific entropy of liquid $s_{4,f} = 0.9439$ kJ / kgK and vapour $s_{4,g} = 7.7686$ kJ / kgK. If expansion through the turbine is isentropic, $s_{4s} = s_3 = 6.5408$ kJ / kgK, and the quality of the mixture in the turbine would be

$$x_{4s} = \frac{s_{4s} - s_{4,f}}{s_{4,g} - s_{4,f}} = \frac{6.5408 \text{ kJ / kgK} - 0.9439 \text{ kJ / kgK}}{7.7686 \text{ kJ / kgK} - 0.9439 \text{ kJ / kgK}} = 0.820 \ 09.$$

The enthalpy at the turbine exit would be

$$h_{4s} = h_{4,f} + x_{4s} \left(h_{4,f} - h_{4,f}\right),$$

$$h_{4s} = 289.23 \text{ kJ / kg} + 0.8201 \times \left(2625.3 \text{ kJ / kg} - 289.23 \text{ kJ / kg}\right) = 2205.0 \text{ kJ / kg}.$$

The enthalpy at the condenser exit is $h_1 = h_{4,f} = 289.23$ kJ / kg.
The enthalpy at the pump exit (from the solution of) is $h_2 = 295.33$ kJ / kg.
The isentropic efficiency of a turbine can be used to find the actual turbine enthalpy:

$$h_4 = h_3 - \eta_t \left(h_3 - h_{4s}\right),$$

$$h_4 = 3177.2 \text{ kJ / kg} - 0.92 \times \left(3177.2 \text{ kJ / kg} - 2205.0 \text{ kJ / kg}\right) = 2282.8 \text{ kJ / kg}.$$

The thermal efficiency of the isentropic Rankine cycle is

$$\eta_{th, \text{Rankine}} = 1 - \frac{\left(h_{4s} - h_1\right)}{\left(h_3 - h_2\right)} = 1 - \frac{2205.0 \text{ kJ / kg} - 289.23 \text{ kJ / kg}}{3177.2 \text{ kJ / kg} - 295.33 \text{ kJ / kg}} = 0.335 \ 23.$$

The thermal efficiency of the Rankine cycle with a turbine with 92% isentropic efficiency is

$$\eta_{th,\text{Rankine}} = 1 - \frac{(h_4 - h_1)}{(h_3 - h_2)} = 1 - \frac{2282.8\,\text{kJ}/\text{kg} - 289.23\,\text{kJ}/\text{kg}}{3177.2\,\text{kJ}/\text{kg} - 295.33\,\text{kJ}/\text{kg}} = 0.308\ 24.$$

**Answer:** The thermal efficiency of the isentropic Rankine cycle is 33.5%, while the thermal efficiency with a turbine with 92% isentropic efficiency is 30.8 %. ∎

If superheating the steam is not sufficient to get rid of moisture in the turbine, it is possible to increase the steam quality further by using *reheat*. Figure 9.4 shows the components of a Rankine cycle with reheat. Steam coming from the boiler expands in two stages, first in a high-pressure and then in a low-pressure turbine. After the steam leaves the high-pressure turbine (stage 4 in Figure 9.4) it is taken through the reheat section of the boiler and more heat added to it, increasing its temperature. It then enters the low-pressure turbine (stage 5 in Figure 9.4) and expands again.

Reheat does not have much effect on the efficiency of the cycle, because while it increases the work output of the turbine it also requires a greater heat input. However, reheat significantly increases the quality at the exit of the low-pressure turbine. Figure 9.5 shows a Rankine cycle with reheat on a *T-s* diagram.

The heat added in the boiler per unit mass of fluid is

$$q_H = \frac{\dot{Q}_H}{\dot{m}} = (h_3 - h_2) + (h_5 - h_4) \tag{9.11}$$

and the total work output from the turbines per unit mass of fluid is

$$w_t = \frac{\dot{W}_t}{\dot{m}} = (h_3 - h_4) + (h_5 - h_6). \tag{9.12}$$

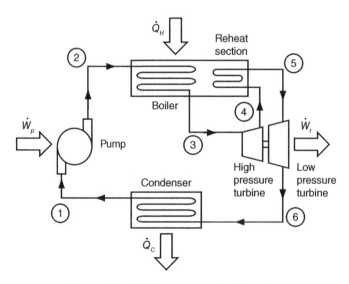

**Figure 9.4**   Rankine power cycle with reheat.

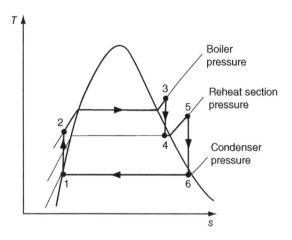

**Figure 9.5**    Rankine cycle with reheat.

## Example 9.3

**Problem:** Superheated steam at a pressure of 6 MPa and temperature of 400 °C enters the first stage of a turbine in a Rankine cycle where it expands to 0.8 MPa. The steam is then reheated to 300 °C and expanded in the second stage of the turbine which it leaves at a condenser pressure of 30 kPa. Find (a) the work done by the turbine per unit mass of fluid, (b) the heat added per unit mass of fluid and (c) the thermal efficiency of the cycle.

**Find:** (a) Work done $w$ per unit mass of fluid, (b) heat added $q$ per unit mass of fluid, (c) thermal efficiency $\eta_{th}$ of the cycle.

**Known:** Turbine inlet pressure $P_2 = P_3 = 6$ MPa, turbine inlet temperature $T_3 = 400$ °C, reheat pressure $P_4 = P_5 = 0.8$ MPa, reheat temperature $T_5 = 300$ °C, condenser pressure $P_1 = P_6 = 30$ kPa.

**Process Diagram:** The Rankine cycle with reheat on a *T-s* diagram (Figure E9.3).

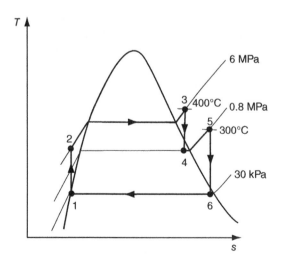

**Figure E9.3**    *T-s* diagram for a Rankine cycle with reheat.

**Assumptions:** Expansion through the turbines is isentropic so $\Delta S_{34} = 0$ and $\Delta S_{56} = 0$.

**Governing Equations:**

Work output from turbines

$$w_t = (h_3 - h_4) + (h_5 - h_6)$$

Heat input to boiler

$$q_H = (h_3 - h_2) + (h_5 - h_4)$$

Thermal efficiency of Rankine cycle

$$\eta_{th, \text{Rankine}} = \frac{w_t - w_p}{q_H}$$

**Solution:**

From Example 9.2 the enthalpy at the condenser exit is $h_1 = 289.23$ kJ / kg.

The enthalpy at the pump exit is $h_2 = 295.33$ kJ / kg.

For superheated steam at $P_3 = 6$ MPa and $T_3 = 400$ °C (Appendix 8c), specific enthalpy $h_3 = 3177.2$ kJ / kg and specific entropy $s_3 = 6.5408$ kJ / kgK.

For saturated water at $P_4 = 0.8$ kPa (Appendix 8b), specific enthalpy of liquid $h_{4,f} = 721.11$ kJ / kg and vapour $h_{4,g} = 2769.1$ kJ / kg, and specific entropy of liquid $s_{4,f} = 2.0462$ kJ / kgK and vapour $s_{4,g} = 6.6628$ kJ / kgK. If expansion through the turbine is isentropic, $s_4 = s_3 = 6.5408$ kJ / kgK and the mixture quality in the high pressure turbine is

$$x_4 = \frac{s_4 - s_{4,f}}{s_{4,g} - s_{4,f}} = \frac{6.5408 \text{ kJ / kgK} - 2.0462 \text{ kJ / kgK}}{6.6628 \text{ kJ / kgK} - 2.0462 \text{ kJ / kgK}} = 0.973\ 57.$$

The enthalpy at the turbine exit is then

$$h_4 = h_{4,f} + x_4 \left(h_{4,g} - h_{4,f}\right),$$

$$h_4 = 721.11 \text{kJ} / \text{kg} + 0.973\ 57 \times \left(2769.1 \text{kJ} / \text{kg} - 721.11 \text{ kJ} / \text{kg}\right) = 2715.0 \text{kJ} / \text{kg}.$$

For superheated steam at $P_5 = 0.8$ MPa and $T_5 = 300$ °C (Appendix 8c), specific enthalpy $h_5 = 3056.5$ kJ / kg and specific entropy $s_5 = 7.2328$ kJ / kgK.

For saturated water at $P_6 = 30$ kPa (Appendix 8b), specific enthalpy of liquid $h_{6,f} = 289.23$ kJ / kg and vapour $h_{6,g} = 2625.3$ kJ / kg, and specific entropy of liquid $s_{6,f} = 0.9439$ kJ / kgK and vapour $s_{6,g} = 7.7686$ kJ / kgK. If expansion through the turbine is isentropic, $s_6 = s_5 = 7.2328$ kJ / kgK and the mixture quality in the low pressure turbine is

$$x_6 = \frac{s_6 - s_{6,f}}{s_{6,g} - s_{6,f}} = \frac{7.2328 \text{ kJ} / \text{kgK} - 0.9439 \text{kJ} / \text{kgK}}{7.7686 \text{ kJ} / \text{kgK} - 0.9439 \text{kJ} / \text{kgK}} = 0.92\ 149.$$

The enthalpy at the turbine exit is then

$$h_6 = h_{6,f} + x_6 \left(h_{6,g} - h_{6,f}\right),$$

$$h_6 = 289.23 \text{kJ} / \text{kg} + 0.92149 \times \left(2625.3 \text{kJ} / \text{kg} - 289.23 \text{ kJ} / \text{kg}\right) = 2441.9 \text{kJ} / \text{kg}.$$

(a) The work done by both turbines is

$$w_t = (h_3 - h_4) + (h_5 - h_6),$$

$$w_t = (3177.2\,\text{kJ}/\text{kg} - 2715.0\,\text{kJ}/\text{kg}) + (3056.5\,\text{kJ}/\text{kg} - 2441.9\,\text{kJ}/\text{kg}) = 1076.8\,\text{kJ}/\text{kg}.$$

(b) The heat input to the boiler is

$$q_H = (h_3 - h_2) + (h_5 - h_4),$$

$$q_H = (3177.2\,\text{kJ}/\text{kg} - 295.33\,\text{kJ}/\text{kg}) + (3056.5\,\text{kJ}/\text{kg} - 2715.0\,\text{kJ}/\text{kg}) = 3223.4\,\text{kJ}/\text{kg}.$$

(c) The work input to the pump is

$$w_p = (h_2 - h_1),$$
$$w_p = 295.33\,\text{kJ}/\text{kg} - 289.23\,\text{kJ}/\text{kg} = 6.1\,\text{kJ}/\text{kg}.$$

Then the thermal efficiency of the Rankine cycle is

$$\eta_{th,\text{Rankine}} = \frac{w_t - w_p}{q_H},$$

$$\eta_{th,\text{Rankine}} = \frac{1076.8\,\text{kJ}/\text{kg} - 6.1\,\text{kJ}/\text{kg}}{3223.4\,\text{kJ}/\text{kg}} = 0.332\,16.$$

**Answer:** (a) The work done by the turbine is 1076.8 kJ / kg, (b) with heat input of 3223.3 kJ / kg of working fluid, (c) the thermal efficiency of the Rankine cycle is 33.2%. ■

## 9.3 Rankine Cycle with Regeneration

The Rankine cycle uses a significant amount of energy to heat subcooled water that enters the boiler at state 2 in Figure 9.6 and bring it to the saturation temperature. The shaded area in Figure 9.6 shows the energy required to heat the water to saturation conditions. We can improve the efficiency of the cycle by increasing the average temperature at which heat is supplied in the boiler if we preheat the water before it enters the boiler using some steam withdrawn from the turbine. This process, known as *regeneration*, increases the temperature of the water entering the boiler and therefore the thermal efficiency of the cycle.

### 9.3.1 Open Feedwater Heater

Figure 9.7 shows how to preheat the water before it enters the boiler by extracting some steam after the high-pressure stage of the turbine (state 5) and supplying it to a heat exchanger known as a *feedwater heater*. The remaining steam expands through the low-pressure stage of the turbine before entering the condenser (state 4). Saturated liquid leaving the condenser

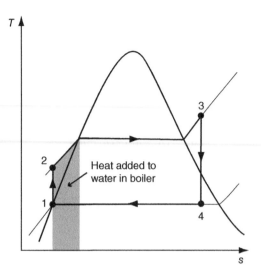

**Figure 9.6**   Heat added to subcooled water in boiler of Rankine cycle to bring it to the saturation temperature.

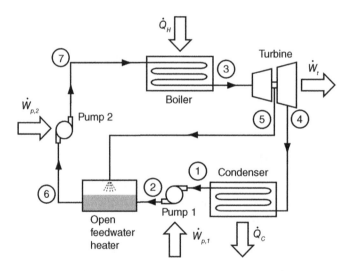

**Figure 9.7**   Rankine cycle with regeneration using an open feedwater heater.

(state 1) is compressed by a pump and fed back into the feedwater heater (state 2). There the subcooled water mixes with the steam extracted from the turbine. This type of heat exchanger, in which the two fluid streams mix, is known as an *open feedwater heater*. The amount of steam added is just enough that the liquid leaving the feedwater heater (state 6) is saturated. A second pump compresses this liquid and supplies it to the boiler (state 7) where it is heated and sent to the turbine (state 3). Figure 9.8 shows a *T-s* diagram of a Rankine cycle with regeneration using an open feedwater heater.

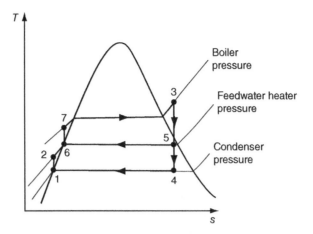

**Figure 9.8**    Rankine cycle with regeneration using an open feedwater heater.

A mass balance around the turbine gives

$$\dot{m}_6 = \dot{m}_4 + \dot{m}_5,$$                                                    (9.13)

where $\dot{m}_6$ is the total rate of mass flow of fluid in the cycle, $\dot{m}_5$ the mass flow rate of steam extracted from the turbine and $\dot{m}_4$ the balance that flows through the second stage of the turbine. Let $f$ denote the fraction of the total mass flow that is extracted for the feedwater heater so that

$$f = \frac{\dot{m}_5}{\dot{m}_6}.$$                                                    (9.14)

An energy balance for the feedwater heater gives

$$h_6 = f h_5 + (1 - f) h_2.$$                                                    (9.15)

Solving for $f$,

$$f = \frac{h_6 - h_2}{h_5 - h_2}.$$                                                    (9.16)

Typically sufficient steam is extracted from the turbine to raise the temperature of the water leaving the condenser about halfway to the temperature of the boiler. The total work output of the turbine per unit mass of fluid is

$$w_t = \frac{\dot{W}_t}{\dot{m}_3} = (h_3 - h_5) + (1 - f)(h_5 - h_4).$$                                                    (9.17)

The total work required for driving both pumps per unit mass of fluid is

$$w_p = \frac{\dot{W}_p}{\dot{m}_3} = (h_7 - h_6) + (1 - f)(h_2 - h_1).$$                                                    (9.18)

The heat added in the boiler is

$$q_H = \frac{\dot{Q}_H}{\dot{m}_3} = h_3 - h_7,$$

(9.19)

and the heat lost in the condenser is

$$q_c = (1-f)(h_4 - h_1).$$

(9.20)

**Example 9.4**

**Problem:** A Rankine cycle with regeneration is supplied superheated steam at a pressure of 6 MPa and temperature of 400 °C. The steam expands to a pressure of 0.8 MPa in the first stage of a turbine, after some of it is diverted into an open feedwater heater. The remaining steam expands in the second stage of the turbine, passes through a condenser at a pressure of 30 kPa, and leaves as saturated liquid that is pumped into the feedwater heater where it mixes with the steam that was removed. Saturated liquid leaves the feedwater heater at 0.8 MPa. Find (a) the fraction of steam extracted and (b) the thermal efficiency of the cycle.

**Find:** (a) Fraction of stream $f$ extracted from the feedwater heater, (b) thermal efficiency $\eta_{th,Rankine}$ of the cycle.

**Known:** Turbine inlet pressure $P_3 = P_7 = 6$ MPa, turbine inlet temperature $T_3 = 400$ °C, feedwater heater pressure $P_5 = P_6 = 0.8$ MPa, condenser pressure $P_1 = P_4 = 30$ kPa.

**Process Diagram:** The Rankine cycle with regeneration on a $T$-$s$ diagram (Figure E9.4).

**Assumptions:** Expansion through the turbines is isentropic so $\Delta S_{35} = 0$ and $\Delta S_{34} = 0$.

**Governing Equations:**

Feedwater mass flow fraction          $f = \dfrac{h_6 - h_2}{h_5 - h_2}$

Work of turbines with regeneration     $w_t = (h_3 - h_5) + (1-f)(h_5 - h_4)$

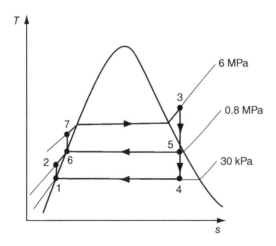

**Figure E9.4**  *T-s* diagram for a Rankine cycle with regeneration using an open feedwater heater.

Work to pumps with regeneration

Heat to boiler with regeneration

Thermal efficiency of Rankine cycle

$$w_p = \left(h_7 - h_6\right) + \left(1 - f\right)\left(h_2 - h_1\right)$$
$$q_H = h_3 - h_7$$
$$\eta_{th,Rankine} = \frac{w_t - w_p}{q_H}$$

**Solution:**

For superheated steam at $P_3 = 6$ MPa and $T_3 = 400\ ^{\circ}$C (Appendix 8c), specific enthalpy $h_3 = 3177.2$ kJ / kg, and specific entropy $s_3 = 6.5408$ kJ / kgK.

For saturated water at $P_5 = 0.8$ MPa (Appendix 8b), specific enthalpy of liquid $h_{5,f} = 721.11$ kJ / kg and vapour $h_{5,g} = 2769.1$ kJ / kg, and specific entropy of liquid $s_{5,f} = 2.0462$ kJ / kgK and vapour $s_{5,g} = 6.6628$ kJ / kgK. If expansion through the first stage of the turbine is isentropic, $s_5 = s_3 = 6.5408$ kJ / kgK, then the quality of the mixture in the first stage is

$$x_5 = \frac{s_5 - s_{5,f}}{s_{5,g} - s_{5,f}} = \frac{6.5408\,\text{kJ}\,/\,\text{kgK} - 2.0462\,\text{kJ}\,/\,\text{kgK}}{6.6628\,\text{kJ}\,/\,\text{kgK} - 2.0462\,\text{kJ}\,/\,\text{kgK}} = 0.973\ 57.$$

The enthalpy at the turbine exit is then

$$h_5 = h_{5,f} + x_5 \left(h_{5,g} - h_{5,f}\right),$$

$$h_5 = 721.11\,\text{kJ}\,/\,\text{kg} + 0.97357 \times \left(2769.1\,\text{kJ}\,/\,\text{kg} - 721.11\ \text{kJ}\,/\,\text{kg}\right) = 2715.0\,\text{kJ}\,/\,\text{kg}.$$

For saturated water at $P_4 = 30$ kPa (Appendix 8b), specific enthalpy of liquid $h_{4,f} = 289.23$ kJ / kg and vapour $h_{4,g} = 2625.3$ kJ / kg, and specific entropy of liquid $s_{4,f} = 0.9439$ kJ / kgK and vapour $s_{4,g} = 7.7686$ kJ / kgK. If expansion through the second stage of the turbine is isentropic, $s_4 = s_3 = 6.5408$ kJ / kgK, and quality in the second turbine is

$$x_4 = \frac{s_4 - s_{4,f}}{s_{4,g} - s_{4,f}} = \frac{6.5408\,\text{kJ}\,/\,\text{kgK} - 0.9439\,\text{kJ}\,/\,\text{kgK}}{7.7686\,\text{kJ}\,/\,\text{kgK} - 0.9439\,\text{kJ}\,/\,\text{kgK}} = 0.820\ 09.$$

The enthalpy at the turbine exit is then

$$h_4 = h_{4,f} + x_4 \left(h_{4,g} - h_{4,f}\right),$$

$$h_4 = 289.23\,\text{kJ}\,/\,\text{kg} + 0.82009 \times \left(2625.3\,\text{kJ}\,/\,\text{kg} - 289.23\ \text{kJ}\,/\,\text{kg}\right) = 2205.0\,\text{kJ}\,/\,\text{kg}.$$

The enthalpy at the condenser exit at 30 kPa is $h_1 = h_f = 289.23$ kJ / kg.

The enthalpy at the exit of pump 1 is

$$h_2 = h_1 + w_{p,1} = h_1 + v_1 \left(P_2 - P_1\right).$$

Using saturated water at $P_1 = 30$ kPa (Appendix 8b), specific volume $v_1 = v_f = 0.001\ 022$ m³ / kg, then

$$h_2 = 289.23\,\text{kJ}\,/\,\text{kgK} + 0.001\ 022\ \text{m}^3\,/\,\text{kg} \times \left(6000\,\text{kPa} - 30\,\text{kPa}\right) = 295.33\,\text{kJ}\,/\,\text{kg}.$$

The enthalpy at the exit of the feedwater heater is $h_6 = h_{5,f} = 721.11$ kJ / kg.
The enthalpy at the exit of pump 2 is

$$h_7 = h_6 + w_{p,1} = h_6 + v_6 \left( P_7 - P_6 \right)$$

For saturated water at $P_6 = 0.8$ MPa (Appendix 8b), specific volume $v_6 = v_f = 0.00115$ m3 / kg, then

$$h_7 = 721.11\,\text{kJ} / \text{kgK} + 0.001115 \ \text{m}^3 / \text{kg} \times \left( 6000\,\text{kPa} - 800\,\text{kPa} \right) = 726.91\,\text{kJ} / \text{kg}.$$

(a)  The mass fraction of steam extracted for the feedwater heater can now be solved for:

$$f = \frac{h_6 - h_2}{h_5 - h_2} = \frac{721.11\,\text{kJ} / \text{kg} - 295.33\,\text{kJ} / \text{kg}}{2715.0\,\text{kJ} / \text{kg} - 295.33\,\text{kJ} / \text{kg}} = 0.175\ 97.$$

(b)  The work done by the turbine is

$$w_t = \left( h_3 - h_5 \right) + \left( 1 - f \right)\left( h_5 - h_4 \right),$$

$$w_t = \left( 3177.2\,\text{kJ} / \text{kg} - 2715.0\,\text{kJ} / \text{kg} \right) + \left( 1 - 0.17597 \right)$$
$$\left( 2715.0\,\text{kJ} / \text{kg} - 2205.0\,\text{kJ} / \text{kg} \right) = 882.46\,\text{kJ} / \text{kg}.$$

The total work done by both pumps is

$$w_p = \left( h_7 - h_6 \right) + \left( 1 - f \right)\left( h_2 - h_1 \right),$$

$$w_p = \left( 726.91\,\text{kJ} / \text{kg} - 721.11\,\text{kJ} / \text{kg} \right) + \left( 1 - 0.175\ 97 \right)$$
$$\left( 295.33\,\text{kJ} / \text{kg} - 289.23\,\text{kJ} / \text{kg} \right) = 10.827\,\text{kJ} / \text{kg}^{\cdot}$$

The heat input to the boiler is

$$q_H = \left( h_3 - h_7 \right) = 3177.2\,\text{kJ} / \text{kg} - 726.91\,\text{kJ} / \text{kg} = 2450.3\,\text{kJ} / \text{kg},$$

The thermal efficiency of the Rankine cycle is then

$$\eta_{th,\text{Rankine}} = \frac{w_t - w_p}{q_H} = \frac{882.46\,\text{kJ} / \text{kg} - 10.827\,\text{kJ} / \text{kg}}{2450.3\,\text{kJ} / \text{kg}} = 0.355\ 73.$$

**Answer:** (a) The mass fraction of steam extracted is 17.6% and (b) the thermal efficiency of the Rankine cycle is 35.6%.                                                                                       ■

## 9.3.2  Closed Feedwater Heater

An open feedwater heater requires the use of a second pump, which may not be desirable since it adds to equipment and maintenance costs. One way to implement regeneration without the use of an additional pump is to use a closed feedwater heater, which is a heat exchanger in which the two fluid streams do not mix with each other. The condensed liquid passes through a tube over which steam flows. Since there is no mixing the two fluids can be at different pressures. Figure 9.9 shows an example of a Rankine cycle with a closed feedwater heater.

A portion of the steam flowing through the high-pressure stage of the turbine is extracted (state 5) and diverted to the closed feedwater heater. The remainder of the steam expands through the low-pressure turbine (state 4) and passes through the condenser. The saturated liquid coming out of the condenser is pumped to the boiler (state 2). The steam passing over the tubes in the feedwater heater condenses and saturated water collects at the bottom of the feedwater heater (state 7). The water flowing through the feedwater heater tube is heated (to state 6 where its temperature is a little below that at state 7) and then enters the boiler. The saturated water in the feedwater heater passes through a steam trap, which is a valve that allows only liquid to pass through to a region of lower pressure. The exit of the steam trap connects back to the condenser (state 8). An energy balance for the steam trap gives, assuming that there is no heat or work transfer in it and changes in kinetic and potential energy are negligible,

$$h_8 = h_7. \tag{9.21}$$

Figure 9.10 shows a $T$-$s$ diagram of a closed feedwater cycle. The expansion of steam from state 7 to 8 is irreversible, so it is shown as a dashed line. An energy balance for the feedwater heater gives

$$\dot{m}_6 \left( h_6 - h_2 \right) = \dot{m}_5 \left( h_5 - h_7 \right). \tag{9.22}$$

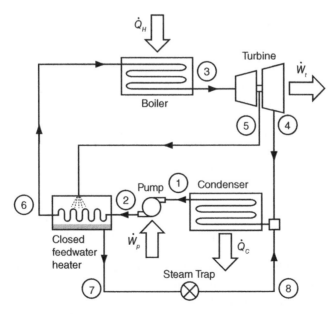

**Figure 9.9**   Regeneration using a closed feedwater heater.

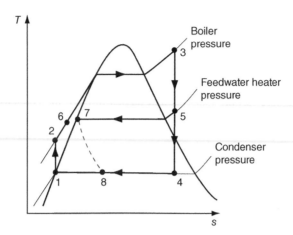

**Figure 9.10** Rankine cycle with regeneration using a closed feedwater heater.

The mass fraction of steam extracted from the turbine,

$$f = \frac{\dot{m}_5}{\dot{m}_6} = \frac{h_6 - h_2}{h_5 - h_7}. \tag{9.23}$$

Energy balances on each component of the cycle can be done to calculate work and heat transfers.

## 9.4  Vapour Refrigeration Cycle

Building a refrigerator based on the Carnot cycle presents the same difficulties that we encountered in trying to construct a Carnot engine. Compressing a two-phase mixture is difficult since liquid droplets can damage the compressor blades. Real refrigerators therefore use a reverse Rankine cycle, in which the compressor takes in saturated vapour and supplies superheated vapour to the condenser. Refrigerators also do not expand vapour in a turbine since the amount of work that obtained from it is very small. Instead, saturated liquid leaving the condenser is expanded through a valve that releases it into the lower pressure evaporator. Figure 9.11 shows a schematic of the components of a refrigerator.

Figure 9.12 shows the reverse Rankine cycle on a $T$-$s$ diagram. There are four processes:

1→2   A saturated liquid–vapour mixture enters the evaporator at state 1. The evaporator pressure is such that the corresponding saturation temperature is lower than the temperature inside the refrigerator. Heat from the refrigerator interior evaporates the refrigerant so that it leaves the condenser as saturated vapour (state 2).

2→3   Saturated vapour enters the compressor at state 2 and is compressed isentropically until it becomes superheated vapour (state 3).

3→4   Superheated vapour is cooled at constant pressure in the condenser until it becomes saturated liquid (state 4). The pressure inside the evaporator is such that the

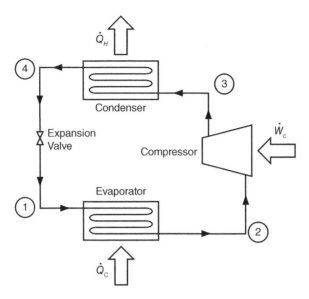

**Figure 9.11**   Vapour refrigeration cycle.

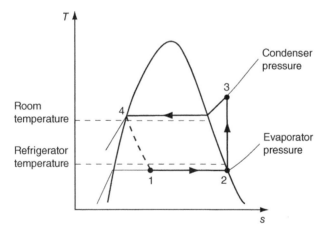

**Figure 9.12**   Reverse Rankine cycle.

corresponding saturation temperature is higher than that of the surrounding air, to which heat is lost.

4→1   Saturated liquid passes through an expansion valve to the lower evaporator pressure. This is an irreversible process in which the liquid flashes into a two-phase mixture. An energy balance for the valve gives that $h_1 = h_4$.

The rate at which heat is removed from the refrigerator per unit mass of fluid is

$$q_C = \frac{\dot{Q}_C}{\dot{m}} = h_2 - h_1.$$                                               (9.24)

The rate at which heat is lost from the condenser to the surroundings per unit mass of fluid is

$$q_H = \frac{\dot{Q}_H}{\dot{m}} = h_3 - h_4. \tag{9.25}$$

The work supplied to the compressor per unit mass of fluid is

$$w_c = \frac{\dot{W}_c}{\dot{m}} = h_3 - h_2. \tag{9.26}$$

The coefficient of performance of the refrigerator is

$$COP_R = \frac{\dot{Q}_C}{\dot{W}_c} = \frac{h_2 - h_1}{h_3 - h_2}. \tag{9.27}$$

Air conditioners also operate on a reverse Rankine cycle, with the evaporator placed on the inside of the house that is being cooled and the condenser on the outside where it can reject heat to the ambient air (Figure 9.13a). Since the evaporator and condenser are similarly sized heat exchangers, it is possible to reverse the direction of refrigerant flow and operate the same device as a heat pump, where it takes heat from the outside air and transfers it to the interior of the house (Figure 9.13b). The coefficient of performance of a heat pump is

$$COP_{HP} = \frac{\dot{Q}_H}{\dot{W}_c} = \frac{h_3 - h_4}{h_3 - h_2}. \tag{9.28}$$

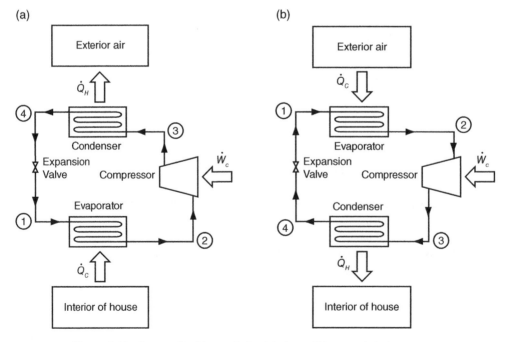

**Figure 9.13**   Reverse Rankine cycle for (a) air conditioner and (b) heat pump.

Air-conditioning and refrigeration capacity is sometimes measured in "tons", a unit that dates back to the days when refrigerators were most likely to be found in ice factories. One ton is the amount of cooling required for freezing 1 ton of ice in 24 h. In SI units 1 ton of refrigeration is defined to be equal to 3.52 kW.

### Example 9.5
**Problem:** A freezer working on a reverse Rankine cycle uses refrigerant 134a with a mass flow rate of 0.10 kg / s. The refrigerant leaves the evaporator as saturated vapour at a temperature of –8 °C. It leaves the condenser as saturated liquid at a pressure of 0.8 MPa. Find the power required to drive the compressor. Calculate the coefficient of performance $(COP_R)$ of the freezer and compare it with the $COP_R$ of a Carnot refrigerator operating between the same temperatures.

**Find:** Power $\dot{W}_c$ required to drive the compressor, coefficient of performance of refrigeration $COP_R$ of the Rankine cycle, the coefficient of performance of refrigeration $COP_R$ of a Carnot cycle operating between the same temperatures.

**Known:** Refrigerant mass flow rate $\dot{m}= 0.10$ kg / s, evaporator temperature $T_2 = T_1 = -8$ °C, condenser pressure $P_3 = P_4 = 0.8$ MPa.

**Process Diagram:** The reverse Rankine cycle on a $T$-$s$ diagram (Figure E9.5).

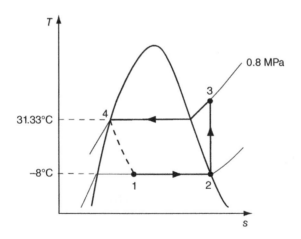

**Figure E9.5**   *T-s* diagram for a reverse Rankine cycle.

**Assumptions:** Expansion through the compressor is isentropic so $\Delta S_{23} = 0$.

**Governing Equations:**

Coefficient of performance Rankine
$$COP_R = \frac{h_2 - h_1}{h_3 - h_2}$$

Power of compressor
$$\dot{W}_c = \dot{m}(h_3 - h_2)$$

### Solution:
We need to find the enthalpy at each of the four states marked in the process diagram.

For saturated refrigerant 134a at $T_2 = -8$ °C (Appendix 9a), specific enthalpy of vapour $h_2$ $= h_g = 242.54$ kJ / kg and specific entropy of vapour $s_2 = s_g = 0.9239$ kJ / kgK.

For saturated refrigerant 134a at $P_4 = 0.8$ MPa (Appendix 9b), specific enthalpy of liquid $h_4 = h_f = 93.42$ kJ / kg.

Expansion through the throttling valve is a constant enthalpy process, so that $h_1 = h_4 = 93.42$ kJ / kg.

Compression is isentropic, so $s_3 = s_2 = 0.9239$ kJ / kgK. At state 3, pressure $P_3 = 0.8$ MPa; with pressure and entropy at state 3, we can interpolate from the tables for superheated refrigerant 134a (Appendix 9c), giving $h_3 = 269.49$ kJ / kg.

The power required by the compressor is

$$\dot{W}_c = \dot{m}(h_3 - h_2) = 0.10\,\text{kg/s} \times (269.49\,\text{kJ/kg} - 242.54\,\text{kJ/kg}) = 2.6950\,\text{kW}.$$

The coefficient of performance of the Rankine cycle is

$$COP_R = \frac{h_2 - h_1}{h_3 - h_2} = \frac{269.49\,\text{kJ/kg} - 93.42\,\text{kJ/kg}}{269.49\,\text{kJ/kg} - 242.54\,\text{kJ/kg}} = 6.5332.$$

Using the saturated refrigerant 134a tables (Appendix 9b), at 0.8 MPa $T_H = T_{\text{sat}} = 31.33$ °C $= 304.48$ K and $T_C = -8$ °C $= 265.15$ K. The coefficient of performance of the related Carnot cycle is

$$COP_{R,\text{Carnot}} = \frac{1}{T_H / T_C - 1} = \frac{1}{(304.48\,\text{K} / 265.15\,\text{K}) - 1} = 6.7417.$$

**Answer:** The compressor requires 2.70 kW of power. The $COP_R$ of the freezer is 6.53, while the $COP_R$ of a Carnot refrigerator operating between the same temperatures is 6.74.  ∎

## Further Reading

1. Y. A. Cengel, M. A. Boles (**2015**) *Thermodynamics – An Engineering Approach*, McGraw Hill, New York.
2. M. J. Moran, H. N. Shapiro, D. D. Boettner, M. B. Bailey (**2014**) *Fundamentals of Engineering Thermodynamics*, John Wiley & Sons, Ltd, London.
3. C. Borgnakke, R. E. Sonntag (**2012**) *Fundamentals of Thermodynamics*, John Wiley & Sons, Ltd, London.

## Summary

In an ideal Rankine cycle the pump work per unit mass of fluid is

$$w_p = \frac{\dot{W}_P}{\dot{m}} = h_2 - h_1.$$

In the boiler the heat input per unit mass of fluid is

$$q_H = \frac{\dot{Q}_H}{\dot{m}} = h_3 - h_2.$$

In the turbine the work done per unit mass of fluid is

$$w_t = \frac{\dot{W}_t}{\dot{m}} = h_3 - h_4.$$

In the condenser the heat removed per unit mass of fluid:

$$q_C = \frac{\dot{Q}_C}{\dot{m}} = h_4 - h_1.$$

The back work ratio is the magnitude of work input to the pump divided by the work output of the turbine:

$$bwr = \frac{w_p}{w_t} = \frac{(h_2 - h_1)}{(h_3 - h_4)}.$$

The thermal efficiency of an ideal Rankine cycle is given by

$$\eta_{th,\text{Rankine}} = \frac{w_{\text{net}}}{q_H} = \frac{w_t - w_p}{q_H} = 1 - \frac{(h_4 - h_1)}{(h_3 - h_2)}.$$

In a Rankine cycle with reheat the heat added in the boiler per unit mass of fluid is

$$q_H = \frac{\dot{Q}_H}{\dot{m}} = (h_3 - h_2) + (h_5 - h_4).$$

The total work output from the turbines per unit mass of fluid is

$$w_t = \frac{\dot{W}_t}{\dot{m}} = (h_3 - h_4) + (h_5 - h_6).$$

In a Rankine cycle with regeneration using an open feedwater heater, the mass flow of steam diverted from the turbine to the feedwater heater is

$$f = \frac{h_6 - h_2}{h_5 - h_2}.$$

The total work output of the turbine per unit mass of fluid is

$$w_t = \frac{\dot{W}_t}{\dot{m}_3} = (h_3 - h_5) + (1 - f)(h_5 - h_4).$$

The total work required for driving both pumps per unit mass of fluid is

$$w_p = \frac{\dot{W}_p}{\dot{m}_3} = (h_7 - h_6) + (1 - f)(h_2 - h_1).$$

The heat added in the boiler is

$$q_H = \frac{\dot{Q}_H}{\dot{m}_3} = h_3 - h_7,$$

and the heat lost in the condenser is

$$q_c = (1-f)(h_4 - h_1).$$

In a refrigerator operating on a reverse Rankine cycle the rate at which heat is removed by the evaporator per unit mass of fluid is

$$q_C = \frac{\dot{Q}_C}{\dot{m}} = h_2 - h_1.$$

The rate at which heat is lost from the condenser per unit mass of fluid is

$$q_H = \frac{\dot{Q}_H}{\dot{m}} = h_3 - h_4.$$

The work supplied to the compressor per unit mass of fluid is

$$w_c = \frac{\dot{W}_c}{\dot{m}} = h_3 - h_2.$$

The coefficient of performance of a Rankine-cycle refrigerator is

$$COP_R = \frac{\dot{Q}_C}{\dot{W}_c} = \frac{h_2 - h_1}{h_3 - h_2}.$$

The coefficient of performance of a Rankine cycle heat pump

$$COP_{HP} = \frac{\dot{Q}_H}{\dot{W}_c} = \frac{h_3 - h_4}{h_3 - h_2}.$$

## Problems

### Ideal Rankine Cycle

9.1   Steam enters the turbine in an ideal Rankine cycle at 8 MPa and leaves at 10 kPa. What is the work done by the turbine per kilogram of steam?

9.2   An ideal Rankine cycle has a condenser operating at a pressure of 50 kPa. If the steam quality at the outlet of the turbine is required to be 80%, what should the boiler pressure be?

9.3   A Rankine cycle uses water as the working fluid. The boiler pressure is 7 MPa and the condenser pressure is 25 kPa. If the isentropic efficiency of the turbine is 90% what is the quality of the steam at the outlet of the turbine?

9.4   A Rankine cycle uses water as the working fluid with a boiler pressure of 6 MPa and a condenser pressure of 40 kPa. If the steam quality at the outlet of the turbine is required to be at least 90%, what should the temperature of steam exiting the boiler be?

9.5    Two thermal reservoirs are available, a high temperature heat source at 350 °C and a low temperature heat sink at 20 °C. Calculate the maximum efficiency of (a) a Carnot cycle and (b) an ideal Rankine cycle using water as a working fluid operating between these two reservoirs.

### *Rankine Cycle with Superheat*

9.6    Steam enters the turbine of a Rankine cycle at a pressure of 10 MPa and temperature of 600 °C with a mass flow rate of 20 kg / s and leaves as saturated vapour. What is the condenser pressure? Calculate the power output of the turbine.

9.7    In a Rankine cycle with superheat, steam enters the turbine at a pressure of 7 MPa and temperature of 500 °C while the condenser has a pressure of 20 kPa. Find the back work ratio for the cycle and the cycle efficiency. If we neglect the work supplied to the pump in the calculation of cycle efficiency, how large an error does it produce?

9.8    Waste heat from the exhaust of an engine is used to heat 0.05 kg / s of refrigerant 134a that is the working fluid of a Rankine cycle. Superheated refrigerant 134a vapour enters the turbine at a pressure of 1.2 MPa and temperature of 80 °C and expands to a condenser pressure of 0.6 MPa. What is the temperature of the refrigerant at the condenser inlet? Find the power output of the cycle.

9.9    A boiler in a Rankine cycle supplies superheated steam at a temperature of 600 °C to a turbine that exhausts a saturated liquid–vapour mixture with 91% quality to a condenser at a pressure of 25 kPa. Calculate the boiler pressure and the cycle efficiency.

9.10   A Rankine cycle has a boiler pressure of 6 MPa while the water that is the working fluid has a maximum temperature of 500 °C and a minimum temperature of 60 °C. If the measured cycle efficiency if 34%, what is the isentropic efficiency of the turbine? Assume that the pump is isentropic.

9.11   A Rankine cycle has a boiler pressure of 3.5 MPa. The highest temperature in the cycle is 500 °C and the lowest temperature is 50 °C. Find the Rankine cycle efficiency and that of a Carnot cycle operating between the same two temperatures.

9.12   In Problem 9.11, what is the mass flow rate of steam required to obtain a 10 MW power output from the turbine if its isentropic efficiency is 80%?

9.13   An ideal Rankine cycle supplies steam to the turbine at a pressure of 8 MPa and a temperature of 500 °C. Steam leaving the turbine is condensed at a pressure of 20 kPa using cooling water taken from a lake at 15 °C. The temperature of the cooling water returned to the lake from the condenser cannot exceed 23 °C. Find the mass flow rate of cooling water required if the cycle has a net power output of 80 MW.

9.14   If the isentropic efficiency of the turbine in Problem 9.13 is 85%, what is the new mass flow rate of cooling water needed?

9.15  A power plant operating on a Rankine cycle has a power output of 50 MW. It has a natural gas-fired boiler that supplies steam at a pressure of 7 MPa and temperature of 500 °C and a condenser that operates at a pressure of 25 kPa. If the boiler transfers 80% of the heat obtained from combustion of natural gas to the steam, how much natural gas (in units of m³ / h) does the plant require? Assume that the gas produce 37 MJ / m³ of heat when it is burned.

### Rankine Cycle with Reheat

9.16  A Rankine cycle with reheat has steam entering the turbine at 8 MPa and 500 °C, which is expanded to a pressure of 1 MPa in the high-pressure stage of the turbine. The steam is then reheated at constant pressure before being expanded in the low-pressure turbine and exhausted to a condenser at a pressure of 20 kPa. Find the reheat temperature such that the steam supplied to the condenser is saturated vapour. What is the work output of the turbine per kilogram of steam passing through it?

9.17  Steam enters the high-pressure stage of a turbine in a Rankine cycle at a pressure of 4 MPa and temperature of 450 °C and expands to a pressure of 300 kPa. The steam is then reheated at constant pressure to 450 °C before entering the low-pressure turbine and expanding to 10 kPa. Calculate the efficiency of the cycle.

9.18  If the reheat pressure in Problem 9.17 is lowered to 200 kPa, how does that affect the work output from the turbines per kilogram of steam and the cycle efficiency?

9.19  Find the work output of the turbine per kilogram of steam in Problem 9.17 if both stages of the turbine have an isentropic efficiency of 90%. What is the temperature of the steam leaving the turbine?

### Rankine Cycle with Regeneration

9.20  In a Rankine cycle with regeneration saturated water from a condenser at a pressure of 20 kPa is pumped at a rate of 15 kg / s into an open feedwater heater operating at a pressure of 500 kPa. The water is heated by mixing it with steam extracted from the turbine at a pressure of 500 kPa and temperature of 350 °C. What is the mass flow rate of steam required if saturated water leaves the feedwater heater?

9.21  A Rankine cycle with regeneration has a boiler pressure of 8 MPa and a condenser pressure of 20 kPa. Steam from the boiler enters the high-pressure turbine at a temper-ature of 450 °C and expands to a pressure of 1 MPa when some of it is extracted and diverted to the feedwater heater. Find the mass fraction of steam extracted if saturated liquid leaves the feedwater heater.

9.22  Find the total work required per unit mass of fluid to drive both pumps in the Rankine cycle described in Problem 9.21.

9.23  A Rankine cycle with regeneration has a boiler pressure of 15 MPa and a condenser pressure of 10 kPa. The open feedwater heater operates at a pressure of 1 MPa and the maximum temperature of the cycle is 600 °C. How much heat is extracted in the condenser per kilogram of steam that flows through it?

9.24 A power plant using the Rankine cycle described in Problem 9.23 is to generate 20 MW of power using heat from a coal-fired furnace. Assuming that the boiler transfers 80% of the energy generated by coal combustion to the steam, how much coal is required in a day? Assume that the coal produces 32 MJ of heat for each kilogram burned.

9.25 Steam enters the high-pressure stage of a turbine in a Rankine cycle with regeneration at a pressure of 6 MPa and temperature of 400 °C. The pressure in the condenser is 10 kPa while that in the feedwater heater is 600 kPa. What percentage of the total work output of the turbine is produced in the high-pressure stage?

9.26 A power plant working on a Rankine cycle with regeneration takes 10 kg / s of sub-cooled water from the condenser and pumps it through a closed feedwater heater to the boiler that is at a pressure of 15 MPa. The water enters at a temperature of 100 °C and leaves at 180 °C. The water is heated by diverting superheated steam at 3.5 MPa and 300 °C from the turbine through the feedwater heater, where it cools and condenses, leaving as saturated liquid. What is the mass flow rate of steam?

9.27 In a Rankine cycle with regeneration superheated steam leaves the boiler at 6 MPa and 400 °C and expands in the first stage of the turbine to a pressure of 1 MPa, after which part of it is diverted to a closed feedwater heater. The remaining steam expands in the low-pressure turbine and is exhausted to the condenser at a pressure of 20 kPa. The saturated liquid leaving the condenser is pumped back to the boiler, first passing through the feedwater heater which it leaves at a temperature of 160 °C and pressure of 6 MPa. The steam condensing in the feedwater heater leaves as saturated liquid that flows back to the condenser through a steam trap. What is the mass fraction of steam extracted in the turbine?

9.28 A Rankine cycle has a boiler pressure of 10 MPa and a condenser pressure of 40 kPa. Steam enters the first stage of the turbine with a mass flow rate of 20 kg / s and expands to a pressure of 800 kPa, at which state it is saturated vapour. Part of the steam is diverted from the turbine to the closed feedwater heater where it is used to heat the saturated liquid leaving the condenser, while the remainder of the steam expands in the second stage of the turbine and is exhausted into the condenser. The saturated liquid leaving the condenser is pumped back to the boiler, first passing through the feedwater heater which it leaves at a temperature of 160 °C and pressure of 10 MPa. The steam condensing in the feedwater heater leaves as saturated liquid that flows back to the condenser through a steam trap. What is the mass flow rate of water diverted from the turbine to the feedwater heater?

9.29 What are the power output of the turbine and the thermal efficiency of the Rankine cycle of Problem 9.28?

9.30 The Rankine cycle plant in Problem 9.28 is fuelled using coal that gives 32 MJ / kg of heat when it is burned. Assuming 85% of the heat produced by coal combustion is transferred to steam in the boiler, how much coal is required each day? If the condenser is cooled with water taken from a lake at a temperature of 18 °C and it is returned at 22 °C, what is the flow rate of cooling water?

## Vapour Refrigeration Cycle

9.31  In an ideal refrigeration cycle saturated refrigerant 134a vapour enters the compressor at –4 °C and leaves at 30 °C. Find the condenser pressure and the coefficient of performance of the refrigerator. What is the coefficient of performance of a Carnot refrigerator operating between the same temperature limits?

9.32  What is the coefficient of performance of the refrigerator in Problem 9.31 if the isentropic efficiency of the compressor is 85%?

9.33  A refrigerator uses refrigerant 134a as the working fluid with a mass flow rate of 0.12 kg / s. Saturated vapour enters the compressor with a pressure of 0.24 MPa and superheated vapour leaves at a pressure of 0.8 MPa. Find the temperature of the vapour at the exit of the compressor. What is the power required to drive the compressor and what is the cooling capacity of the refrigerator?

9.34  An ideal vapour compression refrigerator cycle is used to provide four tons of refrigeration capacity. The compressor takes saturated refrigerant 134a at a pressure of 0.2 MPa and saturated liquid leaves the condenser at a pressure of 1 MPa. How much power is required to drive the compressor?

9.35  A refrigerator using refrigerant 134a as a working fluid has superheated vapour at a pressure of 0.24 MPa and a temperature of 0 °C entering the compressor. Saturated liquid at 0.9 MPa leaves the condenser. Find the coefficient of performance.

9.36  A car air conditioning system takes air at 35 °C from the surrounding atmosphere and cools it down to 15 °C in a heat exchanger from which heat is removed by an ideal refrigeration cycle using refrigerant 134a. The condenser pressure is 1.2 MPa and the evaporator pressure is 0.2 MPa. If the car engine can supply 1.2 kW of power to run the air conditioner what is the maximum mass flow rate of cold air that can be supplied?

9.37  A freezer with an internal temperature of –12 °C is kept in a grocery store where the surrounding air is at 20 °C. The rate of heat transfer due to conduction through the walls of the freezer is estimated to be 8 kW. What is the minimum mass flow rate of refrigerant 134a required if an ideal vapour refrigeration cycle is used in the freezer? How much power is required to drive the compressor?

9.38  To heat a house it is proposed to take atmospheric air at a temperature of –4 °C and pass it through a heat exchanger that raises its temperature to 30 °C. The air can be heated in one of two ways: by using an electrical resistance heater, or by a heat pump that uses refrigerant 134a in an ideal Rankine cycle operating between the same two temperatures. Calculate the work required per kilogram of hot air supplied for each of these methods.

9.39   A heat pump using refrigerant 134a has to supply 10 kW of heat to a high temperature space at 20 °C using a low temperature heat source at –12 °C. Assuming that it works on an ideal Rankine refrigeration cycle find the coefficient of performance of the heat pump and the power required to drive the compressor.

9.40   In a heat pump using refrigerant 134a saturated vapour enters the compressor at a pressure of 0.2 MPa and leaves at a pressure of 1.2 MPa and a temperature of 60 °C. What is the coefficient of performance of the heat pump and the isentropic efficiency of the compressor?

# 10

# Gas Power Cycles

---

**In this chapter you will:**

- Learn how internal combustion engines work.
- Analyse the Otto and Diesel cycles, which are used by reciprocating engines.
- Review the operating principles of gas turbines.
- Study the Brayton cycle, which can be used to evaluate the performance of gas turbines.

---

## 10.1 Internal Combustion Engines

One of the most important applications of heat engines is in transportation, where they are used to power cars, boats and aircraft. Steam engines are not well suited for this purpose because they are large and heavy, requiring boilers, condensers, pumps and other ancillary equipment. Though a vapour cycle is appropriate for use in power plants where size and weight are not major concerns, it is not very practical for use on board a vehicle where it is essential to minimise weight.

Automobiles and aeroplanes use *internal combustion engines*, which can be made light and compact. Internal combustion engines reduce size and weight in a number of ways:

1. Instead of carrying a working fluid that circulates in the engine, air is taken from the atmosphere, heated, expanded and then expelled.
2. There is no separated combustor; instead, fuel is directly mixed with the air that serves as the working fluid and ignited, generating heat.
3. There is no condenser or cooler in which heat is rejected. The hot mixture of air and burned fuel is exhausted to the surrounding air after it has finished expanding.

---

*Energy, Entropy and Engines: An Introduction to Thermodynamics*, First Edition. Sanjeev Chandra.
© 2016 John Wiley & Sons, Ltd. Published 2016 by John Wiley & Sons, Ltd.
Companion website: www.wiley.com/go/chandraSol16

Internal combustion engines can be one of two different types: *reciprocating engines*, in which expansion and compression of air and fuel is carried out in a cylinder fitted with a piston, or *gas turbines*, which use rotating compressors and turbines. In this chapter we will study both types of engines.

## 10.2   Otto Cycle

*Spark ignition engines* are among the most popular types of heat engines, used in every type of vehicle from automobiles to lawnmowers. They are compact and extremely robust, operating reliably in all weather conditions and temperatures. The basic elements of a spark-ignition engine are shown in Figure 10.1. The engine consists of an *engine cylinder* in which a sliding *piston* is fitted. The piston is connected to a *connecting rod* that moves up and down with the piston and makes the *crankshaft* rotate, converting the reciprocating motion of the piston into rotation of the crankshaft. The crankshaft is connected to a flywheel that has a large inertia and keeps the engine rotating at a constant angular velocity so that it can be used to drive the wheels of a vehicle or the propeller of a ship or aircraft. The cylinder has two valves that open alternately, the *intake valve* through which air and fuel enters the cylinder and the *exhaust valve* through which burned gases are expelled.

The position of the piston at which the enclosed volume of gas in the cylinder is minimized is known as the *top dead centre (TDC)* while the position at which the gas volume is a maximum is known as the *bottom dead centre (BDC)*. The distance travelled by the piston as it travels from the TDC to the BDC position is known as the *stroke* of the engine and the internal diameter of the cylinder, which equals the diameter of the piston, is known as the *bore*. When the piston is at the TDC position the volume of gas in the cylinder is known as the *clearance volume* ($V_{min}$) and when the piston is at BDC the volume of gas is maximum ($V_{max}$).

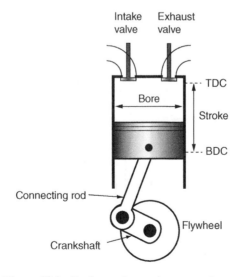

**Figure 10.1**   Reciprocating engine nomenclature.

The difference in gas volume as the piston goes from the BDC to the TDC position is known as the *displacement volume* ($\Delta V = V_{max} - V_{min}$). The ratio of the maximum to the minimum volume is known as the *compression ratio*.

$$r = \frac{V_{max}}{V_{min}}.$$                                                            (10.1)

Figure 10.2 shows the four processes that take place during operation of a spark-ignition engine, which is the reason that it is also known as a *four-stroke engine*. In the *intake* stroke the piston moves from TDC to BDC, drawing in atmospheric air through the open intake valve. Fuel is vaporised and mixed with the air before it enters the engine. The valve then shuts and the piston moves from BDC to TDC in the *compression* stroke, raising the temperature and pressure of the fuel–air mixture. Just before the piston reaches TDC a spark plug is triggered, igniting the fuel. This produces a rapid increase in gas pressure that drives the piston downwards in the *power* or *expansion* stroke, during which the engine does useful work. Finally, the exhaust valve is opened and the piston again moves from BDC to TDC in the *exhaust* stroke, expelling all the combustion products. The engine repeats this cycle, and even though power is generated in only one of four strokes the inertia of the flywheel is sufficient to keep the engine rotating for the remainder of the cycle. Each stroke of the piston corresponds to a half-turn of the crankshaft, so a four-stroke cycle turns the crankshaft through two rotations.

Figure 10.3 shows the pressure variation in the engine cylinder with change in gas volume during the engine cycle. During intake the pressure is almost constant, slightly below atmospheric. The pressure increases during the compression stroke as the piston forces the air–fuel mixture into a smaller volume. Just before the piston reaches TDC ignition occurs, producing a very rapid increase in pressure. The combustion process last for a very brief period, during which the piston moves a very small distance, and once it is over the pressure in the cylinder starts to decrease during the expansion stroke as the piston moves towards BDC and the volume of gas in the cylinder increases. When the exhaust valve is opened the

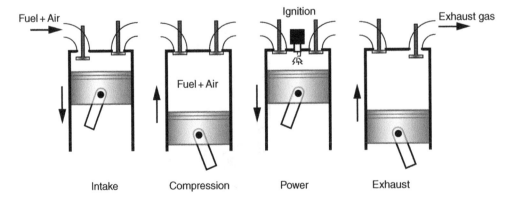

**Figure 10.2**   Four-stroke cycle for a spark-ignition engine.

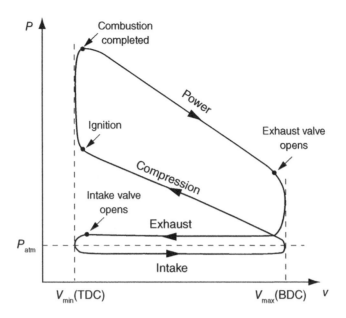

**Figure 10.3** Variation of pressure with cylinder volume during a spark-ignition engine cycle.

pressure in the cylinder drops suddenly and remains slightly above that of the atmosphere during the exhaust stroke.

To model this engine cycle and calculate the work output from it we need to make several simplifying assumptions. The net work done during the exhaust and intake strokes is very small, so we can neglect that portion of the cycle entirely. We make the *air standard assumptions*, that:

1. The working fluid is air, which behaves as an ideal gas.
2. All processes in the cycle are reversible.
3. Fuel ignition can be approximated as heat addition from an external source.
4. Exhausting combustion products and taking in fresh air is equivalent to heat loss to the surroundings.

Sometimes it is also convenient to make the *cold air standard assumption*, in which the specific heats of air are assumed constant, evaluated at the intake temperature. This is not a very realistic assumption since the temperature change of air during combustion can be very large, but it allows us to do a qualitative analysis of engine cycles and understand how changes in engine parameters affect its performance.

With these idealisations we can treat the engine as a control mass to which heat is alternately added and removed and which does net work on the surroundings. Assuming that both the expansion and compression strokes are reversible and adiabatic, we can model them as being isentropic processes. Heat addition (due to ignition) and heat rejection (due to the exhaust valve opening) are very rapid processes, during which the piston moves through a very small distance and we can quite reasonably suppose that both are constant volume processes. With these assumptions, a spark-ignition engine can be approximated by the *Otto cycle*, which is shown in Figure 10.4 on both *P-v* (Figure 10.4a) and *T-s* (Figure 10.4b) diagrams.

The Otto cycle consists of four processes:

1→2    Isentropic compression
2→3    Constant volume heat addition
3→4    Isentropic expansion
4→1    Constant volume heat rejection.

There is no work done by the engine during heat addition or rejection since both are constant volume processes. The heat added to the engine per unit mass of air is, assuming constant specific heat,

$$q_H = u_3 - u_2 = c_v \left( T_3 - T_2 \right), \tag{10.2}$$

and the heat rejected is

$$q_C = u_4 - u_1 = c_v \left( T_4 - T_1 \right). \tag{10.3}$$

The thermal efficiency of an Otto cycle,

$$\eta_{th,\text{Otto}} = \frac{w_{\text{net}}}{q_H} = \frac{q_H - q_C}{q_H} = 1 - \frac{q_C}{q_H}. \tag{10.4}$$

Substituting Equations (10.2) and (10.3) into Equation (10.4) gives

$$\eta_{th,\text{Otto}} = 1 - \frac{T_4 - T_1}{T_3 - T_2} = 1 - \frac{T_1 \left( T_4 / T_1 - 1 \right)}{T_2 \left( T_3 / T_2 - 1 \right)}. \tag{10.5}$$

The processes 2→3 and 4→1 are both constant volume, so $V_1 = V_4 = V_{\text{max}}$ and $V_2 = V_3 = V_{\text{min}}$. Since processes 1→2 and 3→4 are both isentropic, we can write

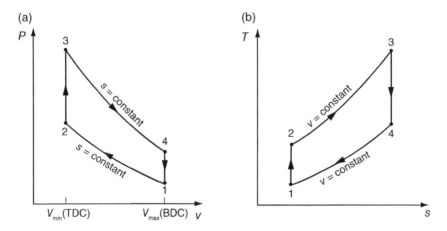

**Figure 10.4**    The Otto cycle on (a) a *P-v* diagram and (b) a *T-s* diagram.

$$\frac{T_1}{T_2} = \left(\frac{V_2}{V_1}\right)^{\gamma-1} = \left(\frac{V_3}{V_4}\right)^{\gamma-1} = \frac{T_4}{T_3},$$

(10.6)

where the specific heat ratio $\gamma = c_p / c_v$. Rearranging Equation (10.6) gives

$$\frac{T_4}{T_1} = \frac{T_3}{T_2}.$$

(10.7)

Substituting Equation (10.7) into Equation (10.5),

$$\eta_{th,Otto} = 1 - \frac{T_1}{T_2} = 1 - \left(\frac{V_2}{V_1}\right)^{\gamma-1} = 1 - \left(\frac{V_{min}}{V_{max}}\right)^{\gamma-1}.$$

(10.8)

Combining with Equation (10.1),

$$\eta_{th,Otto} = 1 - \frac{1}{r^{\gamma-1}}.$$

(10.9)

Figure 10.5 shows the variation of Otto engine efficiency ($\eta_{th,Otto}$) with compression ratio ($r$), assuming that the specific heat ratio for air $\gamma = 1.4$. The efficiency increases continuously with $r$, suggesting that the compression ratio should be made as high as possible. In practice there is a limit to how high the engine compression ratio can be in a spark-ignition engine because the more the gas is compressed the higher its temperature ($T_2$) becomes at the end of the compression stroke. If this exceeds the autoignition temperature of the fuel, above which it ignites spontaneously, the premixed air–fuel mixture in the cylinder may ignite simultaneously in several places even before the spark plug is triggered. Rapid, uncontrolled combustion starting at several locations in the engine cylinder creates shock waves that produce a loud noise

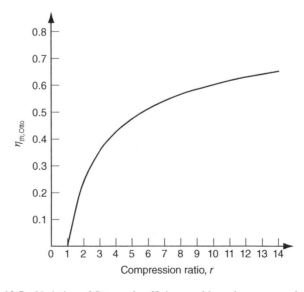

**Figure 10.5**   Variation of Otto cycle efficiency with engine compression ratio.

known as *engine knock* and can damage the piston. Typical spark-ignition engines have compression ratios varying from 8 to 10, which is above the steepest portion of the curve in Figure 10.5 but low enough to prevent knocking.

**Example 10.1**

**Problem:** An engine operating on a cold air standard Otto cycle takes air at 100 kPa and 25 °C and compresses it isentropically to 2.2 MPa. The work output from the cycle is 200 kJ / kg of air. Find the efficiency of the cycle, and the maximum temperature reached in the cycle.

**Find:** Efficiency $\eta_{th,Otto}$ of the Otto cycle, maximum temperature $T_3$ reached.

**Known:** Cold air standard Otto cycle, intake air pressure $P_1 = 100$ kPa, intake temperature $T_1 = 25$ °C, pressure after compression $P_2 = 2.2$ MPa, work output $w_{net} = 200$ kJ / kg.

**Diagram:** Figure E10.1.

**Assumptions:** Constant specific heats evaluated at $T_1 = 25$ °C, air behaves as an ideal gas.

**Governing Equations:**

Otto cycle efficiency (isentropic, ideal gas)
$$\eta_{th,Otto} = \frac{w_{net}}{q_H} = 1 - \frac{1}{r^{\gamma-1}}$$

Heat added in cycle
$$q_H = c_v\left(T_3 - T_2\right)$$

**Properties:** Specific heat ratio of air $\gamma = 1.400$ kJ / kgK (Appendix 1), specific heat of air at constant volume $c_v = 0.717$ kJ / kg (Appendix 1).

**Solution:**

The compression ratio of the cycle can be found, since the process uses an ideal gas and is isentropic:

$$r = \frac{V_1}{V_2} = \left(\frac{P_2}{P_1}\right)^{1/\gamma} = \left(\frac{2.2\,\text{MPa}}{0.1\,\text{MPa}}\right)^{1/1.400} = 9.096\ 48.$$

The cycle efficiency can then be found,

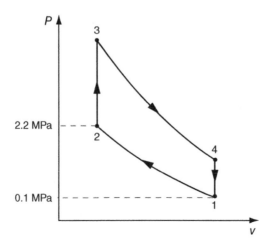

**Figure E10.1**   *P-v* diagram for an Otto cycle.

$$\eta_{th,\text{Otto}} = 1 - \frac{1}{r^{\gamma-1}} = 1 - \frac{1}{9.096\ 48^{0.400}} = 0.586\ 524.$$

The air temperature at the end of compression is

$$T_2 = T_1 \left(\frac{P_2}{P_1}\right)^{(\gamma-1)/\gamma} = 298.15\,\text{K} \times \left(\frac{2.2\,\text{MPa}}{0.1\,\text{MPa}}\right)^{0.400/1.400} = 721.081\,\text{K}.$$

The heat added to the cycle, per unit mass of air is

$$q_H = \frac{w_{\text{net}}}{\eta_{th,\text{Otto}}} = \frac{200\,\text{kJ}/\text{kg}}{0.586\ 524} = 340.992\ \text{kJ}/\text{kg}.$$

Then the maximum temperature in the cycle is

$$T_3 = T_2 + \frac{q_H}{c_v} = 721.081\,\text{K} + \frac{340.992\ \text{kJ}/\text{kg}}{0.717\,\text{kJ}/\text{kgK}} = 1196.66\,\text{K}.$$

**Answer:** The efficiency of the cycle is 58.7% and the maximum air temperature in the cycle is 1197 K. ∎

## 10.3 Diesel Cycle

The compression ratio of spark-ignition engines is limited by the onset of engine knock, which restricts their efficiency. It is possible, however, to circumvent this problem. Engine knock occurs because the cylinder contains a combustible fuel–air mixture that ignites during the compression stroke when the gas temperature is increased. In *compression ignition engines* only air enters the cylinder during the intake stroke, with no fuel mixed with it. Instead, just before the piston approaches TDC at the end of the compression stroke, liquid fuel is sprayed into the high temperature gas in the cylinder. Small fuel droplets evaporate rapidly and the fuel vapour mixes with the hot air and spontaneously ignites, without the need for a spark plug. The rate at which the fuel burns is controlled by the rate at which fuel droplets evaporate, and there is no possibility of engine knock. Compression ignition engines, also known as *diesel engines*, have high compression ratios, typically between 14 and 22.

Combustion in a diesel engine takes place at a much slower rate than it does in a spark-ignition engine since its speed is regulated by the rate at which fuel droplets evaporate and the vapour mixes with air. While the combustion is heating the gas and increasing its pressure the piston has begun the expansion stroke and is expanding the gas. The net result is that the pressure in the cylinder remains approximately constant during fuel combustion. We can model a compression ignition engine using a Diesel cycle, which is shown in Figure 10.6 on both *P-v* and *T-s* diagrams. The only difference between the Diesel and Otto cycles is that heat addition (process 2→3) is at constant pressure in a Diesel cycle while it is at constant volume in an Otto cycle.

The heat added to the engine per unit mass of air in a constant pressure process, assuming constant specific heat is

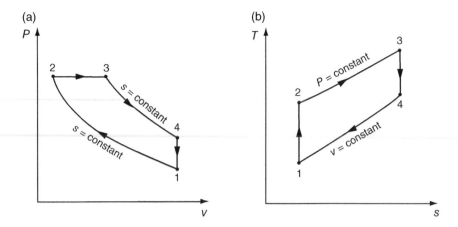

**Figure 10.6**   The Diesel cycle on (a) a *P-v* diagram and (b) on a *T-s* diagram.

$$q_H = h_3 - h_2 = c_p \left( T_3 - T_2 \right).$$
(10.10)

The efficiency of a Diesel cycle is

$$\eta_{th,\text{Diesel}} = 1 - \frac{q_C}{q_H} = 1 - \frac{c_v \left( T_4 - T_1 \right)}{c_p \left( T_3 - T_2 \right)} = 1 - \frac{T_1 \left( T_4 / T_1 - 1 \right)}{\gamma T_2 \left( T_3 / T_2 - 1 \right)}.$$
(10.11)

To simplify this relation, we note that for the isentropic process 1→2,

$$\frac{T_1}{T_2} = \left( \frac{V_2}{V_1} \right)^{\gamma-1} = \frac{1}{r^{\gamma-1}}.$$
(10.12)

Applying the ideal gas equation to the constant pressure process 2→3,

$$\frac{T_3}{T_2} = \frac{P_3 V_3}{P_2 V_2} = \frac{V_3}{V_2},$$
(10.13)

and defining the *cut-off ratio* as the ratio of gas volumes at the end ($V_3$) and start ($V_2$) of heat addition,

$$r_c = \frac{V_3}{V_2} = \frac{T_3}{T_2}.$$
(10.14)

Applying the ideal gas equation to the constant volume process 1→4 and noting that the processes 1→2 and 3→4 are isentropic,

$$\frac{T_4}{T_1} = \frac{P_4 V_4}{P_1 V_1} = \frac{P_4}{P_1} = \frac{P_4 / P_3}{P_1 / P_3} = \frac{P_4 / P_2}{P_1 / P_2} = \frac{\left( V_3 / V_4 \right)^{\gamma}}{\left( V_2 / V_1 \right)^{\gamma}} = \left( \frac{V_3}{V_2} \right)^{\gamma} = r_c^{\gamma}.$$
(10.15)

Substituting Equations (10.12), (10.14) and (10.15) into Equation (10.11) gives

$$\eta_{th,\text{Diesel}} = 1 - \frac{1}{r^{\gamma-1}} \left( \frac{r_c^{\gamma} - 1}{\gamma(r_c - 1)} \right). \tag{10.16}$$

The term in parentheses represents the difference between the efficiencies of Otto and Diesel cycles. This term is always greater than unity, so in principle the Diesel cycle has a lower efficiency than the Otto cycle for the same compression ratio. However $r$ can be much larger in compression ignition engines than it is in spark-ignition engines, and therefore in practice diesel engines are more efficient. In spite of this they have several disadvantages that have prevented them from completely replacing spark-ignition engines. Diesel engines tend to be heavy, since they need to contain the high gas pressures produced by the large compression ratios that they use, making them more suitable for large vehicles such as buses rather than smaller cars. It is also difficult to completely burn all the fuel sprayed into the cylinder. Some small droplets exposed to combustion gases undergo pyrolysis and turn into fine soot particles that are emitted with the engine exhaust. Though diesel engines are more efficient, they require additional equipment such as exhaust filters to control the particulate pollution they create.

**Example 10.2**
**Problem:** An engine operating on a cold air standard Diesel cycle with a compression ratio of 20 and a cut-off ratio of 2 compresses air that is at 100 kPa and 25 °C at the start of the cycle. Find the maximum temperature reached in the cycle and the heat added per kilogram of air.
**Find:** Maximum temperature $T_3$ reached during the cycle, heat added $q$ per kilogram of air.
**Known:** Cold air diesel cycle, compression ratio $r = 20$, cut-off ratio $r_c = 2$, intake air pressure $P_1 = 100$ kPa, intake temperature $T_1 = 25$ °C.
**Diagram:** Figure E10.2.
**Assumptions:** Constant specific heats evaluated at $T_1 = 25$ °C, air behaves as an ideal gas.

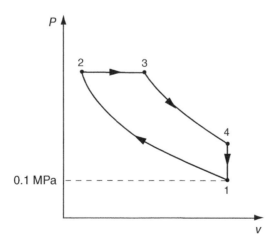

**Figure E10.2**    *P-v* diagram for a Diesel cycle.

**Governing Equations:**

Compression ratio (isentropic, ideal gas)                 $r = \dfrac{V_1}{V_2} = \left(\dfrac{P_2}{P_1}\right)^{1/\gamma}$

Cut-off ratio                                                                 $r_c = \dfrac{V_3}{V_2} = \dfrac{T_3}{T_2}.$

Heat added in cycle                                                     $q_H = c_p\left(T_3 - T_2\right)$

**Properties:** Specific heat ratio of air $\gamma = 1.400$ kJ / kgK (Appendix 1), specific heat of air at constant pressure $c_p = 1.004$ kJ / kg (Appendix 1).

**Solution:**
The pressure after isentropic compression is

$$P_2 = P_1 r^\gamma = 0.1\,\mathrm{MPa} \times (20)^{1.400} = 6.62\ 891\,\mathrm{MPa}.$$

The air temperature after isentropic compression is

$$T_2 = T_1\left(\frac{V_1}{V_2}\right)^{(\gamma-1)/\gamma} = T_1 r^{(\gamma-1)/\gamma} = 298.15\,\mathrm{K} \times (20)^{(0.400)/1.400} = 701.710\,\mathrm{K}.$$

The air temperature at the end of heat addition is

$$T_3 = T_2 r_c = 701.710\,\mathrm{K} \times 2 = 1403.42\,\mathrm{K}.$$

The heat added to the cycle, per unit mass of air, is then

$$q_H = c_p\left(T_3 - T_2\right) = 1.004\,\mathrm{kJ/kgK} \times (1403.42\,\mathrm{K} - 701.710\,\mathrm{K}) = 704.517\,\mathrm{kJ/kg}.$$

**Answer:** The maximum air temperature in the cycle is 1403 K and the heat added is 704.5 kJ / kg. ∎

## 10.4   Gas Turbines

Reciprocating engines are light and extremely versatile, but they also have drawbacks. Fast moving pistons and rotating crankshafts have to be very carefully balanced otherwise they produce vibrations. Pistons that slide in cylinders have to be regularly lubricated and replaced when they wear out. Engines have a large number of moving parts necessary to coordinate the motion of the piston, crankshaft and valves. We could avoid all these problems by making internal combustion engines that do not rely on pistons and cylinders.

*Gas turbines* are heat engines that use rotary compressors and turbines to compress and expand gases without requiring any pistons (see Figure 10.7). Air from the atmosphere is drawn into a diffuser where its pressure increases as it decelerates. The air then enters a rotary compressor where its pressure is increased further until it can be forced into the combustion chamber. Fuel, which can be either gaseous or a liquid spray, is injected into the air and ignited, producing a large increase in its temperature and pressure. Hot combustion gases

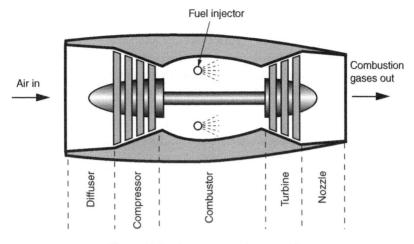

**Figure 10.7**   Components of a gas turbine.

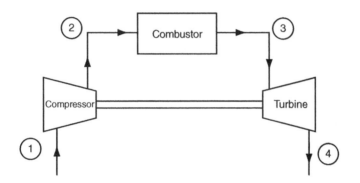

**Figure 10.8**   Schematic diagram of a gas turbine.

enter a turbine where they expand, turning the turbine shaft and producing work. The turbine produces enough power to run the compressor and also do useful work such as turning a ship's propeller or driving an electric generator. It is also possible to make a *turbojet* to propel an aircraft, by expanding the hot gases partially in the turbine, providing just enough power to drive the compressor. Combustion gases that are still at a high temperature and pressure are then accelerated through a nozzle and expelled, providing a large thrust to impel an aircraft forwards. Gas turbines can also be used to rotate a propeller in smaller aircraft, in which case they are known as *turboprops*, or in a hybrid arrangement where they drive a propeller and simultaneously expel some high-pressure combustion gas to provide thrust, which is known as a *turbofan*.

We can simplify the gas turbine for analysis by assuming that it consists of three separate components as shown in Figure 10.8: a compressor driven by the turbine, a combustor in which heat is added to the working fluid and a turbine that does work. Air from the atmosphere flows through these three at steady state so that each component can be modelled as a control volume.

## 10.5  Brayton Cycle

Gas turbines are analysed using the air standard assumptions that we have previously used, in which the working fluid is air, combustion is considered to be a heat addition process, expelling air to the atmosphere is represented by heat rejection from the working fluid, and both the compressor and turbine are assumed to be isentropic. Making these simplifications we can represent a gas turbine by the closed cycle shown in Figure 10.9. It consists of four components: a compressor, a turbine and two heat exchangers, one in which heat is added to the working fluid and one in which heat is rejected. The pressure drop across each heat exchanger is very small, so both heat transfer processes are assumed to be at constant pressure.

This cycle is known as a *Brayton cycle* and is shown in Figure 10.10 on both *P-v* (Figure 10.10a) and *T-s* (Figure 10.10b) diagrams. It consists of four processes:

1→2    Isentropic compression
2→3    Constant pressure heat addition
3→4    Isentropic expansion
4→1    Constant pressure heat rejection.

To find the efficiency and work output of a Brayton cycle we can do a steady-flow energy balance on each of its four components. The heat added at constant pressure in the heat exchanger is

$$q_H = h_3 - h_2 = c_p \left( T_3 - T_2 \right), \qquad (10.17)$$

and the heat rejected from the other heat exchanger is

$$q_C = h_4 - h_1 = c_p \left( T_4 - T_1 \right). \qquad (10.18)$$

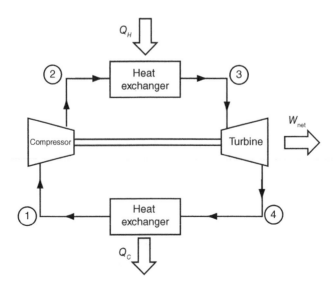

**Figure 10.9**  Gas turbine modelled as a closed cycle.

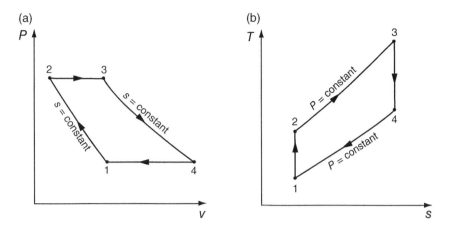

**Figure 10.10**   The Brayton cycle on (a) a $P$-$v$ diagram and (b) a $T$-$s$ diagram.

The thermal efficiency of a Brayton cycle is

$$\eta_{th,\text{Brayton}} = \frac{w_{\text{net}}}{q_H} = \frac{q_H - q_C}{q_H} = 1 - \frac{q_C}{q_H}. \tag{10.19}$$

Substituting Equations (10.17) and (10.18) into Equation (10.19) gives:

$$\eta_{th,\text{Brayton}} = 1 - \frac{T_4 - T_1}{T_3 - T_2} = 1 - \frac{T_1\left(T_4/T_1 - 1\right)}{T_2\left(T_3/T_2 - 1\right)}. \tag{10.20}$$

Processes 1→2 and 3→4 are both isentropic while processes 2→3 and 4→1 are constant pressure, so that $P_2 = P_3$ and $P_1 = P_4$. Therefore,

$$\frac{T_2}{T_1} = \left(\frac{P_2}{P_1}\right)^{(\gamma-1)/\gamma} = \left(\frac{P_3}{P_4}\right)^{(\gamma-1)/\gamma} = \frac{T_3}{T_4}, \tag{10.21}$$

which gives

$$\frac{T_4}{T_1} = \frac{T_3}{T_2}. \tag{10.22}$$

We define the compressor *pressure ratio* as the ratio of the gas pressure at the end and start of compression,

$$r_p = \frac{P_2}{P_1} \tag{10.23}$$

Substituting Equation (10.22) into Equation (10.20) gives

$$\eta_{th,\text{Brayton}} = 1 - \frac{T_1}{T_2} = 1 - \left(\frac{P_1}{P_2}\right)^{(\gamma-1)/\gamma}. \tag{10.24}$$

Combining with Equation (10.23),

$$\eta_{th,Brayton} = 1 - \frac{1}{r_p^{(\gamma-1)/\gamma}}.$$                        (10.25)

The thermal efficiency of a Brayton cycle improves with increasing pressure ratio. The highest temperature of the working fluid is at the exit of the combustion chamber ($T_3$) and this increases with gas pressure. The maximum temperature that the turbine blades can withstand limits the maximum pressure ratio that is possible in a gas turbine. Extensive research has been devoted to developing materials for fabricating turbine blades that are capable of withstanding extreme temperatures.

The work required by the compressor per unit mass of gas flowing through at constant pressure is

$$w_c = h_2 - h_1 = c_p (T_2 - T_1),$$                        (10.26)

and work done by the turbine per unit mass of gas is

$$w_t = h_3 - h_4 = c_p (T_3 - T_4).$$                        (10.27)

The back work ratio for the Brayton cycle is given by

$$bwr = \frac{w_c}{w_t} = \frac{h_2 - h_1}{h_3 - h_4}.$$                        (10.28)

Back work ratios of gas turbines are relatively large as the compressor consumes approximately 40–50% of the work output of the turbine. This is much larger than in a Rankine cycle, where the *bwr* was typically about 1 or 2%, demonstrating that it takes much more work to compress a gas than to pump a liquid.

### Example 10.3

**Problem:** An air standard Brayton cycle with a compressor pressure ratio of 10 takes in air at 100 kPa and 300 K and a mass flow rate of 5 kg / s. The air leaves the combustor at 1260 K. Find the efficiency of the cycle and the net power output. Use air tables to find the properties of air.

**Find:** Efficiency $\eta_{th,Brayton}$ of the cycle, net power output $\dot{W}_{net}$ of the cycle.

**Known:** Air standard Brayton cycle, compressor pressure ratio $r_p = 10$, initial pressure $P_1 = 100$ kPa, initial temperature $T_1 = 300$ K, mass flow rate of air $\dot{m} = 5\,kg/s$, air temperature at combustor exit $T_3 = 1260$ K.

**Diagram:** Figure E10.3.

**Assumptions:** Air is an ideal gas with temperature dependent properties.

**Governing Equation:**

Efficiency of Brayton cycle                    $$\eta_{th,Brayton} = \frac{w_{net}}{q_H} = 1 - \frac{q_C}{q_H} = 1 - \frac{h_4 - h_1}{h_3 - h_2}$$

**Solution:**

From the ideal gas tables (Appendix 7) at $T_1 = 300$ K, specific enthalpy $h_1 = 300.19$ kJ / kg and relative pressure $P_{r1} = 1.3860$.

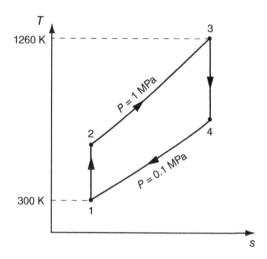

**Figure E10.3** *T-s* diagram for a Brayton cycle.

For the isentropic process 1→2, using the given pressure ratio,

$$P_{r2} = P_{r1}\frac{P_2}{P_1} = 1.3860 \times 10 = 13.860.$$

Interpolating in the ideal gas tables, specific enthalpy $h_2 = 579.8650$ kJ / kg.
From the ideal gas tables (Appendix 7) at $T_3 = 1260$ K, specific enthalpy $h_3 = 1348.55$ kJ / kg, and relative pressure $P_{r3} = 290.8$.
For the isentropic process 3→4, using the inverse of the pressure ratio,

$$P_{r4} = P_{r3}\frac{P_4}{P_3} = \frac{290.8}{10} = 29.080.$$

Interpolating in the ideal gas tables, specific enthalpy $h_4 = 715.1786$ kJ / kg.
The Brayton cycle efficiency is then

$$\eta_{th,\text{Brayton}} = 1 - \frac{h_4 - h_1}{h_3 - h_2} = 1 - \frac{715.1786\,\text{kJ}\,/\,\text{kg} - 300.19\,\text{kJ}\,/\,\text{kg}}{1348.55\,\text{kJ}\,/\,\text{kg} - 579.8650\,\text{kJ}\,/\,\text{kg}} = 0.460\ 131\ 8.$$

The net work output per kilogram of air is

$$w_{\text{net}} = q_H - q_C = (h_3 - h_2) - (h_4 - h_1),$$

$$w_{\text{net}} = (1348.55\,\text{kJ}\,/\,\text{kg} - 579.8650\,\text{kJ}\,/\,\text{kg}) - (715.1786\,\text{kJ}\,/\,\text{kg} - 300.19\,\text{kJ}\,/\,\text{kg})$$
$$= 353.6964\,\text{kJ}\,/\,\text{kg}.$$

Finally, the net power output is

$$\dot{W}_{\text{net}} = \dot{m}w_{\text{net}} = 5\,\text{kg}\,/\,\text{s} \times 353.6964\,\text{kJ}\,/\,\text{kg} = 1768.482\,\text{kW}.$$

**Answer:** The cycle efficiency is 46.0% and the net power output is 1768.5 kW. ∎

## 10.6  Brayton Cycle with Regeneration, Reheat and Intercooling

### 10.6.1  Regeneration

The exhaust gas from a gas turbine is at high temperature, typically hotter than the air that is leaving the compressor. It is therefore possible to preheat air entering the combustor by recovering some of this waste energy, reducing the amount of fuel needed and improving cycle efficiency. Figure 10.11a shows how the exhaust of the turbine is passed through a heat exchanger known as a *regenerator* and heat from the turbine exhaust transferred to air before it enters the combustor, while Figure 10.11b shows the Brayton cycle with regeneration on a *T-s* diagram.

Combustion products at temperature $T_4$ leave the turbine and enter the regenerator where they transfer heat to air leaving the compressor at temperature $T_2$. The air is preheated to temperature $T_5$ by the time it leaves the regenerator and enters the combustor. The heat added in the combustor is

$$q_H = h_3 - h_5 = c_p \left( T_3 - T_5 \right),$$
(10.29)

which is less than it was without the regenerator. Since the work output is not changed the efficiency of the cycle increases. In theory the air can be preheated until $T_5 = T_4$, but in reality it would take a very large heat exchanger to achieve perfect heat transfer. We define *regenerator effectiveness* as the increase in enthalpy of the air divided by the maximum possible enthalpy increase. The effectiveness is given by

$$\varepsilon = \frac{h_5 - h_2}{h_4 - h_2},$$
(10.30)

and assuming constant specific heats,

$$\varepsilon = \frac{T_5 - T_2}{T_4 - T_2}.$$
(10.31)

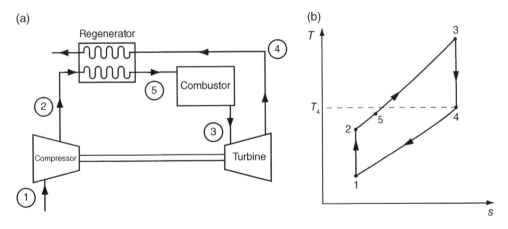

**Figure 10.11**  (a) Gas turbine with a regenerator to preheat air entering the combustor. (b) Brayton cycle with regeneration shown on a *T-s* diagram.

## Example 10.4

**Problem:** A regenerator with an effectiveness of 80% is added to the Brayton cycle of Example 10.3. Find the efficiency of the cycle with the regenerator.

**Find:** The efficiency $\eta_{th,Brayton}$ of the cycle in Example 10.3 with a regenerator.

**Known:** Regenerator effectiveness $\varepsilon = 0.8$, Example 10.3 Brayton cycle with regeneration.

**Diagram:** Figure E10.4.

**Assumptions:** Air is an ideal gas with temperature dependent properties.

**Governing Equation:**

Effectiveness of Brayton cycle
$$\varepsilon = \frac{h_5 - h_2}{h_4 - h_2}$$

Efficiency of Brayton cycle
$$\eta_{th,Brayton} = \frac{w_{net}}{q_H} = 1 - \frac{q_C}{q_H} = 1 - \frac{h_4 - h_1}{h_3 - h_2}$$

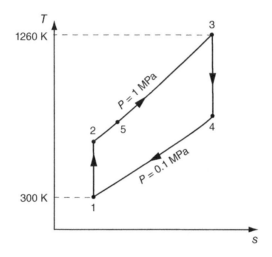

**Figure E10.4**   *T-s* diagram for a Brayton cycle with regeneration.

## Solution:

From the solution to Example 10.3, specific enthalpies $h_1 = 300.19$ kJ / kg, $h_2 = 579.8650$ kJ / kg, $h_3 = 1348.55$ kJ / kg, and $h_4 = 715.1786$ kJ / kg.

   Enthalpy of the air at the exit of the regenerator is

$$h_5 = h_2 + \varepsilon\left(h_4 - h_2\right),$$
$$h_5 = 579.8650 \text{ kJ / kg} + 0.8 \times \left(715.1786 \text{ kJ / kg} - 579.8650 \text{ kJ / kg}\right) = 688.12 \text{ kJ / kg}.$$

The new Brayton cycle efficiency is then

$$\eta_{th,Brayton} = \frac{w_{net}}{q_H} = \frac{w_t - w_c}{q_H} = \frac{\left(h_3 - h_4\right) - \left(h_2 - h_1\right)}{h_3 - h_5},$$

$$\eta_{th,Brayton} = \frac{\left(1348.55 \text{ kJ / kg} - 715.1786 \text{ kJ / kg}\right) - \left(579.8650 \text{ kJ / kg} - 300.19 \text{ kJ / kg}\right)}{1348.55 \text{ kJ / kg} - 688.12 \text{ kJ / kg}} = 0.535 \ 55.$$

**Answer:** The new cycle efficiency is 53.6%.                                           ■

(a)                                                              (b)

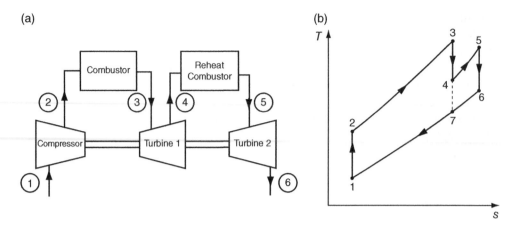

**Figure 10.12** (a) Turbines with reheat combustor. (b) Brayton cycle with reheat shown on a *T-s* diagram.

## 10.6.2 Reheat

The power output of a gas turbine is limited by the maximum temperature the turbine can withstand. If more fuel is added in the combustion chamber, or the pressure ratio of the compressor is increased, the peak temperature in the cycle also increases. One way to solve this problem is to expand the gases in two separate turbines, reheating the gas before it enters the second turbine. Figure 10.12a show a two-stage gas turbine in which the gas leaving the first turbine (stage 4) is directed to a reheat combustor where it is reheated and then expanded in a second turbine. Figure 10.12b shows the corresponding *T-s* diagram for the Brayton cycle. The work done by both turbines per unit mass of gas is

$$w_t = \left(h_3 - h_4\right) + \left(h_5 - h_6\right),\tag{10.32}$$

which is larger than it would have been without reheat. Lines of constant pressure diverge on a *T*-s diagram, so the difference $(h_3 - h_4) + (h_5 - h_6)$ is greater than $h_3 - h_7$, which is the work that would have been done in a single turbine. This increase in power output has been achieved without increasing the maximum temperature in the cycle, which is still $T_3$. The outlet temperature from the second turbine $(T_6)$ is significantly higher than it would have been without reheat $(T_7)$, and this energy can be recovered using regeneration. In practice reheat is usually combined with regeneration.

### Example 10.5
**Problem:** A reheat stage is added to the Brayton cycle of Example 10.3. The air heated in the combustor to 1260 K is expanded in the first turbine to a pressure of 300 kPa, reheated to 1260 K, and then expanded to 100 kPa. Find the power output and efficiency of this cycle.
**Find:** Power output $\dot{W}_{net}$ of new cycle, efficiency $\eta_{th,Brayton}$ of the cycle from Example 10.3 with reheat.
**Known:** Example 10.3 Brayton cycle with reheat, air temperature at combustor exit $T_3 = T_5 = 1260$ K, reheat pressure $P_4 = P_5 = 0.5$ MPa.
**Diagram:** Figure E10.5.

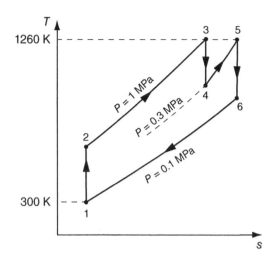

**Figure E10.5** *T-s* diagram for a Brayton cycle with reheat.

**Assumptions:** Air is an ideal gas with temperature dependent properties.
**Governing Equation:**

Efficiency of Brayton cycle

$$\eta_{th,\text{Brayton}} = \frac{w_{net}}{q_H} = \frac{\left(h_3 - h_4\right) + \left(h_5 - h_6\right) - \left(h_2 - h_1\right)}{\left(h_3 - h_2\right) + \left(h_5 - h_4\right)}$$

**Solution:**
From the solution to Example 10.3, specific enthalpies $h_1 = 300.19$ kJ / kg, $h_2 = 579.8650$ kJ / kg and $h_3 = 1348.55$ kJ / kg.

From the ideal gas tables (Appendix 7) at $T_3 = 1260$ K, relative pressure $P_{r3} = 290.8$.
For the isentropic process $3 \rightarrow 4$,

$$P_{r4} = P_{r3} \frac{P_4}{P_3} = 290.8 \times \left(\frac{0.3\,\text{MPa}}{1.0\,\text{MPa}}\right) = 87.24.$$

Interpolating in the ideal gas tables, specific enthalpy $h_4 = 971.5602$ kJ / kg.
From the ideal gas tables (Appendix 7) at $T_5 = 1260$ K, specific enthalpy $h_5 = 1348.55$ kJ / kg, relative pressure $P_{r5} = 290.8$.
For the isentropic process $5 \rightarrow 6$,

$$P_{r6} = P_{r5} \frac{P_6}{P_5} = 290.8 \times \left(\frac{0.1\,\text{MPa}}{0.3\,\text{MPa}}\right) = 96.93.$$

Interpolating in the ideal gas tables, specific enthalpy $h_6 = 1000.3550$ kJ / kg.
The net work output per kilogram of air is

$$w_{net} = \left(h_3 - h_4\right) + \left(h_5 - h_6\right) - \left(h_2 - h_1\right),$$

$$w_{net} = \left(1348.55\,\text{kJ / kg} - 971.5602\,\text{kJ / kg}\right) + \left(1348.55\,\text{kJ / kg} - 1000.3550\,\text{kJ / kg}\right)$$
$$- \left(579.8650\,\text{kJ / kg} - 300.19\,\text{kJ / kg}\right) = 445.5098\,\text{kJ / kg}.$$

Then the net power output is

$$\dot{W}_{net} = \dot{m}w_{net} = 5\,kg\,/\,s \times 445.5098\,kJ\,/\,kg = 2227.549\,kW.$$

The heat added per kilogram of air is

$$q_H = (h_3 - h_2) + (h_5 - h_4),$$

$$q_H = (1348.55\,kJ\,/\,kg - 579.8650\,kJ\,/\,kg) + (1348.55\,kJ\,/\,kg - 1117.917\,kJ\,/\,kg) = 999.3180\,kJ\,/\,kg.$$

The new Brayton cycle efficiency is

$$\eta_{th,\text{Brayton}} = \frac{w_{net}}{q_H} = \frac{445.5098\,kJ\,/\,kg}{999.3180\,kJ\,/\,kg} = 0.4458.$$

**Answer:** The cycle efficiency is 44.6% and the net power output is 2227.5 kW.     ∎

## 10.6.3   *Intercooling*

The work required to compress a gas increases with its specific volume. While a gas is being compressed its temperature rises, which makes it expand and increases the power required by the compressor. It would be more efficient to simultaneously cool the gas as it is being compressed, but this is not very easy to do. A more practical strategy is to partially compress atmospheric air, cool it down in a heat exchanger known as an intercooler, and then use a second compressor to bring it to the desired pressure. Figure 10.13a shows the arrangement of two compressors with an intermediate intercooler and Figure 10.13b shows the process on a *T-s* diagram. Large compressors often use more than one intercooler to further reduce power requirements.

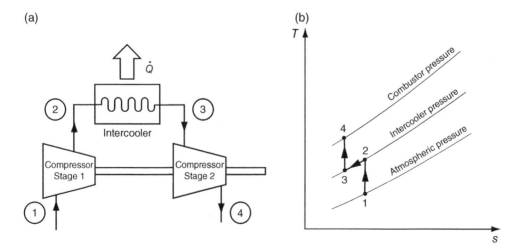

**Figure 10.13**   (a) Two-stage compressor with intercooler. (b) Intercooling shown on a *T-s* diagram.

## Further Reading

1. Y. A. Cengel, M. A. Boles (**2015**) *Thermodynamics – An Engineering Approach*, McGraw Hill, New York.
2. M. J. Moran, H. N. Shapiro, D. D. Boettner, M. B. Bailey (**2014**) *Fundamentals of Engineering Thermodynamics*, John Wiley & Sons, Ltd, London.
3. C. Borgnakke, R. E. Sonntag (**2012**) *Fundamentals of Thermodynamics*, John Wiley & Sons, Ltd, London.
4. J. Heywood (**1988**) *Internal Combustion Engine Fundamentals*, McGraw Hill, New York.

## Summary

The *compression ratio* of a reciprocating engine is the ratio of the maximum to the minimum volume in the cylinder:

$$r = \frac{V_{max}}{V_{min}}.$$

The efficiency of an Otto cycle is

$$\eta_{t,\text{Otto}} = 1 - \frac{1}{r^{\gamma-1}}.$$

The *cut-off ratio* in a diesel engine is the ratio of gas volumes at the end and start of heat addition,

$$r_c = \frac{V_3}{V_2}.$$

The efficiency of a Diesel cycle is

$$\eta_{th,\text{Diesel}} = 1 - \frac{1}{r^{\gamma-1}} \left( \frac{r_c^{\gamma} - 1}{\gamma(r_c - 1)} \right).$$

The compressor *pressure ratio* of a gas turbine is the ratio of the gas pressure at the end and start of compression,

$$r_p = \frac{P_2}{P_1}.$$

The thermal efficiency of a Brayton cycle is

$$\eta_{th,\text{Brayton}} = 1 - \frac{1}{r_p^{(\gamma-1)/\gamma}}.$$

The work required by the compressor in a Brayton cycle per unit mass of gas is

$$w_c = h_2 - h_1 = c_p \left( T_2 - T_1 \right),$$

and work done by the turbine per unit mass of gas is

$$w_t = h_3 - h_4 = c_p \left( T_3 - T_4 \right).$$

The Brayton cycle can be modified with regeneration, reheat, and intercooling to improve cycle efficiency. These techniques can also be combined.

## Problems

*Otto Cycle*

10.1    An engine operating on a cold air standard Otto cycle has a compression ratio of 8.5 and takes in air at a pressure of 100 kPa and temperature of 300 K. The heat added during each cycle is 900 kJ / kg of air in the cylinder. Find the maximum air temperature in the cycle.

10.2    Repeat Problem 10.1 using ideal gas tables to find air properties.

10.3    An engine operating on an Otto cycle has a bore of 90 mm and a stroke of 120 mm. The clearance volume is 12% of the maximum air volume when the piston is at BDC. Making the cold air standard assumption find the efficiency of the engine and the mass of air entering the engine during the intake stroke. The surrounding atmospheric pressure is 100 kPa and temperature is 30 °C.

10.4    An engine working on a cold air standard Otto cycle takes in air at a pressure of 100 kPa and temperature of 295 K. The compression ratio is 9 and the maximum temperature is 2000 K. Find the heat added per kilogram of air during each cycle.

10.5    An air standard Otto cycle has a compression ratio of 8. The minimum air temperature in the cycle is 300 K and the maximum temperature is 1400 K. Calculate the efficiency of the cycle assuming (a) constant specific heats evaluated at 300 K and (b) variable specific heats. Compare these values with that of a Carnot engine operating between the same temperature limits.

10.6    A cold air standard Otto cycle has a compression ratio of 9.5. At the end of isentropic expansion the air temperature is 600 K. The engine rejects 200 kJ / kg of air to the surroundings during each cycle. Determine the maximum temperature of air during the cycle, the work output of the engine per kilogram of air and the cycle efficiency.

10.7    Air in the cylinder of a cold air standard Otto cycle with a compression ratio of 8 has a pressure 100 kPa, temperature of 25 °C and a volume of 300 cm$^3$ at the start of isentropic compression. At the end of isentropic expansion the temperature is 700 K. What is the maximum air temperature and pressure during the cycle? How much heat is added during each cycle?

10.8    An air standard Otto cycle with a compression ratio of 8.5 takes in 1.5 g of air at 22 °C and 100 kPa during the intake stroke. The maximum air temperature during the cycle is 1420 K. Find the mass of fuel added in each cycle assuming that the heat released during combustion is 47 MJ / kg of fuel.

10.9    An engine working on an air standard Otto cycle takes in 500 cm$^3$ of air at a temperature of 300 K and pressure of 100 kPa. The compression ratio is 8 and the engine rotational speed is 3000 RPM. If the heat added in each cycle is 1100 kJ / kg of air what is the power output of the engine?

10.10 A four-cylinder engine operating on a cold air standard Otto cycle takes in air at a temperature of 22 °C and pressure of 100 kPa. Each cylinder has a bore of 100 mm, stroke of 125 mm and a compression ratio of 9. If the maximum air temperature during the cycle is 1200 K, what is the power output of the engine when it is operating at 3200 RPM?

## Diesel Cycle

10.11 The air in a cold air standard Diesel cycle is at a pressure of 95 kPa and a temperature of 300 K at the start of compression. The maximum pressure in the cycle is 6500 kPa and the maximum temperature is 2200 K. Find the compression ratio and the cut-off ratio for the cycle. Use the properties of air at 300 K.

10.12 A cold air standard Diesel cycle has air at a pressure of 90 kPa and a temperature of 300 K at the start of compression. The compression ratio is 18 and the heat added due to combustion is 900 kJ / kg of air. What are the maximum air pressure and temperature in the cycle?

10.13 The temperature in an air standard Diesel cycle is 320 K at the start of isentropic compression. If the maximum temperature in the cycle is 2200 K and the compression ratio is 20, what is the cut-off ratio?

10.14 A cold air standard Diesel cycle compresses air that is initially at a pressure of 90 kPa and temperature of 300 K. The maximum pressure in the cycle is 8 MPa and the maximum temperature is 1900 K. Find the compression ratio, the cut-off ratio and the thermal efficiency of the cycle. Use the properties of air at 300 K.

10.15 Solve Problem 10.14 using the ideal gas tables to find the properties of air.

10.16 In a cold air standard Diesel cycle the compression ratio is 18 and the cut-off ratio is 2.0. Air is at a pressure of 100 kPa, and a temperature of 300 K at the start of compression. Find the maximum temperature and pressure in the cycle, the heat added and the net work done during the cycle.

10.17 An air standard Diesel cycle begins the compression stroke with a 0.01 m$^3$ volume of air at a pressure 120 kPa and a temperature of 310 K. The compression ratio is 18 and the maximum temperature in the cycle is 1620 K. Find the heat added and rejected in each cycle and the thermal efficiency of the cycle.

10.18 A six-cylinder diesel engine has a cylinder bore of 95 mm, a stroke of 110 mm, a compression ratio of 21 and a cut-off ratio of 2. During each cycle the mass of diesel fuel injected into each cylinder is 18 mg. The fuel releases 44.8 MJ / kg of heat when it is burned completely. Assuming that the intake air is at a pressure of 100 kPa and a temperature of 300 K, find the power output of the engine when it is running at 3500 RPM. Use the properties of air at 300 K.

10.19   The engine in Problem 10.18 is "turbocharged", in which air entering the engine during the intake stroke is compressed to a pressure of 200 kPa. Assuming that the inlet air temperature and that mass of fuel injected remain the same, how does turbocharging affect the peak air temperature and pressure during the cycle? What effect does it have on the work output? How can we increase the work output from the turbocharged engine?

10.20   An air standard Diesel cycle compresses air initially at a temperature of 300 K and pressure of 100 kPa. The compression ratio is 20 and the heat added during combustion is 1100 kJ / kg of air. Find the work done by the cycle per kilogram of air and its thermal efficiency.

### Brayton Cycle

10.21   A cold air standard Brayton cycle has a minimum temperature of 320 K, a maximum temperature of 1400 K, and a compressor pressure ratio of 10. Find the back work ratio and the thermal efficiency of the cycle. Assume constant specific heats.

10.22   Solve Problem 10.21 using ideal gas tables to find the properties of air.

10.23   A gas turbine operating on an air standard Brayton cycle has air entering the compressor with a temperature of 300 K, a pressure of 100 kPa and a mass flow rate of 4 kg / s. The compressor pressure ratio is 8 and air enters the turbine with a temperature of 1200 K. Find the thermal efficiency of the cycle and the net power supplied by it. Assume constant gas properties.

10.24   If the isentropic efficiency of the turbine in Problem 10.23 is 85% while all other parameters remain the same, what is the thermal efficiency of the Brayton cycle?

10.25   Air is at a temperature of 300 K and a pressure of 100 kPa at the start of compression in a Brayton cycle. The combustor adds 900 kJ of heat per kilogram of air flowing through it. What is the highest compression ratio possible if the turbine blades can withstand a maximum temperature of 1400 K?

10.26   Show that if gas properties are assumed constant in a cold air standard Brayton cycle the back work ratio $bwr = T_1/T_4$, where $T_1$ is the air temperature at the entrance of the compressor and $T_4$ is the air temperature at the exit of the turbine.

10.27   Air enters the compressor of a gas turbine with a velocity of 20 m/s through a circular opening that is 0.5 m in diameter. The air pressure at the compressor inlet is 90 kPa and the temperature is 280 K. The air temperature at the inlet to the turbine is 1400 K and the pressure is 1 MPa. Assuming that the gas turbine operates on an air standard Brayton cycle find the rate of heat addition, the net power output, and the thermal efficiency of the cycle.

10.28   Heat is added at the rate of 200 kW to an air standard Brayton cycle that has a compressor pressure ratio of 12. Air enters the compressor at a temperature of 320 K and the turbine at a temperature of 1340 K. Find the mass flow rate of air and the power output of the cycle.

10.29    The air in a Brayton cycle is at a temperature of 320 K at the start of compression. The compressor pressure ratio is 8 and 800 kJ of heat are added per kilogram of air flowing through the combustor. If the temperature of the air entering the turbine is 1300 K, what is the isentropic efficiency of the compressor?

10.30    In a gas turbine operating on a Brayton cycle air enters the compressor with a temperature of 300 K and pressure of 100 kPa. Measurements show that at the exit of the compressor the temperature is 640 K and the pressure is 1 MPa while the air temperature at the exit of the turbine is 840 K. The combustor adds 800 kJ of heat for every kilogram of air flowing through it. Find the isentropic efficiencies of the compressor and turbine and the thermal efficiency of the Brayton cycle.

### Brayton Cycle with Regeneration, Reheat and Intercooling

10.31    An air standard Brayton cycle with regeneration has a minimum temperature of 320 K, a maximum temperature of 1400 K, and a regenerator effectiveness of 90%. What is the thermal efficiency of the cycle? Assume constant specific heats.

10.32    Air at a temperature of 285 K enters the compressor of an air standard Brayton cycle with regeneration at a rate of 4 kg/s. The compressor pressure ratio is 16, the maximum air temperature in the cycle is 1600 K, and the regenerator effectiveness is 90%. Find the net power output of the cycle and its thermal efficiency.

10.33    Solve Problem 10.32 assuming that both the turbine and compressor have isentropic efficiencies of 85%.

10.34    Air at a temperature of 300 K enters the compressor of an air standard Brayton cycle. The compressor pressure ratio is 12, the maximum air temperature in the cycle is 1200 K. Calculate the heat supplied per kilogram of air for (a) the ideal Brayton cycle (b) the cycle with a regenerator with 80% effectiveness and assuming the turbine is isentropic (c) the cycle with a regenerator with 80% effectiveness and assuming the turbine isentropic efficiency is 85%.

10.35    In a Brayton cycle with reheat air at pressure $P_1$ and temperature $T_1$ enters the first turbine stage and expands to $P_x$, the pressure of the reheat combustor, where it is reheated at constant pressure to temperature $T_1$. It then enters the second turbine stage and expands to pressure $P_2$ where it is exhausted. Show that to maximise the combined work output of the two turbine stages the reheat combustor pressure should be $P_x = \sqrt{P_1 P_2}$. Assume that the specific heats of air are constant.

10.36    In a gas turbine with reheat air leaves the combustor at a temperature of 1500 K and pressure of 1.2 MPa. It expands in the first turbine stage to the pressure of the reheat combustor, where it is reheated at constant pressure to 1500 K. It then enters the second turbine stage where it expands to 100 kPa and is exhausted. Using the results of Problem 10.35 to determine the reheat combustor pressure, calculate the maximum work output per kilogram of air that can be obtained from the two turbine stages combined. How much heat is added in the reheat combustor per kilogram of air?

10.37   Air in a cold air standard Brayton cycle enters the turbine at a temperature of 1000 K
         and pressure of 1 MPa and expands in a single stage to 100 kPa. Calculate the work
         done per kilogram of air. How much work per kilogram is done if the air is expanded
         in the first part of a two-stage turbine to a pressure of 300 kPa, heated to a temperature
         of 1000 K in a reheat combustor, and then expanded in the second turbine stage to
         100 kPa? Assume air has constant specific heats evaluated at 300 K.

10.38   A cold air standard Brayton cycle with regeneration and reheat has air at a pressure of
         100 kPa and temperature of 300 K at the start of compression. The compressor
         pressure ratio is 8, the reheat combustor pressure is 280 kPa and the regenerator effec-
         tiveness is 80%. The temperature at the entrance to both turbine stages is 1200 K. Find
         the net work done per kilogram of air and the thermal efficiency of the cycle. Assume
         air has constant specific heats evaluated at 300 K.

10.39   Solve Problem 10.38 given that the isentropic efficiencies of the compressor and both
         stages of the turbine are 85%.

10.40   Air enters the compressor of an air standard Brayton cycle at a temperature of 300 K
         and pressure of 100 kPa with a mass flow rate of 4 kg / s. It is compressed to 1 MPa in
         a two-stage compressor, with an intercooler at a pressure of 320 kPa that cools the air
         to 300 K before it enters the second stage. The air enters the first turbine stage at a
         temperature of 1200 K and expands to a pressure of 320 kPa. It is reheated at this
         pressure in a reheat combustor to a temperature of 1200 K after which it enters the
         second turbine stage and expands to 100 kPa. Find the net power output of the cycle.

# Appendices

## Appendix 1: Properties of Gases

| Gas | Chemical formula | Molar mass, kg/kmol | $R$, kJ/kgK | $c_p$, kJ/kgK | $c_v$, kJ/kgK | $\gamma = \dfrac{c_p}{c_v}$ |
|---|---|---|---|---|---|---|
| Air | — | 28.97 | 0.2870 | 1.004 | 0.717 | 1.400 |
| Ammonia | $NH_3$ | 17.03 | 0.4882 | 2.130 | 1.642 | 1.297 |
| Argon | Ar | 39.95 | 0.2081 | 0.520 | 0.312 | 1.667 |
| Butane | $C_4H_{10}$ | 58.12 | 0.1430 | 1.716 | 1.573 | 1.091 |
| Carbon dioxide | $CO_2$ | 44.01 | 0.1889 | 0.842 | 0.653 | 1.289 |
| Carbon monoxide | CO | 28.01 | 0.2968 | 1.041 | 0.744 | 1.400 |
| Ethane | $C_2H_6$ | 30.07 | 0.2765 | 1.766 | 1.490 | 1.186 |
| Ethylene | $C_2H_4$ | 28.05 | 0.2964 | 1.548 | 1.252 | 1.237 |
| Helium | He | 4.00 | 2.0770 | 5.193 | 3.116 | 1.667 |
| Hydrogen | $H_2$ | 2.02 | 4.1242 | 14.209 | 10.085 | 1.409 |
| Methane | $CH_4$ | 16.04 | 0.5184 | 2.254 | 1.736 | 1.299 |
| Neon | Ne | 20.18 | 0.4120 | 1.030 | 0.618 | 1.667 |
| Nitrogen | $N_2$ | 28.01 | 0.2968 | 1.042 | 0.745 | 1.400 |
| Octane | $C_8H_{18}$ | 114.23 | 0.0728 | 1.711 | 1.638 | 1.044 |
| Oxygen | $O_2$ | 32.00 | 0.2598 | 0.922 | 0.662 | 1.393 |
| Propane | $C_3H_8$ | 44.10 | 0.1886 | 1.679 | 1.490 | 1.126 |
| Water (steam) | $H_2O$ | 18.02 | 0.4615 | 1.872 | 1.410 | 1.327 |

Specific heats evaluated at 25 °C, 100 kPa

*Energy, Entropy and Engines: An Introduction to Thermodynamics*, First Edition. Sanjeev Chandra.
© 2016 John Wiley & Sons, Ltd. Published 2016 by John Wiley & Sons, Ltd.
Companion website: www.wiley.com/go/chandraSol16

# Appendix 2: Properties of Solids

## Metals

| Substance | Density, $\rho$ (kg/m³) | Specific heat, $c$ (kJ/kg°C) | Molar mass, $M$ (kg/kmol) |
|---|---|---|---|
| Aluminum | 2700 | 0.902 | 26.98 |
| Copper | 8900 | 0.386 | 63.55 |
| Iron | 7840 | 0.450 | 55.85 |
| Lead | 11310 | 0.128 | 207.20 |
| Magnesium | 1730 | 1.000 | 24.31 |
| Nickel | 8890 | 0.440 | 58.69 |
| Silver | 10470 | 0.235 | 107.87 |
| Steel, mild | 7830 | 0.500 | 55.71 |
| Tungsten | 19400 | 0.130 | 183.85 |

## Non-Metals

| Substance | Density, $\rho$ (kg/m³) | Specific heat, $c$ (kJ/kg°C) | Molar mass, $M$ (kg/kmol) |
|---|---|---|---|
| Asphalt | 2110 | 0.920 | 200 |
| Brick, common | 1922 | 0.790 | 59.49 |
| Concrete | 2300 | 0.653 | 270.1 |
| Clay | 1000 | 0.920 | 258.16 |
| Diamond | 2420 | 0.616 | 12.01 |
| Glass, window | 2700 | 0.800 | 60.08 |
| Graphite | 2500 | 0.711 | 12.01 |
| Granite | 2700 | 1.017 | 62.44 |
| Ice (0°C) | 921 | 2.110 | 18.02 |
| Marble | 2600 | 0.880 | 100.09 |
| Plywood (Douglas fir) | 545 | 1.210 | 162.14 |
| Sand | 1520 | 0.800 | 60.08 |
| Stone | 1500 | 0.800 | 66.42 |

# Appendix 3:  Properties of Liquids

| Substance | Temperature, $T$ (°C) | Density, $\rho$ (kg/m$^3$) | Specific heat, $c$ (kJ/kg°C) | Molar mass, $M$ (kg/kmol) |
|---|---|---|---|---|
| Ammonia | 25 | 602 | 4.8 | 17.03 |
| Argon | −185.6 | 1,394 | 1.14 | 39.95 |
| Benzene | 20 | 879 | 1.72 | 78.11 |
| n-Butane | −0.5 | 601 | 2.31 | 58.12 |
| Carbon dioxide | 0 | 298 | 0.59 | 44.01 |
| Ethanol | 25 | 783 | 2.46 | 46.07 |
| Ethylene glycol | 20 | 1,109 | 2.84 | 62.07 |
| Glycerine | 20 | 1,261 | 2.32 | 92.09 |
| Helium | −268.9 | 146 | 22.8 | 4.00 |
| Hydrogen | −252.8 | 71 | 10 | 2.02 |
| Kerosene | 20 | 820 | 2 | 170.34 |
| Mercury | 25 | 13,560 | 0.139 | 200.59 |
| Methane | −161.5 | 423 | 3.49 | 16.04 |
| Methanol | 25 | 787 | 2.55 | 32.04 |
| Nitrogen | −195.8 | 809 | 2.06 | 28.01 |
| Octane | 20 | 703 | 2.1 | 114.23 |
| Oil (light) | 25 | 910 | 1.8 | 114.23 |
| Oxygen | −183 | 1,141 | 1.71 | 32.00 |
| Petroleum | 20 | 640 | 2 | 95 |
| Propane | −42.1 | 581 | 2.25 | 44.1 |
|  | 0 | 529 | 2.53 | 44.1 |
|  | 50 | 449 | 3.13 | 44.1 |
| Refrigerant-134a | −50 | 1,443 | 1.23 | 102.03 |
|  | −26.1 | 1,374 | 1.27 | 102.03 |
|  | 0 | 1,294 | 1.34 | 102.03 |
|  | 25 | 1,206 | 1.42 | 102.03 |
| Water | 0 | 1,000 | 4.23 | 18.02 |
|  | 25 | 997 | 4.18 | 18.02 |
|  | 50 | 988 | 4.18 | 18.02 |
|  | 75 | 975 | 4.19 | 18.02 |
|  | 100 | 958 | 4.22 | 18.02 |

# Appendix 4: Specific Heats of Gases

| | Air | | | Carbon dioxide, $CO_2$ | | | Carbon monoxide, CO | | |
|---|---|---|---|---|---|---|---|---|---|
| Temp. (K) | $c_p$ (kJ/ kgK) | $c_v$ (kJ/ kgK) | $\gamma = c_p/c_v$ | $c_p$ (kJ/ kgK) | $c_v$ (kJ/ kgK) | $\gamma = c_p/c_v$ | $c_p$ (kJ/ kgK) | $c_v$ (kJ/ kgK) | $\gamma = c_p/c_v$ |
| 250 | 1.003 | 0.716 | 1.401 | 0.791 | 0.602 | 1.314 | 1.039 | 0.743 | 1.400 |
| 300 | 1.005 | 0.718 | 1.400 | 0.846 | 0.657 | 1.288 | 1.040 | 0.744 | 1.399 |
| 350 | 1.008 | 0.721 | 1.398 | 0.895 | 0.706 | 1.268 | 1.043 | 0.746 | 1.398 |
| 400 | 1.013 | 0.726 | 1.395 | 0.939 | 0.750 | 1.252 | 1.047 | 0.751 | 1.395 |
| 450 | 1.020 | 0.733 | 1.391 | 0.978 | 0.790 | 1.239 | 1.054 | 0.757 | 1.392 |
| 500 | 1.029 | 0.742 | 1.387 | 1.014 | 0.825 | 1.229 | 1.063 | 0.767 | 1.387 |
| 550 | 1.040 | 0.753 | 1.381 | 1.046 | 0.857 | 1.220 | 1.075 | 0.778 | 1.382 |
| 600 | 1.051 | 0.764 | 1.376 | 1.075 | 0.886 | 1.213 | 1.087 | 0.790 | 1.376 |
| 650 | 1.063 | 0.776 | 1.370 | 1.102 | 0.913 | 1.207 | 1.100 | 0.803 | 1.370 |
| 700 | 1.075 | 0.788 | 1.364 | 1.126 | 0.937 | 1.202 | 1.113 | 0.816 | 1.364 |
| 750 | 1.087 | 0.800 | 1.359 | 1.148 | 0.959 | 1.197 | 1.126 | 0.829 | 1.358 |
| 800 | 1.099 | 0.812 | 1.354 | 1.169 | 0.980 | 1.193 | 1.139 | 0.842 | 1.353 |
| 900 | 1.121 | 0.834 | 1.344 | 1.204 | 1.015 | 1.186 | 1.163 | 0.866 | 1.343 |
| 1000 | 1.142 | 0.855 | 1.336 | 1.234 | 1.045 | 1.181 | 1.185 | 0.888 | 1.335 |

| | Hydrogen, $H_2$ | | | Nitrogen, $N_2$ | | | Oxygen, $O_2$ | | |
|---|---|---|---|---|---|---|---|---|---|
| Temp. (K) | $c_p$ (kJ/ kgK) | $c_v$ (kJ/ kgK) | $\gamma = c_p/c_v$ | $c_p$ (kJ/ kgK) | $c_v$ (kJ/ kgK) | $\gamma = c_p/c_v$ | $c_p$ (kJ/ kgK) | $c_v$ (kJ/ kgK) | $\gamma = c_p/c_v$ |
| 250 | 14.051 | 9.927 | 1.416 | 1.039 | 0.742 | 1.400 | 0.913 | 0.653 | 1.398 |
| 300 | 14.307 | 10.183 | 1.405 | 1.039 | 0.743 | 1.400 | 0.918 | 0.658 | 1.395 |
| 350 | 14.427 | 10.302 | 1.400 | 1.041 | 0.744 | 1.399 | 0.928 | 0.668 | 1.389 |
| 400 | 14.476 | 10.352 | 1.398 | 1.044 | 0.747 | 1.397 | 0.941 | 0.681 | 1.382 |
| 450 | 14.501 | 10.377 | 1.398 | 1.049 | 0.752 | 1.395 | 0.956 | 0.696 | 1.373 |
| 500 | 14.513 | 10.389 | 1.397 | 1.056 | 0.759 | 1.391 | 0.972 | 0.712 | 1.365 |
| 550 | 14.530 | 10.405 | 1.396 | 1.065 | 0.768 | 1.387 | 0.988 | 0.728 | 1.358 |
| 600 | 14.546 | 10.422 | 1.396 | 1.075 | 0.778 | 1.382 | 1.003 | 0.743 | 1.350 |
| 650 | 14.571 | 10.447 | 1.395 | 1.086 | 0.789 | 1.376 | 1.017 | 0.758 | 1.343 |
| 700 | 14.604 | 10.480 | 1.394 | 1.098 | 0.801 | 1.371 | 1.031 | 0.771 | 1.337 |
| 750 | 14.645 | 10.521 | 1.392 | 1.110 | 0.813 | 1.365 | 1.043 | 0.783 | 1.332 |
| 800 | 14.695 | 10.570 | 1.390 | 1.121 | 0.825 | 1.360 | 1.054 | 0.794 | 1.327 |
| 900 | 14.822 | 10.698 | 1.385 | 1.145 | 0.849 | 1.349 | 1.074 | 0.814 | 1.319 |
| 1000 | 14.983 | 10.859 | 1.380 | 1.167 | 0.870 | 1.341 | 1.090 | 0.830 | 1.313 |

# Appendix 5: Polynomial Relations for Ideal Gas Specific Heat as a Function of Temperature

$$\overline{c}_p = Mc_p = a + bT + cT^2 + dT^3$$

Valid for temperatures ranging from 300 to 1500 K, with a typical accuracy of ±1%.

| Substance | | a | b | c | d |
|---|---|---|---|---|---|
| Nitrogen | $N_2$ | 28.9 | $-0.1571 \times 10^{-2}$ | $0.8081 \times 10^{-5}$ | $-2.873 \times 10^{-9}$ |
| Oxygen | $O_2$ | 25.48 | $1.52 \times 10^{-2}$ | $-0.7155 \times 10^{-5}$ | $1.312 \times 10^{-9}$ |
| Air | — | 28.11 | $0.1967 \times 10^{-2}$ | $0.4802 \times 10^{-5}$ | $-1.966 \times 10^{-9}$ |
| Hydrogen | $H_2$ | 29.11 | $-0.1916 \times 10^{-2}$ | $0.4003 \times 10^{-5}$ | $-0.8704 \times 10^{-9}$ |
| Carbon monoxide | CO | 28.16 | $0.1675 \times 10^{-2}$ | $0.5372 \times 10^{-5}$ | $-2.222 \times 10^{-9}$ |
| Carbon dioxide | $CO_2$ | 22.26 | $5.981 \times 10^{-2}$ | $-3.501 \times 10^{-5}$ | $7.469 \times 10^{-9}$ |
| Water vapor | $H_2O$ | 32.24 | $0.1923 \times 10^{-2}$ | $1.055 \times 10^{-5}$ | $-3.595 \times 10^{-9}$ |
| Nitric oxide | NO | 29.34 | $-0.09395 \times 10^{-2}$ | $0.9747 \times 10^{-5}$ | $-4.187 \times 10^{-9}$ |
| Nitrous oxide | $N_2O$ | 24.11 | $5.8632 \times 10^{-2}$ | $-3.562 \times 10^{-5}$ | $10.58 \times 10^{-9}$ |
| Nitrogen dioxide | $NO_2$ | 22.9 | $5.715 \times 10^{-2}$ | $-3.52 \times 10^{-5}$ | $7.87 \times 10^{-9}$ |
| Ammonia | $NH_3$ | 27.568 | $2.563 \times 10^{-2}$ | $0.99072 \times 10^{-5}$ | $-6.6909 \times 10^{-9}$ |
| Sulfur | $S_2$ | 27.21 | $2.218 \times 10^{-2}$ | $-1.628 \times 10^{-5}$ | $3.986 \times 10^{-9}$ |
| Sulfur dioxide | $SO_2$ | 25.78 | $5.795 \times 10^{-2}$ | $-3.812 \times 10^{-5}$ | $8.612 \times 10^{-9}$ |
| Sulfur trioxide | $SO_3$ | 16.4 | $14.58 \times 10^{-2}$ | $-11.2 \times 10^{-5}$ | $32.42 \times 10^{-9}$ |
| Acetylene | $C_2H_2$ | 21.8 | $9.2143 \times 10^{-2}$ | $-6.527 \times 10^{-5}$ | $18.21 \times 10^{-9}$ |
| Benzene | $C_6H_6$ | −36.22 | $48.475 \times 10^{-2}$ | $-31.57 \times 10^{-5}$ | $77.62 \times 10^{-9}$ |
| Methanol | $CH_4O$ | 19 | $9.152 \times 10^{-2}$ | $-1.22 \times 10^{-5}$ | $-8.039 \times 10^{-9}$ |
| Ethanol | $C_2H_6O$ | 19.9 | $20.96 \times 10^{-2}$ | $-10.38 \times 10^{-5}$ | $20.05 \times 10^{-9}$ |
| Hydrogen chloride | HCl | 30.33 | $-0.762 \times 10^{-2}$ | $1.327 \times 10^{-5}$ | $-4.338 \times 10^{-9}$ |
| Methane | $CH_4$ | 19.89 | $5.024 \times 10^{-2}$ | $1.269 \times 10^{-5}$ | $-11.01 \times 10^{-9}$ |
| Ethane | $C_2H_6$ | 6.9 | $17.27 \times 10^{-2}$ | $-6.406 \times 10^{-5}$ | $7.285 \times 10^{-9}$ |
| Propane | $C_3H_8$ | −4.04 | $30.48 \times 10^{-2}$ | $-15.72 \times 10^{-5}$ | $31.74 \times 10^{-9}$ |
| n-Butane | $C_4H_{10}$ | 3.96 | $37.15 \times 10^{-2}$ | $-18.34 \times 10^{-5}$ | $35 \times 10^{-9}$ |
| i-Butane | $C_4H_{10}$ | −7.913 | $41.6 \times 10^{-2}$ | $-23.01 \times 10^{-5}$ | $49.91 \times 10^{-9}$ |
| n-Pentane | $C_5H_{12}$ | 6.774 | $45.43 \times 10^{-2}$ | $-22.46 \times 10^{-5}$ | $42.29 \times 10^{-9}$ |
| n-Hexane | $C_6H_{14}$ | 6.938 | $55.22 \times 10^{-2}$ | $-28.65 \times 10^{-5}$ | $57.69 \times 10^{-9}$ |
| Ethylene | $C_2H_4$ | 3.95 | $15.64 \times 10^{-2}$ | $-8.344 \times 10^{-5}$ | $17.67 \times 10^{-9}$ |
| Propylene | $C_3H_6$ | 3.15 | $23.83 \times 10^{-2}$ | $-12.18 \times 10^{-5}$ | $24.62 \times 10^{-9}$ |

# Appendix 6: Critical Properties of Fluids

| Substance | Formula | Molar mass, $M$, kg/kmol | Temp., $T_c$, K | Press., $P_c$, MPa | Volume, $\bar{v}_c$, m³/kmol |
|---|---|---|---|---|---|
| Air | — | 28.97 | 132.5 | 3.77 | 0.0883 |
| Ammonia | $NH_3$ | 17.03 | 405.5 | 11.28 | 0.0724 |
| Argon | Ar | 39.948 | 151 | 4.86 | 0.0749 |
| Benzene | $C_6H_6$ | 78.115 | 562 | 4.92 | 0.2603 |
| Bromine | $Br_2$ | 159.808 | 584 | 10.34 | 0.1355 |
| n-Butane | $C_4H_{10}$ | 58.124 | 425.2 | 3.8 | 0.2547 |
| Carbon dioxide | $CO_2$ | 44.01 | 304.2 | 7.39 | 0.0943 |
| Carbon monoxide | CO | 28.011 | 133 | 3.5 | 0.093 |
| Carbon tetrachloride | $CCl_4$ | 153.82 | 556.4 | 4.56 | 0.2759 |
| Chlorine | $Cl_2$ | 70.906 | 417 | 7.71 | 0.1242 |
| Chloroform | $CHCl_3$ | 119.38 | 536.6 | 5.47 | 0.2403 |
| Dichlorodifluoromethane (R-12) | $CCl_2F_2$ | 120.91 | 384.7 | 4.01 | 0.2179 |
| Dichlorofluoromethane (R-21) | $CHCl_2F$ | 102.92 | 451.7 | 5.17 | 0.1973 |
| Ethane | $C_2H_6$ | 30.07 | 305.5 | 4.48 | 0.148 |
| Ethyl alcohol | $C_2H_5OH$ | 46.07 | 516 | 6.38 | 0.1673 |
| Ethylene | $C_2H_4$ | 28.054 | 282.4 | 5.12 | 0.1242 |
| Helium | He | 4.003 | 5.3 | 0.23 | 0.0578 |
| n-Hexane | $C_6H_{14}$ | 86.179 | 507.9 | 3.03 | 0.3677 |
| Hydrogen | $H_2$ | 2.016 | 33.3 | 1.3 | 0.0649 |
| Krypton | Kr | 83.8 | 209.4 | 5.5 | 0.0924 |
| Methane | $CH_4$ | 16.043 | 191.1 | 4.64 | 0.0993 |
| Methyl alcohol | $CH_3OH$ | 32.042 | 513.2 | 7.95 | 0.118 |
| Methyl chloride | $CH_3Cl$ | 50.488 | 416.3 | 6.68 | 0.143 |
| Neon | Ne | 20.183 | 44.5 | 2.73 | 0.0417 |
| Nitrogen | $N_2$ | 28.013 | 126.2 | 3.39 | 0.0899 |
| Nitrous oxide | $N_2O$ | 44.013 | 309.7 | 7.27 | 0.0961 |
| Oxygen | $O_2$ | 31.999 | 154.8 | 5.08 | 0.078 |
| Propane | $C_3H_8$ | 44.097 | 370 | 4.26 | 0.1998 |
| Propylene | $C_3H_6$ | 42.081 | 365 | 4.62 | 0.181 |
| Sulfur dioxide | $SO_2$ | 64.063 | 430.7 | 7.88 | 0.1217 |
| Tetrafluoroethane (R-134a) | $CF_3CH_2F$ | 102.03 | 374.3 | 4.067 | 0.1847 |
| Trichlorofluoromethane (R-11) | $CCl_3F$ | 137.37 | 471.2 | 4.38 | 0.2478 |
| Water | $H_2O$ | 18.015 | 647.3 | 22.09 | 0.0568 |
| Xenon | Xe | 131.3 | 289.8 | 5.88 | 0.1186 |

# Appendix 7: Ideal Gas Tables for Air

| $T$, K | $h$, kJ/kg | $P_r$ | $u$, kJ/kg | $v_r$ | $s^o$, kJ/kgK |
|---|---|---|---|---|---|
| 200 | 199.97 | 0.3363 | 142.56 | 1707.0 | 1.29559 |
| 210 | 209.97 | 0.3987 | 149.69 | 1512.0 | 1.34444 |
| 220 | 219.97 | 0.4690 | 156.82 | 1346.0 | 1.39105 |
| 230 | 230.02 | 0.5477 | 164.00 | 1205.0 | 1.43557 |
| 240 | 240.02 | 0.6355 | 171.13 | 1084.0 | 1.47824 |
| 250 | 250.05 | 0.7329 | 178.28 | 979.0 | 1.51917 |
| 260 | 260.09 | 0.8405 | 185.45 | 887.8 | 1.55848 |
| 270 | 270.11 | 0.9590 | 192.60 | 808.0 | 1.59634 |
| 280 | 280.13 | 1.0889 | 199.75 | 738.0 | 1.63279 |
| 285 | 285.14 | 1.1584 | 203.33 | 706.1 | 1.65055 |
| 290 | 290.16 | 1.2311 | 206.91 | 676.1 | 1.66802 |
| 295 | 295.17 | 1.3068 | 210.49 | 647.9 | 1.68515 |
| 300 | 300.19 | 1.3860 | 214.07 | 621.2 | 1.70203 |
| 305 | 305.22 | 1.4686 | 217.67 | 596.0 | 1.71865 |
| 310 | 310.24 | 1.5546 | 221.25 | 572.3 | 1.73498 |
| 315 | 315.27 | 1.6442 | 224.85 | 549.8 | 1.75106 |
| 320 | 320.29 | 1.7375 | 228.42 | 528.6 | 1.76690 |
| 325 | 325.31 | 1.8345 | 232.02 | 508.4 | 1.78249 |
| 330 | 330.34 | 1.9352 | 235.61 | 489.4 | 1.79783 |
| 340 | 340.42 | 2.149 | 242.82 | 454.1 | 1.82790 |
| 350 | 350.49 | 2.379 | 250.02 | 422.2 | 1.85708 |
| 360 | 360.58 | 2.626 | 257.24 | 393.4 | 1.88543 |
| 370 | 370.67 | 2.892 | 264.46 | 367.2 | 1.91313 |
| 380 | 380.77 | 3.176 | 271.69 | 343.4 | 1.94001 |
| 390 | 390.88 | 3.481 | 278.93 | 321.5 | 1.96633 |
| 400 | 400.98 | 3.806 | 286.16 | 301.6 | 1.99194 |
| 410 | 411.12 | 4.153 | 293.43 | 283.3 | 2.01699 |
| 420 | 421.26 | 4.522 | 300.69 | 266.6 | 2.04142 |
| 430 | 431.43 | 4.915 | 307.99 | 251.1 | 2.06533 |
| 440 | 441.61 | 5.332 | 315.30 | 236.8 | 2.08870 |
| 450 | 451.80 | 5.775 | 322.62 | 223.6 | 2.11161 |
| 460 | 462.02 | 6.245 | 329.97 | 211.4 | 2.13407 |
| 470 | 472.24 | 6.742 | 337.32 | 200.1 | 2.15604 |
| 480 | 482.49 | 7.268 | 344.70 | 189.5 | 2.17760 |
| 490 | 492.74 | 7.824 | 352.08 | 179.7 | 2.19876 |
| 500 | 503.02 | 8.411 | 359.49 | 170.6 | 2.21952 |

(*continued*)

(*continued*)

| $T$, K | $h$, kJ/kg | $P_r$ | $u$, kJ/kg | $v_r$ | $s^o$, kJ/kgK |
|---|---|---|---|---|---|
| 510 | 513.32 | 9.031 | 366.92 | 162.1 | 2.23993 |
| 520 | 523.63 | 9.684 | 374.36 | 154.1 | 2.25997 |
| 530 | 533.98 | 10.37 | 381.84 | 146.7 | 2.27967 |
| 540 | 544.35 | 11.10 | 389.34 | 139.7 | 2.29906 |
| 550 | 555.74 | 11.86 | 396.86 | 133.1 | 2.31809 |
| 560 | 565.17 | 12.66 | 404.42 | 127.0 | 2.33685 |
| 570 | 575.59 | 13.50 | 411.97 | 121.2 | 2.35531 |
| 580 | 586.04 | 14.38 | 419.55 | 115.7 | 2.37348 |
| 590 | 596.52 | 15.31 | 427.15 | 110.6 | 2.39140 |
| 600 | 607.02 | 16.28 | 434.78 | 105.8 | 2.40902 |
| 610 | 617.53 | 17.30 | 442.42 | 101.2 | 2.42644 |
| 620 | 628.07 | 18.36 | 450.09 | 96.92 | 2.44356 |
| 630 | 638.63 | 19.84 | 457.78 | 92.84 | 2.46048 |
| 640 | 649.22 | 20.64 | 465.50 | 88.99 | 2.47716 |
| 650 | 659.84 | 21.86 | 473.25 | 85.34 | 2.49364 |
| 660 | 670.47 | 23.13 | 481.01 | 81.89 | 2.50985 |
| 670 | 681.14 | 24.46 | 488.81 | 78.61 | 2.52589 |
| 680 | 691.82 | 25.85 | 496.62 | 75.50 | 2.54175 |
| 690 | 702.52 | 27.29 | 504.45 | 72.56 | 2.55731 |
| 700 | 713.27 | 28.80 | 512.33 | 69.76 | 2.57277 |
| 710 | 724.04 | 30.38 | 520.23 | 67.07 | 2.58810 |
| 720 | 734.82 | 32.02 | 528.14 | 64.53 | 2.60319 |
| 730 | 745.62 | 33.72 | 536.07 | 62.13 | 2.61803 |
| 740 | 756.44 | 35.50 | 544.02 | 59.82 | 2.63280 |
| 750 | 767.29 | 37.35 | 551.99 | 57.63 | 2.64737 |
| 760 | 778.18 | 39.27 | 560.01 | 55.54 | 2.66176 |
| 780 | 800.03 | 43.35 | 576.12 | 51.64 | 2.69013 |
| 800 | 821.95 | 47.75 | 592.30 | 48.08 | 2.71787 |
| 820 | 843.98 | 52.59 | 608.59 | 44.84 | 2.74504 |
| 840 | 866.08 | 57.60 | 624.95 | 41.85 | 2.77170 |
| 860 | 888.27 | 63.09 | 641.40 | 39.12 | 2.79783 |
| 880 | 910.56 | 68.98 | 657.95 | 36.61 | 2.82344 |
| 900 | 932.93 | 75.29 | 674.58 | 34.31 | 2.84856 |
| 920 | 955.38 | 82.05 | 691.28 | 32.18 | 2.87324 |
| 940 | 977.92 | 89.28 | 708.08 | 30.22 | 2.89748 |
| 960 | 1000.55 | 97.00 | 725.02 | 28.40 | 2.92128 |
| 980 | 1023.25 | 105.2 | 741.98 | 26.73 | 2.94468 |
| 1000 | 1046.04 | 114.0 | 758.94 | 25.17 | 2.96770 |
| 1020 | 1068.89 | 123.4 | 776.10 | 23.72 | 2.99034 |
| 1040 | 1091.85 | 133.3 | 793.36 | 23.29 | 3.01260 |
| 1060 | 1114.86 | 143.9 | 810.62 | 21.14 | 3.03449 |
| 1080 | 1137.89 | 155.2 | 827.88 | 19.98 | 3.05608 |
| 1100 | 1161.07 | 167.1 | 845.33 | 18.896 | 3.07732 |
| 1120 | 1184.28 | 179.7 | 862.79 | 17.886 | 3.09825 |
| 1140 | 1207.57 | 193.1 | 880.35 | 16.946 | 3.11883 |
| 1160 | 1230.92 | 207.2 | 897.91 | 16.064 | 3.13916 |
| 1180 | 1254.34 | 222.2 | 915.57 | 15.241 | 3.15916 |

*(continued)*

| $T$, K | $h$, kJ/kg | $P_r$ | $u$, kJ/kg | $v_r$ | $s^o$, kJ/kgK |
|---|---|---|---|---|---|
| 1200 | 1277.79 | 238.0 | 933.33 | 14.470 | 3.17888 |
| 1220 | 1301.31 | 254.7 | 951.09 | 13.747 | 3.19834 |
| 1240 | 1324.93 | 272.3 | 968.95 | 13.069 | 3.21751 |
| 1260 | 1348.55 | 290.8 | 986.90 | 12.435 | 3.23638 |
| 1280 | 1372.24 | 310.4 | 1004.76 | 11.835 | 3.25510 |
| 1300 | 1395.97 | 330.9 | 1022.82 | 11.275 | 3.27345 |
| 1320 | 1419.76 | 352.5 | 1040.88 | 10.747 | 3.29160 |
| 1340 | 1443.60 | 375.3 | 1058.94 | 10.247 | 3.30959 |
| 1360 | 1467.49 | 399.1 | 1077.10 | 9.780 | 3.32724 |
| 1380 | 1491.44 | 424.2 | 1095.26 | 9.337 | 3.34474 |
| 1400 | 1515.42 | 450.5 | 1113.52 | 8.919 | 3.36200 |
| 1420 | 1539.44 | 478.0 | 1131.77 | 8.526 | 3.37901 |
| 1440 | 1563.51 | 506.9 | 1150.13 | 8.153 | 3.39586 |
| 1460 | 1587.63 | 537.1 | 1168.49 | 7.801 | 3.41247 |
| 1480 | 1611.79 | 568.8 | 1186.95 | 7.468 | 3.42892 |
| 1500 | 1635.97 | 601.9 | 1205.41 | 7.152 | 3.44516 |
| 1520 | 1660.23 | 636.5 | 1223.87 | 6.854 | 3.46120 |
| 1540 | 1684.51 | 672.8 | 1242.43 | 6.569 | 3.47712 |
| 1560 | 1708.82 | 710.5 | 1260.99 | 6.301 | 3.49276 |
| 1580 | 1733.17 | 750.0 | 1279.65 | 6.046 | 3.50829 |
| 1600 | 1757.57 | 791.2 | 1298.30 | 5.804 | 3.52364 |
| 1620 | 1782.00 | 834.1 | 1316.96 | 5.574 | 3.53879 |
| 1640 | 1806.46 | 878.9 | 1335.72 | 5.355 | 3.55381 |
| 1660 | 1830.96 | 925.6 | 1354.48 | 5.147 | 3.56867 |
| 1680 | 1855.50 | 974.2 | 1373.24 | 4.949 | 3.58335 |
| 1700 | 1880.1 | 1025 | 1392.7 | 4.761 | 3.5979 |
| 1750 | 1941.6 | 1161 | 1439.8 | 4.328 | 3.6336 |
| 1800 | 2003.3 | 1310 | 1487.2 | 3.994 | 3.6684 |
| 1850 | 2065.3 | 1475 | 1534.9 | 3.601 | 3.7023 |
| 1900 | 2127.4 | 1655 | 1582.6 | 3.295 | 3.7354 |
| 1950 | 2189.7 | 1852 | 1630.6 | 3.022 | 3.7677 |
| 2000 | 2252.1 | 2068 | 1678.7 | 2.776 | 3.7994 |
| 2050 | 2314.6 | 2303 | 1726.8 | 2.555 | 3.8303 |
| 2100 | 2377.7 | 2559 | 1775.3 | 2.356 | 3.8605 |
| 2150 | 2440.3 | 2837 | 1823.8 | 2.175 | 3.8901 |
| 2200 | 2503.2 | 3138 | 1872.4 | 2.012 | 3.9191 |
| 2250 | 2566.4 | 3464 | 1921.3 | 1.864 | 3.9474 |

# Appendix 8: Properties of Water

## Appendix 8a: Properties of Saturated Water (Temperature Table)

| Temp., °C | Pressure, MPa | Specific volume, m³/kg | | Internal energy, kJ/kg | | Enthalpy, kJ/kg | | Entropy, kJ/kgK | |
|---|---|---|---|---|---|---|---|---|---|
| $T_{sat}$ | $P$ | $v_f$ | $v_g$ | $u_f$ | $u_g$ | $h_f$ | $h_g$ | $s_f$ | $s_g$ |
| 0.01 | 0.0006113 | 0.001000 | 206.14 | 0.00 | 2375.3 | 0.00 | 2501.4 | 0.0000 | 9.1562 |
| 5 | 0.0008721 | 0.001000 | 147.12 | 20.97 | 2382.3 | 20.98 | 2510.6 | 0.0761 | 9.0257 |
| 10 | 0.0012276 | 0.001000 | 106.38 | 42.00 | 2389.2 | 42.01 | 2519.8 | 0.1510 | 8.9008 |
| 15 | 0.0017051 | 0.001001 | 77.93 | 62.99 | 2396.1 | 62.99 | 2528.9 | 0.2245 | 8.7814 |
| 20 | 0.002339 | 0.001002 | 57.79 | 83.95 | 2402.9 | 83.96 | 2538.1 | 0.2966 | 8.6672 |
| 25 | 0.003169 | 0.001003 | 43.36 | 104.88 | 2409.8 | 104.89 | 2547.2 | 0.3674 | 8.5580 |
| 30 | 0.004246 | 0.001004 | 32.89 | 125.78 | 2416.6 | 125.79 | 2556.3 | 0.4369 | 8.4533 |
| 35 | 0.005628 | 0.001006 | 25.22 | 146.67 | 2423.4 | 146.68 | 2565.3 | 0.5053 | 8.3531 |
| 40 | 0.007384 | 0.001008 | 19.52 | 167.56 | 2430.1 | 167.57 | 2574.3 | 0.5725 | 8.2570 |
| 45 | 0.009593 | 0.001010 | 15.26 | 188.44 | 2436.8 | 188.45 | 2583.2 | 0.6387 | 8.1648 |
| 50 | 0.012349 | 0.001012 | 12.03 | 209.32 | 2443.5 | 209.33 | 2592.1 | 0.7038 | 8.0763 |
| 55 | 0.015758 | 0.001015 | 9.568 | 230.21 | 2450.1 | 230.23 | 2600.9 | 0.7679 | 7.9913 |
| 60 | 0.019940 | 0.001017 | 7.671 | 251.11 | 2456.6 | 251.13 | 2609.6 | 0.8312 | 7.9096 |
| 65 | 0.02503 | 0.001020 | 6.197 | 272.02 | 2463.1 | 272.06 | 2618.3 | 0.8935 | 7.8310 |
| 70 | 0.03119 | 0.001023 | 5.042 | 292.95 | 2469.6 | 292.98 | 2626.8 | 0.9549 | 7.7553 |
| 75 | 0.03858 | 0.001026 | 4.131 | 313.90 | 2475.9 | 313.93 | 2643.7 | 1.0155 | 7.6824 |
| 80 | 0.04739 | 0.001029 | 3.407 | 334.86 | 2482.2 | 334.91 | 2635.3 | 1.0753 | 7.6122 |
| 85 | 0.05783 | 0.001033 | 2.828 | 355.84 | 2488.4 | 355.90 | 2651.9 | 1.1343 | 7.5445 |
| 90 | 0.07014 | 0.001036 | 2.361 | 376.85 | 2494.5 | 376.92 | 2660.1 | 1.1925 | 7.4791 |
| 95 | 0.08455 | 0.001040 | 1.982 | 397.88 | 2500.6 | 397.96 | 2668.1 | 1.2500 | 7.4159 |
| 100 | 0.10135 | 0.001044 | 1.6729 | 418.94 | 2506.5 | 419.04 | 2676.1 | 1.3069 | 7.3549 |
| 105 | 0.12082 | 0.001048 | 1.4194 | 440.02 | 2512.4 | 440.15 | 2683.8 | 1.3630 | 7.2958 |
| 110 | 0.14327 | 0.001052 | 1.2102 | 461.14 | 2518.1 | 461.30 | 2691.5 | 1.4185 | 7.2387 |
| 115 | 0.16906 | 0.001056 | 1.0366 | 482.30 | 2523.7 | 482.48 | 2699.0 | 1.4734 | 7.1833 |
| 120 | 0.19853 | 0.001060 | 0.8919 | 503.50 | 2529.3 | 503.71 | 2706.3 | 1.5276 | 7.1296 |
| 125 | 0.2321 | 0.001065 | 0.7706 | 524.74 | 2534.6 | 524.99 | 2713.5 | 1.5813 | 7.0775 |
| 130 | 0.2701 | 0.001070 | 0.6685 | 546.02 | 2539.9 | 546.31 | 2720.5 | 1.6344 | 7.0269 |
| 135 | 0.3130 | 0.001075 | 0.5822 | 567.35 | 2545.0 | 567.69 | 2727.3 | 1.6870 | 6.9777 |
| 140 | 0.3613 | 0.001080 | 0.5089 | 588.74 | 2550.0 | 589.13 | 2733.9 | 1.7391 | 6.9299 |
| 145 | 0.4154 | 0.001085 | 0.4463 | 610.18 | 2554.9 | 610.63 | 2740.3 | 1.7907 | 6.8833 |
| 150 | 0.4758 | 0.001091 | 0.3928 | 631.68 | 2559.5 | 632.20 | 2746.5 | 1.8418 | 6.8379 |
| 155 | 0.5431 | 0.001096 | 0.3468 | 653.24 | 2564.1 | 653.84 | 2752.4 | 1.8925 | 6.7935 |
| 160 | 0.6178 | 0.001102 | 0.3071 | 674.87 | 2568.4 | 675.55 | 2758.1 | 1.9427 | 6.7502 |
| 165 | 0.7005 | 0.001108 | 0.2727 | 696.56 | 2572.5 | 697.34 | 2763.5 | 1.9925 | 6.7078 |

(*continued*)

| Temp., °C | Pressure, MPa | Specific volume, m³/kg | | Internal energy, kJ/kg | | Enthalpy, kJ/kg | | Entropy, kJ/kgK | |
|---|---|---|---|---|---|---|---|---|---|
| $T_{sat}$ | $P$ | $v_f$ | $v_g$ | $u_f$ | $u_g$ | $h_f$ | $h_g$ | $s_f$ | $s_g$ |
| 170 | 0.7917 | 0.001114 | 0.2428 | 718.33 | 2576.5 | 719.21 | 2768.7 | 2.0419 | 6.6663 |
| 175 | 0.8920 | 0.001121 | 0.2168 | 740.17 | 2580.2 | 741.17 | 2773.6 | 2.0909 | 6.6256 |
| 180 | 1.0021 | 0.001127 | 0.19405 | 762.09 | 2583.7 | 763.22 | 2778.2 | 2.1396 | 6.5857 |
| 185 | 1.1227 | 0.001134 | 0.17409 | 784.10 | 2587.0 | 785.37 | 2782.4 | 2.1879 | 6.5465 |
| 190 | 1.2544 | 0.001141 | 0.15654 | 806.19 | 2590.0 | 807.62 | 2786.4 | 2.2359 | 6.5079 |
| 195 | 1.3978 | 0.001149 | 0.14105 | 828.37 | 2592.8 | 829.98 | 2790.0 | 2.2835 | 6.4698 |
| 200 | 1.5538 | 0.001157 | 0.12736 | 850.65 | 2595.3 | 852.45 | 2793.2 | 2.3309 | 6.4323 |
| 205 | 1.7230 | 0.001164 | 0.11521 | 873.04 | 2597.5 | 875.04 | 2796.0 | 2.3780 | 6.3952 |
| 210 | 1.9062 | 0.001173 | 0.10441 | 895.53 | 2599.5 | 897.76 | 2798.5 | 2.4248 | 6.3585 |
| 215 | 2.104 | 0.001181 | 0.09479 | 918.14 | 2601.1 | 920.62 | 2800.5 | 2.4714 | 6.3221 |
| 220 | 2.318 | 0.001190 | 0.08619 | 940.87 | 2602.4 | 943.62 | 2802.1 | 2.5178 | 6.2861 |
| 225 | 2.548 | 0.001199 | 0.07849 | 963.73 | 2603.3 | 966.78 | 2803.3 | 2.5639 | 6.2503 |
| 230 | 2.795 | 0.001209 | 0.07158 | 986.74 | 2603.9 | 990.12 | 2804.0 | 2.6099 | 6.2146 |
| 235 | 3.060 | 0.001219 | 0.06537 | 1009.89 | 2604.1 | 1013.62 | 2804.2 | 2.6558 | 6.1791 |
| 240 | 3.344 | 0.001229 | 0.05976 | 1033.21 | 2604.0 | 1037.32 | 2803.8 | 2.7015 | 6.1437 |
| 245 | 3.648 | 0.001240 | 0.05471 | 1056.71 | 2603.4 | 1061.23 | 2803.0 | 2.7472 | 6.1083 |
| 250 | 3.973 | 0.001251 | 0.05013 | 1080.39 | 2602.4 | 1085.36 | 2801.5 | 2.7927 | 6.0730 |
| 255 | 4.319 | 0.001263 | 0.04598 | 1104.28 | 2600.9 | 1109.73 | 2799.5 | 2.8383 | 6.0375 |
| 260 | 4.688 | 0.001276 | 0.04221 | 1128.39 | 2599.0 | 1134.37 | 2796.9 | 2.8838 | 6.0019 |
| 265 | 5.081 | 0.001289 | 0.03877 | 1152.74 | 2596.6 | 1159.28 | 2793.6 | 2.9294 | 5.9662 |
| 270 | 5.499 | 0.001302 | 0.03564 | 1177.36 | 2593.7 | 1184.51 | 2789.7 | 2.9751 | 5.9301 |
| 275 | 5.942 | 0.001317 | 0.03279 | 1202.25 | 2590.2 | 1210.07 | 2785.0 | 3.0208 | 5.8938 |
| 280 | 6.412 | 0.001332 | 0.03017 | 1227.46 | 2586.1 | 1235.99 | 2779.6 | 3.0668 | 5.8571 |
| 285 | 6.909 | 0.001348 | 0.02777 | 1253.00 | 2581.4 | 1262.31 | 2773.3 | 3.1130 | 5.8199 |
| 290 | 7.436 | 0.001366 | 0.02557 | 1278.92 | 2576.0 | 1289.07 | 2766.2 | 3.1594 | 5.7821 |
| 295 | 7.993 | 0.001384 | 0.02354 | 1305.20 | 2569.9 | 1316.30 | 2758.1 | 3.2062 | 5.7437 |
| 300 | 8.581 | 0.001404 | 0.02167 | 1332.00 | 2563.0 | 1344.00 | 2749.0 | 3.2534 | 5.7045 |
| 305 | 9.202 | 0.001425 | 0.019948 | 1359.30 | 2555.2 | 1372.40 | 2738.7 | 3.3010 | 5.6643 |
| 310 | 9.856 | 0.001447 | 0.018350 | 1387.10 | 2546.4 | 1401.30 | 2727.3 | 3.3493 | 5.6230 |
| 315 | 10.547 | 0.001472 | 0.016867 | 1415.50 | 2536.6 | 1431.00 | 2714.5 | 3.3982 | 5.5804 |
| 320 | 11.274 | 0.001499 | 0.015488 | 1444.60 | 2525.5 | 1461.50 | 2700.1 | 3.4480 | 5.5362 |
| 330 | 12.845 | 0.001561 | 0.012996 | 1505.30 | 2498.9 | 1525.30 | 2665.9 | 3.5507 | 5.4417 |
| 340 | 14.586 | 0.001638 | 0.010797 | 1570.30 | 2464.6 | 1594.20 | 2622.0 | 3.6594 | 5.3357 |
| 350 | 16.513 | 0.001740 | 0.008813 | 1641.90 | 2418.4 | 1670.60 | 2563.9 | 3.7777 | 5.2112 |
| 360 | 18.651 | 0.001893 | 0.006945 | 1725.20 | 2351.5 | 1760.50 | 2481.0 | 3.9147 | 5.0526 |
| 370 | 21.03 | 0.002213 | 0.004925 | 1844.00 | 2228.5 | 1890.50 | 2332.1 | 4.1106 | 4.7971 |
| 374.14 | 22.09 | 0.003155 | 0.003155 | 2029.60 | 2029.6 | 2099.30 | 2099.3 | 4.4298 | 4.4298 |

## Appendix 8b: Properties of Saturated Water (Pressure Table)

| Pressure, MPa | Temp., °C | Specific volume, m³/kg | | Internal energy, kJ/kg | | Enthalpy, kJ/kg | | Entropy, kJ/kgK | |
|---|---|---|---|---|---|---|---|---|---|
| $P$ | $T_{sat}$ | $v_f$ | $v_g$ | $u_f$ | $u_g$ | $h_f$ | $h_g$ | $s_f$ | $s_g$ |
| 0.0006113 | 0.01 | 0.001000 | 206.14 | 0 | 2375.3 | 0.00 | 2501.4 | 0.0000 | 9.1562 |
| 0.0010 | 6.98 | 0.001000 | 129.21 | 29.3 | 2385.0 | 29.30 | 2514.2 | 0.1059 | 8.9756 |
| 0.0015 | 13.03 | 0.001001 | 87.98 | 54.71 | 2393.3 | 54.71 | 2525.3 | 0.1957 | 8.8279 |
| 0.0020 | 17.50 | 0.001001 | 67.00 | 73.48 | 2399.5 | 73.48 | 2533.5 | 0.2607 | 8.7237 |
| 0.0025 | 21.08 | 0.001002 | 54.25 | 88.48 | 2404.4 | 88.49 | 2540.0 | 0.3120 | 8.6432 |
| 0.0030 | 24.08 | 0.001003 | 45.67 | 101.04 | 2408.5 | 101.05 | 2545.5 | 0.3545 | 8.5776 |
| 0.0040 | 28.96 | 0.001004 | 34.80 | 121.45 | 2415.2 | 121.46 | 2554.4 | 0.4226 | 8.4746 |
| 0.0050 | 32.88 | 0.001005 | 28.19 | 137.81 | 2420.5 | 137.82 | 2561.5 | 0.4764 | 8.3951 |
| 0.0075 | 40.29 | 0.001008 | 19.24 | 168.78 | 2430.5 | 168.79 | 2574.8 | 0.5764 | 8.2515 |
| 0.010 | 45.81 | 0.001010 | 14.67 | 191.82 | 2437.9 | 191.83 | 2584.7 | 0.6493 | 8.1502 |
| 0.015 | 53.97 | 0.001014 | 10.02 | 225.92 | 2448.7 | 225.94 | 2599.1 | 0.7549 | 8.0085 |
| 0.020 | 60.06 | 0.001017 | 7.649 | 251.38 | 2456.7 | 251.40 | 2609.7 | 0.8320 | 7.9085 |
| 0.025 | 64.97 | 0.001020 | 6.204 | 271.90 | 2463.1 | 271.93 | 2618.2 | 0.8931 | 7.8314 |
| 0.030 | 69.10 | 0.001022 | 5.229 | 289.20 | 2468.4 | 289.23 | 2625.3 | 0.9439 | 7.7686 |
| 0.040 | 75.87 | 0.001027 | 3.993 | 317.53 | 2477.0 | 317.58 | 2636.8 | 1.0259 | 7.6700 |
| 0.050 | 81.33 | 0.001030 | 3.240 | 340.44 | 2483.9 | 340.49 | 2645.9 | 1.0910 | 7.5939 |
| 0.075 | 91.78 | 0.001037 | 2.217 | 384.31 | 2496.7 | 384.39 | 2663.0 | 1.2130 | 7.4564 |
| 0.100 | 99.63 | 0.001043 | 1.694 | 417.36 | 2506.1 | 417.46 | 2675.5 | 1.3026 | 7.3594 |
| 0.125 | 105.99 | 0.001048 | 1.3749 | 444.19 | 2513.5 | 444.32 | 2685.4 | 1.3740 | 7.2844 |
| 0.150 | 111.37 | 0.001053 | 1.1593 | 466.94 | 2519.7 | 467.11 | 2693.6 | 1.4336 | 7.2233 |
| 0.175 | 116.06 | 0.001057 | 1.0036 | 486.80 | 2524.9 | 486.99 | 2700.6 | 1.4849 | 7.1717 |
| 0.200 | 120.23 | 0.001061 | 0.8857 | 504.49 | 2529.5 | 504.70 | 2706.7 | 1.5301 | 7.1271 |
| 0.225 | 124.00 | 0.001064 | 0.7933 | 520.47 | 2533.6 | 520.72 | 2712.1 | 1.5706 | 7.0878 |
| 0.250 | 127.44 | 0.001067 | 0.7187 | 535.10 | 2537.2 | 535.37 | 2716.9 | 1.6072 | 7.0527 |
| 0.275 | 130.60 | 0.001070 | 0.6573 | 548.59 | 2540.5 | 548.89 | 2721.3 | 1.6408 | 7.0209 |
| 0.300 | 133.55 | 0.001073 | 0.6058 | 561.15 | 2543.6 | 561.47 | 2725.3 | 1.6718 | 6.9919 |
| 0.325 | 136.30 | 0.001076 | 0.5620 | 572.90 | 2546.4 | 573.25 | 2729.0 | 1.7006 | 6.9652 |
| 0.350 | 138.88 | 0.001079 | 0.5243 | 583.95 | 2548.9 | 584.33 | 2732.4 | 1.7275 | 6.9405 |
| 0.375 | 141.32 | 0.001081 | 0.4914 | 594.40 | 2551.3 | 594.81 | 2735.6 | 1.7528 | 6.9175 |
| 0.40 | 143.63 | 0.001084 | 0.4625 | 604.31 | 2553.6 | 604.74 | 2738.6 | 1.7766 | 6.8959 |
| 0.45 | 147.93 | 0.001088 | 0.4140 | 622.77 | 2557.6 | 623.25 | 2743.9 | 1.8207 | 6.8565 |
| 0.50 | 151.86 | 0.001093 | 0.3749 | 639.68 | 2561.2 | 640.23 | 2748.7 | 1.8607 | 6.8213 |
| 0.55 | 155.48 | 0.001097 | 0.3427 | 655.32 | 2564.5 | 665.93 | 2753.0 | 1.8973 | 6.7893 |
| 0.60 | 158.85 | 0.001101 | 0.3157 | 669.90 | 2567.4 | 670.56 | 2756.8 | 1.9312 | 6.7600 |
| 0.65 | 162.01 | 0.001104 | 0.2927 | 683.56 | 2570.1 | 684.28 | 2760.3 | 1.9627 | 6.7331 |
| 0.70 | 164.97 | 0.001108 | 0.2729 | 696.44 | 2572.5 | 697.22 | 2763.5 | 1.9922 | 6.7080 |
| 0.75 | 167.78 | 0.001112 | 0.2556 | 708.64 | 2574.7 | 709.47 | 2766.4 | 2.0200 | 6.6847 |
| 0.80 | 170.43 | 0.001115 | 0.2404 | 720.22 | 2576.8 | 721.11 | 2769.1 | 2.0462 | 6.6628 |
| 0.85 | 172.96 | 0.001118 | 0.2270 | 731.27 | 2578.7 | 732.22 | 2771.6 | 2.0710 | 6.6421 |
| 0.90 | 175.38 | 0.001121 | 0.2150 | 741.83 | 2580.5 | 742.83 | 2773.9 | 2.0946 | 6.6226 |
| 0.95 | 177.69 | 0.001124 | 0.2042 | 751.95 | 2582.1 | 753.02 | 2776.1 | 2.1172 | 6.6041 |
| 1.00 | 179.91 | 0.001127 | 0.19444 | 761.68 | 2583.6 | 762.81 | 2778.1 | 2.1387 | 6.5865 |

(*continued*)

| Pressure, MPa | Temp., °C | Specific volume, m³/kg | | Internal energy, kJ/kg | | Enthalpy, kJ/kg | | Entropy, kJ/kgK | |
|---|---|---|---|---|---|---|---|---|---|
| $P$ | $T_{sat}$ | $v_f$ | $v_g$ | $u_f$ | $u_g$ | $h_f$ | $h_g$ | $s_f$ | $s_g$ |
| 1.10 | 184.09 | 0.001133 | 0.17753 | 780.09 | 2586.4 | 781.34 | 2871.7 | 2.1792 | 6.5536 |
| 1.20 | 187.99 | 0.001139 | 0.16333 | 797.29 | 2588.8 | 798.65 | 2784.8 | 2.2166 | 6.5233 |
| 1.30 | 191.64 | 0.001144 | 0.15125 | 813.44 | 2591.0 | 814.93 | 2787.6 | 2.2515 | 6.4953 |
| 1.40 | 195.07 | 0.001149 | 0.14084 | 828.70 | 2592.8 | 830.30 | 2790.0 | 2.2842 | 6.4693 |
| 1.50 | 198.32 | 0.001154 | 0.13177 | 843.16 | 2594.5 | 844.89 | 2792.2 | 2.3150 | 6.4448 |
| 1.75 | 205.76 | 0.001166 | 0.11349 | 876.46 | 2597.8 | 878.50 | 2796.4 | 2.3851 | 6.3896 |
| 2.00 | 212.42 | 0.001177 | 0.09963 | 906.44 | 2600.3 | 908.79 | 2799.5 | 2.4474 | 6.3409 |
| 2.25 | 218.45 | 0.001187 | 0.08875 | 933.83 | 2602.0 | 936.49 | 2801.7 | 2.5035 | 6.2972 |
| 2.50 | 223.99 | 0.001197 | 0.07998 | 959.11 | 2603.1 | 962.11 | 2803.1 | 2.5547 | 6.2575 |
| 3.00 | 233.90 | 0.001217 | 0.06668 | 1004.78 | 2604.1 | 1008.42 | 2804.2 | 2.6457 | 6.1869 |
| 3.50 | 242.60 | 0.001235 | 0.05707 | 1045.43 | 2603.7 | 1049.75 | 2803.4 | 2.7253 | 6.1253 |
| 4 | 250.40 | 0.001252 | 0.04978 | 1082.31 | 2602.3 | 1087.31 | 2801.4 | 2.7964 | 6.0701 |
| 5 | 263.99 | 0.001286 | 0.03944 | 1147.81 | 2597.1 | 1154.23 | 2794.3 | 2.9202 | 5.9734 |
| 6 | 275.64 | 0.001319 | 0.03244 | 1205.44 | 2589.7 | 1213.35 | 2784.3 | 3.0267 | 5.8892 |
| 7 | 285.88 | 0.001351 | 0.02737 | 1257.55 | 2580.5 | 1267.00 | 2772.1 | 3.1211 | 5.8133 |
| 8 | 295.06 | 0.001384 | 0.02352 | 1305.57 | 2569.8 | 1316.64 | 2758.0 | 3.2068 | 5.7432 |
| 9 | 303.40 | 0.001418 | 0.02048 | 1350.51 | 2557.8 | 1363.26 | 2742.1 | 3.2858 | 5.6722 |
| 10 | 311.06 | 0.001452 | 0.018026 | 1393.04 | 2544.4 | 1407.56 | 2724.7 | 3.3596 | 5.6141 |
| 11 | 318.15 | 0.001489 | 0.015987 | 1433.7 | 2529.8 | 1450.1 | 2705.6 | 3.4295 | 5.5527 |
| 12 | 324.75 | 0.001527 | 0.014263 | 1473.0 | 2513.7 | 1491.3 | 2684.9 | 3.4962 | 5.4924 |
| 13 | 330.93 | 0.001567 | 0.012780 | 1511.1 | 2496.1 | 1531.5 | 2662.2 | 3.5606 | 5.4323 |
| 14 | 336.75 | 0.001611 | 0.011485 | 1548.6 | 2476.8 | 1571.1 | 2637.6 | 3.6232 | 5.3717 |
| 15 | 342.24 | 0.001658 | 0.010337 | 1585.6 | 2455.5 | 1610.5 | 2610.5 | 3.6848 | 5.3098 |
| 16 | 347.44 | 0.001711 | 0.009306 | 1622.7 | 2431.7 | 1650.1 | 2580.6 | 3.7461 | 5.2455 |
| 17 | 352.37 | 0.001770 | 0.008364 | 1660.2 | 2405.0 | 1690.3 | 2547.2 | 3.8079 | 5.1777 |
| 18 | 357.06 | 0.001840 | 0.007489 | 1698.9 | 2374.3 | 1732.0 | 2509.1 | 3.8715 | 5.1044 |
| 19 | 361.54 | 0.001924 | 0.006657 | 1739.9 | 2338.1 | 1776.5 | 2464.5 | 3.9388 | 5.0228 |
| 20 | 365.81 | 0.002036 | 0.005834 | 1785.6 | 2293.0 | 1826.3 | 2409.7 | 4.0139 | 4.9269 |
| 21 | 369.89 | 0.002207 | 0.004952 | 1842.1 | 2230.6 | 1888.4 | 2334.6 | 4.1075 | 4.8013 |
| 22 | 373.80 | 0.002742 | 0.003568 | 1961.9 | 2087.1 | 2022.2 | 2165.6 | 4.3110 | 4.5327 |
| 22.09 | 374.14 | 0.003155 | 0.003155 | 2029.6 | 2029.6 | 2099.3 | 2099.3 | 4.4298 | 4.4298 |

# Appendix 8c: Properties of Superheated Steam

| T, °C | P=0.01 MPa ($T_{sat}$=45.81 °C) | | | | P=0.05 MPa ($T_{sat}$=81.33 °C) | | | | P=0.10 MPa ($T_{sat}$=99.63 °C) | | | |
|---|---|---|---|---|---|---|---|---|---|---|---|---|
| | v, m³/kg | u, kJ/kg | h, kJ/kg | s, kJ/kgK | v, m³/kg | u, kJ/kg | h, kJ/kg | s, kJ/kgK | v, m³/kg | u, kJ/kg | h, kJ/kg | s, kJ/kgK |
| $T_{sat}$ | 14.674 | 2437.9 | 2584.7 | 8.1502 | 3.24 | 2483.9 | 2645.9 | 7.5939 | 1.694 | 2506.1 | 2675.5 | 7.3594 |
| 50 | 14.869 | 2443.9 | 2592.6 | 8.1749 | | | | | | | | |
| 100 | 17.196 | 2515.5 | 2687.5 | 8.4479 | 3.418 | 2511.6 | 2682.5 | 7.6947 | 1.6958 | 2506.7 | 2676.2 | 7.3614 |
| 150 | 19.512 | 2587.9 | 2783.0 | 8.6882 | 3.889 | 2585.6 | 2780.1 | 7.9401 | 1.9364 | 2582.8 | 2776.4 | 7.6143 |
| 200 | 21.825 | 2661.3 | 2879.5 | 8.9038 | 4.356 | 2659.9 | 2877.7 | 8.1580 | 2.172 | 2658.1 | 2875.3 | 7.8343 |
| 250 | 24.136 | 2736.0 | 2977.3 | 9.1002 | 4.820 | 2735.0 | 2976.0 | 8.3556 | 2.406 | 2733.7 | 2974.3 | 8.0333 |
| 300 | 26.445 | 2812.1 | 3076.5 | 9.2813 | 5.284 | 2811.3 | 3075.5 | 8.5373 | 2.639 | 2810.4 | 3074.3 | 8.2158 |
| 400 | 31.063 | 2968.9 | 3279.6 | 9.6077 | 6.209 | 2968.5 | 3278.9 | 8.8642 | 3.103 | 2967.9 | 3278.2 | 8.5435 |
| 500 | 35.679 | 3132.3 | 3489.1 | 9.8978 | 7.134 | 3132.0 | 3488.7 | 9.1546 | 3.565 | 3131.6 | 3488.1 | 8.8342 |
| 600 | 40.295 | 3302.5 | 3705.4 | 10.1608 | 8.057 | 3302.2 | 3705.1 | 9.4178 | 4.028 | 3301.9 | 3704.4 | 9.0976 |
| 700 | 44.911 | 3479.6 | 3928.7 | 10.4028 | 8.981 | 3479.4 | 3928.5 | 9.6599 | 4.490 | 3479.2 | 3928.2 | 9.3398 |
| 800 | 49.526 | 3663.8 | 4159.0 | 10.6281 | 9.904 | 3663.6 | 4158.9 | 9.8852 | 4.952 | 3663.5 | 4158.6 | 9.5652 |
| 900 | 54.141 | 3855.0 | 4396.4 | 10.8396 | 10.828 | 3854.9 | 4396.3 | 10.0967 | 5.414 | 3854.8 | 4396.1 | 9.7767 |
| 1000 | 58.757 | 4053.0 | 4640.6 | 11.0393 | 11.751 | 4052.9 | 4640.5 | 10.2964 | 5.875 | 4052.8 | 4640.3 | 9.9764 |
| 1100 | 63.372 | 4257.5 | 4891.2 | 11.2287 | 12.674 | 4257.4 | 4891.1 | 10.4859 | 6.337 | 4257.3 | 4891.0 | 10.1659 |
| 1200 | 67.987 | 4467.9 | 5147.8 | 11.4091 | 13.597 | 4467.8 | 5147.7 | 10.6662 | 6.799 | 4467.7 | 5147.6 | 10.3463 |
| 1300 | 72.602 | 4683.7 | 5409.7 | 11.5811 | 14.521 | 4683.6 | 5409.6 | 10.8382 | 7.260 | 4683.5 | 5409.5 | 10.5183 |

| T, °C | P=0.20 MPa ($T_{sat}$=120.23 °C) | | | | P=0.30 MPa ($T_{sat}$=133.35 °C) | | | | P=0.40 MPa ($T_{sat}$=143.63 °C) | | | |
|---|---|---|---|---|---|---|---|---|---|---|---|---|
| | v, m³/kg | u, kJ/kg | h, kJ/kg | s, kJ/kgK | v, m³/kg | u, kJ/kg | h, kJ/kg | s, kJ/kgK | v, m³/kg | u, kJ/kg | h, kJ/kg | s, kJ/kgK |
| $T_{sat}$ | 0.8857 | 2529.5 | 2706.7 | 7.1272 | 0.6058 | 2543.6 | 2725.3 | 6.9919 | 0.4625 | 2553.6 | 2738.6 | 6.8959 |
| 150 | 0.9596 | 2576.9 | 2768.8 | 7.2795 | 0.6339 | 2570.8 | 2761.0 | 7.0778 | 0.4708 | 2564.5 | 2752.8 | 6.9299 |
| 200 | 1.0803 | 2654.4 | 2870.5 | 7.5066 | 0.7163 | 2650.7 | 2865.6 | 7.3115 | 0.5342 | 2646.8 | 2860.5 | 7.1706 |
| 250 | 1.1988 | 2731.2 | 2971.0 | 7.7086 | 0.7964 | 2728.7 | 2967.6 | 7.5166 | 0.5951 | 2726.1 | 2964.2 | 7.3789 |
| 300 | 1.3162 | 2808.6 | 3071.8 | 7.8926 | 0.8753 | 2806.7 | 3069.3 | 7.7022 | 0.6548 | 2804.8 | 3066.8 | 7.5662 |

| T, °C | v, m³/kg | u, kJ/kg | h, kJ/kg | s, kJ/kgK | v, m³/kg | u, kJ/kg | h, kJ/kg | s, kJ/kgK | v, m³/kg | u, kJ/kg | h, kJ/kg | s, kJ/kgK |
|---|---|---|---|---|---|---|---|---|---|---|---|---|
| 400 | 1.5493 | 2966.7 | 3276.6 | 8.2218 | 1.0315 | 2965.6 | 3275.0 | 8.0330 | 0.7726 | 2964.4 | 3273.4 | 7.8985 |
| 500 | 1.7814 | 3130.8 | 3487.1 | 8.5133 | 1.1867 | 3130.0 | 3486.0 | 8.3251 | 0.8893 | 3129.2 | 3484.9 | 8.1913 |
| 600 | 2.013 | 3301.4 | 3704.0 | 8.7770 | 1.3414 | 3300.8 | 3703.2 | 8.5892 | 1.0055 | 3300.2 | 3702.4 | 8.4558 |
| 700 | 2.244 | 3478.8 | 3927.6 | 9.0194 | 1.4957 | 3478.4 | 3927.1 | 8.8319 | 1.1215 | 3477.9 | 3926.5 | 8.6987 |
| 800 | 2.475 | 3663.1 | 4158.2 | 9.2449 | 1.6499 | 3662.9 | 4157.8 | 9.0576 | 1.2372 | 3662.4 | 4157.3 | 8.9244 |
| 900 | 2.705 | 3854.5 | 4395.8 | 9.4566 | 1.8041 | 3854.2 | 4395.4 | 9.2692 | 1.3529 | 3853.9 | 4395.1 | 9.1362 |
| 1000 | 2.937 | 4052.5 | 4640.0 | 9.6563 | 1.9581 | 4052.3 | 4639.7 | 9.4690 | 1.4685 | 4052.0 | 4639.4 | 9.3360 |
| 1100 | 3.168 | 4257.0 | 4890.7 | 9.8458 | 2.1121 | 4256.8 | 4890.4 | 9.6585 | 1.5840 | 4256.5 | 4890.2 | 9.5256 |
| 1200 | 3.399 | 4467.5 | 5147.5 | 10.0262 | 2.2661 | 4467.2 | 5147.1 | 9.8389 | 1.6996 | 4467.0 | 5146.8 | 9.7060 |
| 1300 | 3.630 | 4683.2 | 5409.3 | 10.1982 | 2.4201 | 4683.0 | 5409.0 | 10.0110 | 1.8151 | 4682.8 | 5408.8 | 9.8780 |

| | P = 0.50 MPa (T_sat = 151.86 °C) | | | | P = 0.60 MPa (T_sat = 158.85 °C) | | | | P = 0.80 MPa (T_sat = 170.43 °C) | | | |
|---|---|---|---|---|---|---|---|---|---|---|---|---|
| T, °C | v, m³/kg | u, kJ/kg | h, kJ/kg | s, kJ/kgK | v, m³/kg | u, kJ/kg | h, kJ/kg | s, kJ/kgK | v, m³/kg | u, kJ/kg | h, kJ/kg | s, kJ/kgK |
| T_sat | 0.3749 | 2561.2 | 2748.7 | 6.8213 | 0.3175 | 2567.4 | 2756.8 | 6.7600 | 0.2404 | 2576.8 | 2769.1 | 6.6628 |
| 200 | 0.4249 | 2642.9 | 2855.4 | 7.0592 | 0.3520 | 2638.9 | 2850.1 | 6.9665 | 0.2608 | 2630.6 | 2839.3 | 6.8158 |
| 250 | 0.4744 | 2723.5 | 2960.7 | 7.2709 | 0.3938 | 2720.9 | 2957.2 | 7.1816 | 0.2931 | 2715.5 | 2950.0 | 7.0384 |
| 300 | 0.5226 | 2802.9 | 3064.2 | 7.4599 | 0.4344 | 2801.0 | 3061.6 | 7.3724 | 0.3241 | 2797.2 | 3056.5 | 7.2328 |
| 350 | 0.5701 | 2882.6 | 3167.7 | 7.6329 | 0.4742 | 2881.2 | 3165.7 | 7.5464 | 0.3544 | 2878.2 | 3161.7 | 7.4089 |
| 400 | 0.6173 | 2963.2 | 3271.9 | 7.7938 | 0.5137 | 2962.1 | 3270.3 | 7.7079 | 0.3843 | 2959.7 | 3267.1 | 7.5716 |
| 500 | 0.7109 | 3128.4 | 3483.9 | 8.0873 | 0.5920 | 3127.6 | 3482.8 | 8.0021 | 0.4433 | 3126.0 | 3480.6 | 7.8673 |
| 600 | 0.8041 | 3299.6 | 3701.7 | 8.3522 | 0.6697 | 3299.1 | 3700.9 | 8.2674 | 0.5018 | 3297.9 | 3699.4 | 8.1333 |
| 700 | 0.8969 | 3477.5 | 3925.9 | 8.5952 | 0.7472 | 3477.0 | 3925.3 | 8.5107 | 0.5601 | 3476.2 | 3924.2 | 8.3770 |
| 800 | 0.9896 | 3662.1 | 4156.9 | 8.8211 | 0.8245 | 3661.8 | 4156.6 | 8.7367 | 0.6181 | 3661.1 | 4155.6 | 8.6033 |
| 900 | 1.0822 | 3853.6 | 4394.7 | 9.0329 | 0.9017 | 3853.4 | 4394.4 | 8.9486 | 0.6761 | 3852.8 | 4393.7 | 8.8153 |
| 1000 | 1.1747 | 4051.8 | 4639.1 | 9.2328 | 0.9788 | 4051.5 | 4638.8 | 9.1485 | 0.7340 | 4051.0 | 4638.2 | 9.0153 |
| 1100 | 1.2672 | 4256.3 | 4889.9 | 9.4224 | 1.0559 | 4256.1 | 4889.6 | 9.3381 | 0.7919 | 4255.6 | 4889.1 | 9.2050 |
| 1200 | 1.3596 | 4466.8 | 5146.6 | 9.6029 | 1.1330 | 4466.5 | 5146.3 | 9.5185 | 0.8497 | 4466.1 | 5145.9 | 9.3855 |
| 1300 | 1.4521 | 4682.5 | 5408.6 | 9.7749 | 1.2101 | 4682.3 | 5408.3 | 9.6906 | 0.9076 | 4681.8 | 5407.9 | 9.5575 |

(continued)

(continued)

| | P=1.00 MPa (T_sat =179.91°C) | | | | | P=1.20 MPa (T_sat =187.99°C) | | | | | P=1.40 MPa (T_sat =195.07°C) | | | |
|---|---|---|---|---|---|---|---|---|---|---|---|---|---|---|
| T, °C | v, m³/kg | u, kJ/kg | h, kJ/kg | s, kJ/kgK | | v, m³/kg | u, kJ/kg | h, kJ/kg | s, kJ/kgK | | v, m³/kg | u, kJ/kg | h, kJ/kg | s, kJ/kgK |
| T_sat | 0.19444 | 2583.6 | 2778.1 | 6.5865 | | 0.16333 | 2588.8 | 2784.4 | 6.5233 | | 0.14084 | 2592.8 | 2790.0 | 6.4693 |
| 200 | 0.2060 | 2621.9 | 2827.9 | 6.6940 | | 0.16930 | 2612.8 | 2815.9 | 6.5898 | | 0.14302 | 2603.1 | 2803.3 | 6.4975 |
| 250 | 0.2327 | 2709.9 | 2942.6 | 6.9247 | | 0.19234 | 2704.2 | 2935.0 | 6.8294 | | 0.16350 | 2698.3 | 2927.2 | 6.7467 |
| 300 | 0.2579 | 2793.2 | 3051.2 | 7.1229 | | 0.2138 | 2789.2 | 3045.8 | 7.0317 | | 0.18228 | 2785.2 | 3040.4 | 6.9534 |
| 350 | 0.2825 | 2875.2 | 3157.7 | 7.3011 | | 0.2345 | 2872.2 | 3153.6 | 7.2121 | | 0.2003 | 2869.2 | 3149.5 | 7.1360 |
| 400 | 0.3066 | 2957.3 | 3263.9 | 7.4651 | | 0.2548 | 2954.9 | 3260.7 | 7.3774 | | 0.2178 | 2952.5 | 3257.5 | 7.3026 |
| 500 | 0.3541 | 3124.4 | 3478.5 | 7.7622 | | 0.2946 | 3122.8 | 3476.3 | 7.6759 | | 0.2521 | 3121.1 | 3474.1 | 7.6027 |
| 600 | 0.4011 | 3296.8 | 3697.9 | 8.0290 | | 0.3339 | 3295.6 | 3696.3 | 7.9435 | | 0.2860 | 3294.4 | 3694.8 | 7.8710 |
| 700 | 0.4478 | 3475.3 | 3923.1 | 8.2731 | | 0.3729 | 3474.4 | 3922.0 | 8.1881 | | 0.3195 | 3473.6 | 3920.8 | 8.1160 |
| 800 | 0.4943 | 3660.4 | 4154.7 | 8.4996 | | 0.4118 | 3659.7 | 4153.8 | 8.4148 | | 0.3528 | 3659.0 | 4153.0 | 8.3431 |
| 900 | 0.5407 | 3852.2 | 4392.9 | 8.7118 | | 0.4505 | 3851.6 | 4392.2 | 8.6272 | | 0.3861 | 3851.1 | 4391.5 | 8.5556 |
| 1000 | 0.5871 | 4050.5 | 4637.6 | 8.9119 | | 0.4892 | 4050.0 | 4637.0 | 8.8274 | | 0.4192 | 4049.5 | 4636.4 | 8.7559 |
| 1100 | 0.6335 | 4255.1 | 4888.6 | 9.1017 | | 0.5278 | 4254.6 | 4888.0 | 9.0172 | | 0.4524 | 4254.1 | 4887.5 | 8.9457 |
| 1200 | 0.6798 | 4465.6 | 5145.4 | 9.2822 | | 0.5665 | 4465.1 | 5144.9 | 9.1977 | | 0.4855 | 4464.7 | 5144.4 | 9.1262 |
| 1300 | 0.7261 | 4681.3 | 5407.4 | 9.4543 | | 0.6051 | 4680.9 | 5407.0 | 9.3698 | | 0.5186 | 4680.4 | 5406.5 | 9.2984 |

| | P=1.60 MPa (T_sat =201.41°C) | | | | | P=1.80 MPa (T_sat =207.15°C) | | | | | P=2.00 MPa (T_sat =212.42°C) | | | |
|---|---|---|---|---|---|---|---|---|---|---|---|---|---|---|
| T, °C | v, m³/kg | u, kJ/kg | h, kJ/kg | s, kJ/kgK | | v, m³/kg | u, kJ/kg | h, kJ/kg | s, kJ/kgK | | v, m³/kg | u, kJ/kg | h, kJ/kg | s, kJ/kgK |
| T_sat | 0.12380 | 2596.0 | 2794.0 | 6.4218 | | 0.11042 | 2598.4 | 2797.1 | 6.3794 | | 0.09963 | 2600.3 | 2799.5 | 6.3409 |
| 225 | 0.13287 | 2644.7 | 2857.3 | 6.5518 | | 0.11673 | 2636.6 | 2846.7 | 6.4808 | | 0.10377 | 2628.3 | 2835.8 | 6.4147 |
| 250 | 0.14184 | 2692.3 | 2919.2 | 6.6732 | | 0.12497 | 2686.0 | 2911.0 | 6.6066 | | 0.11144 | 2679.6 | 2902.5 | 6.5453 |
| 300 | 0.15862 | 2781.1 | 3034.8 | 6.8844 | | 0.14021 | 2776.9 | 3029.2 | 6.8226 | | 0.12547 | 2772.6 | 3023.5 | 6.7664 |
| 350 | 0.17456 | 2866.1 | 3145.4 | 7.0694 | | 0.15457 | 2863.0 | 3141.2 | 7.0100 | | 0.13857 | 2859.8 | 3137.0 | 6.9563 |
| 400 | 0.19005 | 2950.1 | 3254.2 | 7.2374 | | 0.16847 | 2947.7 | 3250.9 | 7.1794 | | 0.15120 | 2945.2 | 3247.6 | 7.1271 |
| 500 | 0.2203 | 3119.5 | 3472.0 | 7.5390 | | 0.19550 | 3117.9 | 3469.8 | 7.4825 | | 0.17568 | 3116.2 | 3467.6 | 7.4317 |
| 600 | 0.2500 | 3293.3 | 3693.2 | 7.8080 | | 0.2220 | 3292.1 | 3691.7 | 7.7523 | | 0.19960 | 3290.9 | 3690.1 | 7.7024 |
| 700 | 0.2794 | 3472.7 | 3919.7 | 8.0535 | | 0.2482 | 3471.8 | 3918.5 | 7.9983 | | 0.2232 | 3470.9 | 3917.4 | 7.9487 |
| 800 | 0.3086 | 3658.3 | 4152.1 | 8.2808 | | 0.2742 | 3657.6 | 4151.2 | 8.2258 | | 0.2467 | 3657.0 | 4150.3 | 8.1765 |

| | v, m³/kg | u, kJ/kg | h, kJ/kg | s, kJ/kgK | v, m³/kg | u, kJ/kg | h, kJ/kg | s, kJ/kgK | v, m³/kg | u, kJ/kg | h, kJ/kg | s, kJ/kgK |
|---|---|---|---|---|---|---|---|---|---|---|---|---|
| 900 | 0.3377 | 3850.5 | 4390.8 | 8.4935 | 0.3001 | 3849.9 | 4390.1 | 8.4386 | 0.2700 | 3849.3 | 4389.4 | 8.3895 |
| 1000 | 0.3668 | 4049.0 | 4635.8 | 8.6938 | 0.3260 | 4048.5 | 4635.2 | 8.6391 | 0.2933 | 4048.0 | 4634.6 | 8.5901 |
| 1100 | 0.3958 | 4253.7 | 4887.0 | 8.8837 | 0.3518 | 4253.2 | 4886.4 | 8.8290 | 0.3166 | 4252.7 | 4885.9 | 8.7800 |
| 1200 | 0.4248 | 4464.2 | 5143.9 | 9.0643 | 0.3776 | 4463.7 | 5143.4 | 9.0096 | 0.3398 | 4463.3 | 5142.9 | 8.9607 |
| 1300 | 0.4538 | 4679.9 | 5406.0 | 9.2364 | 0.4034 | 4679.5 | 5405.6 | 9.1818 | 0.3631 | 4679.0 | 5405.1 | 9.1329 |

| | P=2.50 MPa ($T_{sat}$=223.99°C) | | | | P=3.00 MPa ($T_{sat}$=233.90°C) | | | | P=3.50 MPa ($T_{sat}$=242.60°C) | | | |
|---|---|---|---|---|---|---|---|---|---|---|---|---|
| T, °C | v, m³/kg | u, kJ/kg | h, kJ/kg | s, kJ/kgK | v, m³/kg | u, kJ/kg | h, kJ/kg | s, kJ/kgK | v, m³/kg | u, kJ/kg | h, kJ/kg | s, kJ/kgK |
| $T_{sat}$ | 0.07998 | 2603.1 | 2803.1 | 6.2575 | 0.06668 | 2604.1 | 2804.2 | 6.1869 | 0.05070 | 2603.7 | 2803.4 | 6.1253 |
| 225 | 0.08027 | 2605.6 | 2806.3 | 6.2639 | | | | | | | | |
| 250 | 0.08700 | 2662.6 | 2880.1 | 6.4085 | 0.07058 | 2644.0 | 2855.8 | 6.2872 | 0.05872 | 2623.7 | 2829.2 | 6.1749 |
| 300 | 0.09890 | 2761.6 | 3008.8 | 6.6438 | 0.08114 | 2750.1 | 2993.5 | 6.5390 | 0.06842 | 2738 | 2977.5 | 6.4461 |
| 350 | 0.10976 | 2851.9 | 3126.3 | 6.8403 | 0.09053 | 2843.7 | 3115.3 | 6.7428 | 0.07678 | 2835.3 | 3104.0 | 6.6579 |
| 400 | 0.12010 | 2939.1 | 3239.3 | 7.0148 | 0.09936 | 2932.8 | 3230.9 | 6.9212 | 0.08453 | 2926.4 | 3222.3 | 6.8405 |
| 450 | 0.13014 | 3025.5 | 3350.8 | 7.1746 | 0.10787 | 3020.4 | 3344.0 | 7.0834 | 0.09196 | 3015.3 | 3337.2 | 7.0052 |
| 500 | 0.13993 | 3112.1 | 3462.1 | 7.3234 | 0.11619 | 3108.0 | 3456.5 | 7.2338 | 0.09918 | 3103.0 | 3450.9 | 7.1572 |
| 600 | 0.15930 | 3288.0 | 3686.3 | 7.5960 | 0.13243 | 3285.0 | 3682.3 | 7.5085 | 0.11324 | 3282.1 | 3678.4 | 7.4339 |
| 700 | 0.17832 | 3468.7 | 3914.5 | 7.8435 | 0.14838 | 3466.5 | 3911.7 | 7.7571 | 0.12699 | 3464.3 | 3908.8 | 7.6837 |
| 800 | 0.19716 | 3655.3 | 4148.2 | 8.0720 | 0.16414 | 3653.5 | 4145.9 | 7.9862 | 0.14056 | 3651.8 | 4143.7 | 7.9134 |
| 900 | 0.21590 | 3847.9 | 4387.6 | 8.2853 | 0.17980 | 3846.5 | 4385.9 | 8.1999 | 0.15402 | 3845.0 | 4384.1 | 8.1276 |
| 1000 | 0.2346 | 4046.7 | 4633.1 | 8.4861 | 0.19541 | 4045.4 | 4631.6 | 8.4009 | 0.16743 | 4044.1 | 4630.1 | 8.3288 |
| 1100 | 0.2532 | 4251.5 | 4884.6 | 8.6762 | 0.21098 | 4250.3 | 4883.3 | 8.5912 | 0.18080 | 4249.2 | 4881.9 | 8.5192 |
| 1200 | 0.2718 | 4462.1 | 5141.7 | 8.8569 | 0.22652 | 4460.9 | 5140.5 | 8.7720 | 0.19415 | 4459.8 | 5139.3 | 8.7000 |
| 1300 | 0.2905 | 4677.8 | 5404.0 | 9.0291 | 0.24206 | 4676.6 | 5402.8 | 8.9442 | 0.20749 | 4675.5 | 5401.7 | 8.8723 |

(continued)

(continued)

| P=4.0 MPa ($T_{sat}$ =250.40°C) | | | | P=4.5 MPa ($T_{sat}$ =257.49°C) | | | | P=5.0 MPa ($T_{sat}$ =263.99°C) | | | |
|---|---|---|---|---|---|---|---|---|---|---|---|
| T, °C | v, m³/kg | u, kJ/kg | h, kJ/kg | s, kJ/kgK | v, m³/kg | u, kJ/kg | h, kJ/kg | s, kJ/kgK | v, m³/kg | u, kJ/kg | h, kJ/kg | s, kJ/kgK |
| $T_{sat}$ | 0.04978 | 2602.3 | 2801.4 | 6.0701 | 0.04406 | 2600.1 | 2798.3 | 6.0198 | 0.03944 | 2597.1 | 2794.3 | 5.9734 |
| 275 | 0.05457 | 2667.9 | 2886.2 | 6.2285 | 0.04730 | 2650.3 | 2863.2 | 6.1401 | 0.04141 | 2631.3 | 2838.3 | 6.0544 |
| 300 | 0.05884 | 2725.3 | 2960.7 | 6.3615 | 0.05135 | 2712.0 | 2943.1 | 6.2828 | 0.04532 | 2698.0 | 2924.5 | 6.2084 |
| 350 | 0.06645 | 2826.7 | 3092.5 | 6.5821 | 0.05840 | 2817.8 | 3080.6 | 6.5131 | 0.05194 | 2808.7 | 3068.4 | 6.4493 |
| 400 | 0.07341 | 2919.9 | 3213.6 | 6.7690 | 0.06475 | 2913.3 | 3204.7 | 6.7047 | 0.05781 | 2906.6 | 3195.7 | 6.6459 |
| 450 | 0.08002 | 3010.2 | 3330.3 | 6.9363 | 0.07074 | 3005.0 | 3323.3 | 6.8746 | 0.06330 | 2999.7 | 3316.2 | 6.8186 |
| 500 | 0.08643 | 3099.5 | 3445.3 | 7.0901 | 0.07651 | 3095.3 | 3439.6 | 7.0301 | 0.06857 | 3091.0 | 3433.8 | 6.9759 |
| 600 | 0.09885 | 3279.1 | 3674.4 | 7.3688 | 0.08765 | 3276.0 | 3670.5 | 7.3110 | 0.07869 | 3273.0 | 3666.5 | 7.2589 |
| 700 | 0.11095 | 3462.1 | 3905.9 | 7.6198 | 0.09847 | 3459.9 | 3903.0 | 7.5631 | 0.08849 | 3457.6 | 3900.1 | 7.5122 |
| 800 | 0.12287 | 3650.0 | 4141.5 | 7.8502 | 0.10911 | 3648.3 | 4139.3 | 7.7942 | 0.09811 | 3646.6 | 4137.1 | 7.7440 |
| 900 | 0.13469 | 3843.6 | 4382.3 | 8.0647 | 0.11965 | 3842.2 | 4380.6 | 8.0091 | 0.10762 | 3840.7 | 4378.8 | 7.9593 |
| 1000 | 0.14645 | 4042.9 | 4628.7 | 8.2662 | 0.13013 | 4041.6 | 4627.2 | 8.2108 | 0.11707 | 4040.4 | 4625.7 | 8.1612 |
| 1100 | 0.15817 | 4248.0 | 4880.6 | 8.4567 | 0.14056 | 4246.8 | 4879.3 | 8.4015 | 0.12648 | 4245.6 | 4878.0 | 8.3520 |
| 1200 | 0.16987 | 4458.6 | 5138.1 | 8.6376 | 0.15098 | 4457.5 | 5136.9 | 8.5825 | 0.13587 | 4456.3 | 5135.7 | 8.5331 |
| 1300 | 0.18156 | 4674.3 | 5400.5 | 8.8100 | 0.16139 | 4673.1 | 5399.4 | 8.7549 | 0.14526 | 4672.0 | 5398.2 | 8.705 |

| P=6.0 MPa ($T_{sat}$ =257.64°C) | | | | P=7.0 MPa ($T_{sat}$ =285.88°C) | | | | P=8.0 MPa ($T_{sat}$ =295.06°C) | | | |
|---|---|---|---|---|---|---|---|---|---|---|---|
| T, °C | v, m³/kg | u, kJ/kg | h, kJ/kg | s, kJ/kgK | v, m³/kg | u, kJ/kg | h, kJ/kg | s, kJ/kgK | v, m³/kg | u, kJ/kg | h, kJ/kg | s, kJ/kgK |
| $T_{sat}$ | 0.03244 | 2589.7 | 2784.3 | 5.8892 | 0.02737 | 2580.5 | 2772.1 | 5.8133 | 0.02352 | 2569.8 | 2758.0 | 5.7432 |
| 300 | 0.03616 | 2667.2 | 2884.2 | 6.0674 | 0.02947 | 2632.2 | 2838.4 | 5.9305 | 0.02426 | 2590.9 | 2785.0 | 5.7906 |
| 350 | 0.04223 | 2789.6 | 3043.0 | 6.3335 | 0.03524 | 2769.4 | 3016.0 | 6.2283 | 0.02995 | 2747.7 | 2987.3 | 6.1301 |
| 400 | 0.04739 | 2892.9 | 3177.2 | 6.5408 | 0.03993 | 2878.6 | 3158.1 | 6.4478 | 0.03432 | 2863.8 | 3138.3 | 6.3634 |
| 450 | 0.05214 | 2988.9 | 3301.8 | 6.7193 | 0.04416 | 2978.0 | 3287.1 | 6.6327 | 0.03817 | 2966.7 | 3272.0 | 6.5551 |
| 500 | 0.05665 | 3082.2 | 3422.2 | 6.8803 | 0.04814 | 3073.4 | 3410.3 | 6.7975 | 0.04175 | 3064.3 | 3398.3 | 6.7240 |
| 550 | 0.06101 | 3174.6 | 3540.6 | 7.0288 | 0.05195 | 3167.2 | 3530.9 | 6.9486 | 0.04516 | 3159.8 | 3521.0 | 6.8778 |
| 600 | 0.06525 | 3266.9 | 3658.4 | 7.1677 | 0.05565 | 3260.7 | 3650.3 | 7.0894 | 0.04845 | 3254.4 | 3642.0 | 7.0206 |
| 700 | 0.07352 | 3453.1 | 3894.2 | 7.4234 | 0.06283 | 3448.5 | 3888.3 | 7.3476 | 0.05481 | 3443.9 | 3882.4 | 7.2812 |
| 800 | 0.0816 | 3643.1 | 4132.7 | 7.6566 | 0.06981 | 3639.5 | 4128.2 | 7.5822 | 0.06097 | 3636.0 | 4123.8 | 7.5173 |
| 900 | 0.08958 | 3837.8 | 4375.3 | 7.8727 | 0.07669 | 3835.0 | 4371.8 | 7.7991 | 0.06702 | 3832.1 | 4368.3 | 7.7351 |
| 1000 | 0.09749 | 4037.8 | 4622.7 | 8.0751 | 0.08350 | 4035.3 | 4619.8 | 8.0020 | 0.07301 | 4032.8 | 4616.9 | 7.9384 |

| T | v, m³/kg | u, kJ/kg | h, kJ/kg | s, kJ/kgK | v, m³/kg | u, kJ/kg | h, kJ/kg | s, kJ/kgK | v, m³/kg | u, kJ/kg | h, kJ/kg | s, kJ/kgK |
|---|---|---|---|---|---|---|---|---|---|---|---|---|
| 1100 | 0.10536 | 4243.3 | 4875.4 | 8.2661 | 0.09027 | 4240.9 | 4872.8 | 8.1933 | 0.07896 | 4238.6 | 4870.3 | 8.1300 |
| 1200 | 0.11321 | 4454.0 | 5133.3 | 8.4474 | 0.09703 | 4451.7 | 5130.9 | 8.3747 | 0.08489 | 4449.5 | 5128.5 | 8.3115 |
| 1300 | 0.12106 | 4669.6 | 5396.0 | 8.6199 | 0.10377 | 4667.3 | 5393.7 | 8.5475 | 0.09080 | 4665.0 | 5391.5 | 8.4842 |

|  | P = 9.0 MPa ($T_{sat}$ = 303.4°C) | | | | P = 10.0 MPa ($T_{sat}$ = 311.06°C) | | | | P = 12.5 MPa ($T_{sat}$ = 327.89°C) | | | |
|---|---|---|---|---|---|---|---|---|---|---|---|---|
| T, °C | v, m³/kg | u, kJ/kg | h, kJ/kg | s, kJ/kgK | v, m³/kg | u, kJ/kg | h, kJ/kg | s, kJ/kgK | v, m³/kg | u, kJ/kg | h, kJ/kg | s, kJ/kgK |
| $T_{sat}$ | 0.02048 | 2557.8 | 2742.1 | 5.6772 | 0.018026 | 2544.4 | 2724.7 | 5.6141 | 0.013495 | 2505.1 | 2673.8 | 5.4624 |
| 325 | 0.02327 | 2646.6 | 2856.0 | 5.8712 | 0.019861 | 2610.4 | 2809.1 | 5.7568 |  |  |  |  |
| 350 | 0.02580 | 2724.4 | 2956.6 | 6.0361 | 0.02242 | 2699.2 | 2923.4 | 5.9443 | 0.016126 | 2624.6 | 2826.2 | 5.7118 |
| 400 | 0.02993 | 2848.4 | 3117.8 | 6.2854 | 0.02641 | 2832.4 | 3096.5 | 6.2120 | 0.02000 | 2789.3 | 3039.3 | 6.0417 |
| 450 | 0.03350 | 2955.2 | 3256.6 | 6.4844 | 0.02975 | 2943.4 | 3240.9 | 6.4190 | 0.02299 | 2912.5 | 3199.8 | 6.2719 |
| 500 | 0.03677 | 3055.2 | 3386.1 | 6.6576 | 0.03279 | 3045.8 | 3373.7 | 6.5966 | 0.02560 | 3021.7 | 3341.8 | 6.4618 |
| 550 | 0.03987 | 3152.2 | 3511.0 | 6.8142 | 0.03564 | 3144.6 | 3500.9 | 6.7561 | 0.02801 | 3125.0 | 3475.2 | 6.6290 |
| 600 | 0.04285 | 3248.1 | 3633.7 | 6.9589 | 0.03837 | 3241.7 | 3625.3 | 6.9029 | 0.03029 | 3225.4 | 3604.0 | 6.7810 |
| 650 | 0.04574 | 3343.6 | 3755.3 | 7.0943 | 0.04101 | 3338.2 | 3748.2 | 7.0398 | 0.03248 | 3324.4 | 3730.4 | 6.9218 |
| 700 | 0.04857 | 3439.3 | 3876.5 | 7.2221 | 0.04358 | 3434.7 | 3870.5 | 7.1687 | 0.03460 | 3422.9 | 3855.3 | 7.0536 |
| 800 | 0.05409 | 3632.5 | 4119.3 | 7.4596 | 0.04859 | 3628.9 | 4114.8 | 7.4077 | 0.03869 | 3620.0 | 4103.6 | 7.2965 |
| 900 | 0.05950 | 3829.2 | 4364.8 | 7.6783 | 0.05349 | 3826.3 | 4361.2 | 7.6272 | 0.04267 | 3819.1 | 4352.5 | 7.5182 |
| 1000 | 0.06485 | 4030.3 | 4614.0 | 7.8821 | 0.05832 | 4027.8 | 4611.0 | 7.8315 | 0.04658 | 4021.6 | 4603.8 | 7.7237 |
| 1100 | 0.07016 | 4236.3 | 4867.7 | 8.0740 | 0.06312 | 4234.0 | 4865.1 | 8.0237 | 0.05045 | 4228.2 | 4858.8 | 7.9165 |
| 1200 | 0.07544 | 4447.2 | 5126.2 | 8.2556 | 0.06789 | 4444.9 | 5123.8 | 8.2055 | 0.05430 | 4439.3 | 5118.0 | 8.0937 |
| 1300 | 0.08072 | 4662.7 | 5389.2 | 8.4284 | 0.07265 | 4660.5 | 5387.0 | 8.3783 | 0.05813 | 4654.8 | 5381.4 | 8.2717 |

|  | P = 15.0 MPa ($T_{sat}$ = 342.24°C) | | | | P = 17.5 MPa ($T_{sat}$ = 354.75°C) | | | | P = 20.0 MPa ($T_{sat}$ = 365.81°C) | | | |
|---|---|---|---|---|---|---|---|---|---|---|---|---|
| T, °C | v, m³/kg | u, kJ/kg | h, kJ/kg | s, kJ/kgK | v, m³/kg | u, kJ/kg | h, kJ/kg | s, kJ/kgK | v, m³/kg | u, kJ/kg | h, kJ/kg | s, kJ/kgK |
| $T_{sat}$ | 0.010337 | 2455.5 | 2610.5 | 5.3098 | 0.007920 | 2390.2 | 2528.8 | 5.1419 | 0.005834 | 2293.0 | 2409.7 | 4.9269 |
| 350 | 0.011470 | 2520.4 | 2692.4 | 5.4421 |  |  |  |  |  |  |  |  |
| 400 | 0.015649 | 2740.7 | 2975.5 | 5.8811 | 0.012447 | 2685.0 | 2902.9 | 5.7213 | 0.009942 | 2619.3 | 2818.1 | 5.5540 |
| 450 | 0.018445 | 2879.5 | 3156.2 | 6.1404 | 0.015174 | 2844.2 | 3109.7 | 6.0184 | 0.012695 | 2806.2 | 3060.1 | 5.9017 |

(continued)

(continued)

| T, °C | P=15.0 MPa ($T_{sat}$ =342.24 °C) | | | | P=17.5 MPa ($T_{sat}$ =354.75 °C) | | | | P=20.0 MPa ($T_{sat}$ =365.81 °C) | | | |
|---|---|---|---|---|---|---|---|---|---|---|---|---|
| | v, m³/kg | u, kJ/kg | h, kJ/kg | s, kJ/kgK | v, m³/kg | u, kJ/kg | h, kJ/kg | s, kJ/kgK | v, m³/kg | u, kJ/kg | h, kJ/kg | s, kJ/kgK |
| 500 | 0.02080 | 2996.6 | 3308.6 | 6.3443 | 0.017358 | 2970.3 | 3274.1 | 6.2383 | 0.014768 | 2942.9 | 3238.2 | 6.1401 |
| 550 | 0.02293 | 3104.7 | 3448.6 | 6.5199 | 0.019288 | 3083.9 | 3421.4 | 6.4230 | 0.016555 | 3062.4 | 3393.5 | 6.3348 |
| 600 | 0.02491 | 3208.6 | 3582.3 | 6.6776 | 0.02106 | 3191.5 | 3560.1 | 6.5866 | 0.018178 | 3174.0 | 3537.6 | 6.5048 |
| 650 | 0.02680 | 3310.3 | 3712.3 | 6.8224 | 0.02274 | 3296.0 | 3693.9 | 6.7357 | 0.019693 | 3281.4 | 3675.3 | 6.6582 |
| 700 | 0.02861 | 3410.9 | 3840.1 | 6.9572 | 0.02434 | 3398.7 | 3824.6 | 6.8736 | 0.02113 | 3386.4 | 3809.0 | 6.7993 |
| 800 | 0.03210 | 3610.9 | 4092.4 | 7.2040 | 0.02738 | 3601.8 | 4081.1 | 7.1244 | 0.02385 | 3592.7 | 4069.7 | 7.0544 |
| 900 | 0.03546 | 3811.9 | 4343.8 | 7.4279 | 0.03031 | 3804.7 | 4335.1 | 7.3507 | 0.02645 | 3797.5 | 4326.4 | 7.2830 |
| 1000 | 0.03875 | 4015.4 | 4596.6 | 7.6348 | 0.03316 | 4009.3 | 4589.5 | 7.5589 | 0.02897 | 4003.1 | 4582.5 | 7.4925 |
| 1100 | 0.04200 | 4222.6 | 4852.6 | 7.8283 | 0.03597 | 4216.9 | 4846.4 | 7.7531 | 0.03145 | 4211.3 | 4840.2 | 7.6874 |
| 1200 | 0.04523 | 4433.8 | 5112.3 | 8.0108 | 0.03876 | 4428.3 | 5106.6 | 7.9360 | 0.03391 | 4422.8 | 5101.0 | 7.8707 |
| 1300 | 0.04845 | 4649.1 | 5376.0 | 8.1840 | 0.04154 | 4643.5 | 5370.5 | 8.1093 | 0.03636 | 4638.0 | 5365.1 | 8.0442 |

| T, °C | P=25.0 MPa | | | | P=30.0 MPa | | | | P=35.0 MPa | | | |
|---|---|---|---|---|---|---|---|---|---|---|---|---|
| | v, m³/kg | u, kJ/kg | h, kJ/kg | s, kJ/kgK | v, m³/kg | u, kJ/kg | h, kJ/kg | s, kJ/kgK | v, m³/kg | u, kJ/kg | h, kJ/kg | s, kJ/kgK |
| 375 | 0.0019731 | 1798.7 | 1848.0 | 4.0320 | 0.0017892 | 1737.8 | 1791.5 | 3.9305 | 0.0017003 | 1702.9 | 1762.4 | 3.8722 |
| 400 | 0.006004 | 2430.1 | 2580.2 | 5.1418 | 0.002790 | 2067.4 | 2151.1 | 4.4728 | 0.002100 | 1914.1 | 1987.6 | 4.2126 |
| 425 | 0.007881 | 2609.2 | 2806.3 | 5.4723 | 0.005303 | 2455.1 | 2614.2 | 5.1504 | 0.003428 | 2253.4 | 2373.4 | 4.7747 |
| 450 | 0.009162 | 2720.7 | 2949.7 | 5.6744 | 0.006735 | 2619.3 | 2821.4 | 5.4424 | 0.004961 | 2498.7 | 2672.4 | 5.1962 |
| 500 | 0.011123 | 2884.3 | 3162.4 | 5.9592 | 0.008678 | 2820.7 | 3081.1 | 5.7905 | 0.006927 | 2751.9 | 2994.4 | 5.6282 |
| 550 | 0.012724 | 3017.5 | 3335.6 | 6.1765 | 0.010168 | 2970.3 | 3275.4 | 6.0342 | 0.008345 | 2921.0 | 3213.0 | 5.9026 |
| 600 | 0.014137 | 3137.9 | 3491.4 | 6.3602 | 0.011446 | 3100.5 | 3443.9 | 6.2331 | 0.009527 | 3062.0 | 3395.5 | 6.1179 |
| 650 | 0.015433 | 3251.6 | 3637.4 | 6.5229 | 0.012596 | 3221.0 | 3598.9 | 6.4058 | 0.010575 | 3189.8 | 3559.9 | 6.3010 |
| 700 | 0.016646 | 3361.3 | 3777.5 | 6.6707 | 0.013661 | 3335.8 | 3745.6 | 6.5606 | 0.011533 | 3309.8 | 3713.5 | 6.4631 |
| 800 | 0.018912 | 3574.3 | 4047.1 | 6.9345 | 0.015623 | 3555.5 | 4024.2 | 6.8332 | 0.013278 | 3536.7 | 4001.5 | 6.7450 |
| 900 | 0.021045 | 3783.0 | 4309.1 | 7.1680 | 0.017448 | 3768.5 | 4291.9 | 7.0718 | 0.014883 | 3754.0 | 4274.9 | 6.9386 |
| 1000 | 0.02310 | 3990.9 | 4568.5 | 7.3802 | 0.019196 | 3978.8 | 4554.7 | 7.2867 | 0.016410 | 3966.7 | 4541.1 | 7.2064 |
| 1100 | 0.02512 | 4200.2 | 4828.2 | 7.5765 | 0.020903 | 4189.2 | 4816.3 | 7.4845 | 0.017895 | 4178.3 | 4804.6 | 7.4037 |
| 1200 | 0.02711 | 4412.0 | 5089.9 | 7.7605 | 0.022589 | 4401.3 | 5079.0 | 7.6692 | 0.019360 | 4390.7 | 5068.3 | 7.5910 |
| 1300 | 0.02910 | 4626.9 | 5354.4 | 7.9342 | 0.024266 | 4616.0 | 5344.0 | 7.8432 | 0.020815 | 4605.1 | 5333.6 | 7.7653 |

## P=40.0 MPa

| T, °C | v, m³/kg | u, kJ/kg | h, kJ/kg | s, kJ/kgK |
|---|---|---|---|---|
| 375 | 0.0016407 | 1677.1 | 1742.8 | 3.8290 |
| 400 | 0.0019077 | 1854.6 | 1930.9 | 4.1135 |
| 425 | 0.002532 | 2096.9 | 2198.1 | 4.5029 |
| 450 | 0.003693 | 2365.1 | 2512.8 | 4.9459 |
| 500 | 0.005622 | 2678.4 | 2903.3 | 5.4700 |
| 550 | 0.006984 | 2869.7 | 3149.1 | 5.7785 |
| 600 | 0.008094 | 3022.6 | 3346.4 | 6.0144 |
| 650 | 0.009063 | 3158.0 | 3520.6 | 6.2054 |
| 700 | 0.009941 | 3283.6 | 3681.2 | 6.3750 |
| 800 | 0.011523 | 3517.8 | 3978.7 | 6.6662 |
| 900 | 0.012962 | 3739.4 | 4257.9 | 6.9150 |
| 1000 | 0.014324 | 3954.6 | 4527.6 | 7.1356 |
| 1100 | 0.015642 | 4167.4 | 4793.1 | 7.3364 |
| 1200 | 0.016940 | 4380.1 | 5057.7 | 7.5224 |
| 1300 | 0.018229 | 4594.3 | 5323.5 | 7.6969 |

## P=50.0 MPa

| T, °C | v, m³/kg | u, kJ/kg | h, kJ/kg | s, kJ/kgK |
|---|---|---|---|---|
| 375 | 0.0015594 | 1638.6 | 1716.6 | 3.7639 |
| 400 | 0.0017309 | 1788.1 | 1874.6 | 4.0031 |
| 425 | 0.002007 | 1959.7 | 2060.0 | 4.2734 |
| 450 | 0.002486 | 2159.6 | 2284.0 | 4.5884 |
| 500 | 0.003892 | 2525.5 | 2720.1 | 5.1726 |
| 550 | 0.005118 | 2763.6 | 3019.5 | 5.5485 |
| 600 | 0.006112 | 2942.0 | 3247.6 | 5.8178 |
| 650 | 0.006966 | 3093.5 | 3441.8 | 6.0342 |
| 700 | 0.007727 | 3230.5 | 3616.8 | 6.2189 |
| 800 | 0.009076 | 3479.8 | 3933.6 | 6.5290 |
| 900 | 0.010283 | 3710.3 | 4224.4 | 6.7882 |
| 1000 | 0.011411 | 3930.5 | 4501.1 | 7.0146 |
| 1100 | 0.012496 | 4145.7 | 4770.5 | 7.2184 |
| 1200 | 0.013561 | 4359.1 | 5037.2 | 7.4058 |
| 1300 | 0.014616 | 4572.8 | 5303.6 | 7.5808 |

## P=60.0 MPa

| T, °C | v, m³/kg | u, kJ/kg | h, kJ/kg | s, kJ/kgK |
|---|---|---|---|---|
| 375 | 0.0015028 | 1609.4 | 1699.5 | 3.7141 |
| 400 | 0.0016335 | 1745.4 | 1843.4 | 3.9318 |
| 425 | 0.0018165 | 1892.7 | 2001.7 | 4.1626 |
| 450 | 0.002085 | 2053.9 | 2179.0 | 4.4121 |
| 500 | 0.002956 | 2390.6 | 2567.9 | 4.9321 |
| 550 | 0.003956 | 2658.8 | 2896.2 | 5.3441 |
| 600 | 0.004834 | 2861.1 | 3151.2 | 5.6452 |
| 650 | 0.005595 | 3028.8 | 3364.5 | 5.8829 |
| 700 | 0.006272 | 3177.2 | 3553.5 | 6.0824 |
| 800 | 0.007459 | 3441.5 | 3889.1 | 6.4109 |
| 900 | 0.008505 | 3681.0 | 4191.5 | 6.6805 |
| 1000 | 0.009480 | 3906.4 | 4475.2 | 6.9127 |
| 1100 | 0.010409 | 4124.1 | 4748.6 | 7.1195 |
| 1200 | 0.011317 | 4338.2 | 5017.2 | 7.3083 |
| 1300 | 0.012215 | 4551.4 | 5284.3 | 7.483 |

## Appendix 8d: Properties of Subcooled Water

| | P=5.0 MPa (T$_{sat}$=263.99°C) | | | | P=10.0 MPa (T$_{sat}$=311.06°C) | | | | P=15.0 MPa (T$_{sat}$=342.24°C) | | | |
|---|---|---|---|---|---|---|---|---|---|---|---|---|
| T, °C | v, m³/kg | u, kJ/kg | h, kJ/kg | s, kJ/kgK | v, m³/kg | u, kJ/kg | h, kJ/kg | s, kJ/kgK | v, m³/kg | u, kJ/kg | h, kJ/kg | s, kJ/kgK |
| Sat. | 0.0012859 | 1147.8 | 1154.2 | 2.9202 | 0.0014524 | 1393.0 | 1407.6 | 3.3596 | 0.0016581 | 1585.60 | 1610.5 | 3.6848 |
| 0 | 0.0009977 | 0.0 | 5.0 | 0.0001 | 0.0009952 | 0.1 | 10.0 | 0.0002 | 0.0009928 | 0.15 | 15.1 | 0.0004 |
| 20 | 0.0009995 | 83.7 | 88.7 | 0.2956 | 0.0009972 | 83.4 | 93.3 | 0.2945 | 0.0009950 | 83.06 | 98.0 | 0.2934 |
| 40 | 0.0010056 | 167.0 | 172.0 | 0.5705 | 0.0010034 | 166.4 | 176.4 | 0.5686 | 0.0010013 | 165.76 | 180.8 | 0.5666 |
| 60 | 0.0010149 | 250.2 | 255.3 | 0.8285 | 0.0010127 | 249.4 | 259.5 | 0.8258 | 0.0010105 | 248.51 | 263.7 | 0.8232 |
| 80 | 0.0010268 | 333.7 | 338.9 | 1.0720 | 0.0010245 | 332.6 | 342.8 | 1.0688 | 0.0010222 | 331.48 | 346.8 | 1.0656 |
| 100 | 0.0010410 | 417.5 | 422.7 | 1.3030 | 0.0010385 | 416.1 | 426.5 | 1.2992 | 0.0010361 | 414.74 | 430.3 | 1.2955 |
| 120 | 0.0010576 | 501.8 | 507.1 | 1.5233 | 0.0010549 | 500.1 | 510.6 | 1.5189 | 0.0010522 | 498.40 | 514.2 | 1.5145 |
| 140 | 0.0010768 | 586.8 | 592.2 | 1.7343 | 0.0010737 | 584.7 | 595.4 | 1.7292 | 0.0010707 | 582.66 | 598.7 | 1.7242 |
| 160 | 0.0010988 | 672.6 | 678.1 | 1.9375 | 0.0010953 | 670.1 | 681.1 | 1.9317 | 0.0010918 | 667.71 | 684.1 | 1.9260 |
| 180 | 0.0011240 | 759.6 | 765.3 | 2.1341 | 0.0011199 | 756.7 | 767.8 | 2.1275 | 0.0011159 | 753.76 | 770.5 | 2.1210 |
| 200 | 0.0011530 | 848.1 | 853.9 | 2.3255 | 0.0011480 | 844.5 | 856.0 | 2.3178 | 0.0011433 | 841.00 | 858.2 | 2.3104 |

(Continued)

(continued)

| | P=5.0 MPa (T_sat =263.99°C) | | | | P=10.0 MPa (T_sat =311.06°C) | | | | P=15.0 MPa (T_sat =342.24°C) | | | |
|---|---|---|---|---|---|---|---|---|---|---|---|---|
| T, °C | v, m³/kg | u, kJ/kg | h, kJ/kg | s, kJ/kgK | v, m³/kg | u, kJ/kg | h, kJ/kg | s, kJ/kgK | v, m³/kg | u, kJ/kg | h, kJ/kg | s, kJ/kgK |
| 220 | 0.0011866 | 938.4 | 944.4 | 2.5128 | 0.0011805 | 934.1 | 945.9 | 2.5039 | 0.0011748 | 929.90 | 947.5 | 2.4953 |
| 240 | 0.0012264 | 1031.4 | 1037.5 | 2.6979 | 0.0012187 | 1026.0 | 1038.1 | 2.6872 | 0.0012114 | 1020.80 | 1039.0 | 2.6771 |
| 260 | 0.0012749 | 1127.9 | 1134.3 | 2.8830 | 0.0012645 | 1121.1 | 1133.7 | 2.8699 | 0.0012550 | 1114.60 | 1133.4 | 2.8576 |
| 280 | | | | | 0.0013216 | 1220.9 | 1234.1 | 3.0548 | 0.0013084 | 1212.50 | 1232.1 | 3.0393 |
| 300 | | | | | 0.0013972 | 1328.4 | 1342.3 | 3.2469 | 0.0013770 | 1316.60 | 1337.3 | 3.2260 |
| 320 | | | | | | | | | 0.0014724 | 1431.10 | 1453.2 | 3.4247 |
| 340 | | | | | | | | | 0.0016311 | 1567.50 | 1591.9 | 3.6546 |

| | P=20.0 MPa (T_sat =365.81°C) | | | | P=30.0 MPa | | | | P=50.0 MPa | | | |
|---|---|---|---|---|---|---|---|---|---|---|---|---|
| T, °C | v, m³/kg | u, kJ/kg | h, kJ/kg | s, kJ/kgK | v, m³/kg | u, kJ/kg | h, kJ/kg | s, kJ/kgK | v, m³/kg | u, kJ/kg | h, kJ/kg | s, kJ/kgK |
| Sat. | 0.002036 | 1785.6 | 1826.3 | 4.0139 | | | | | | | | |
| 0 | 0.0009904 | 0.2 | 20.0 | 0.0004 | 0.0009856 | 0.3 | 29.8 | 0.0001 | 0.0009766 | 0.20 | 49.0 | 0.0014 |
| 20 | 0.0009928 | 82.8 | 102.6 | 0.2923 | 0.0009886 | 82.2 | 111.8 | 0.2899 | 0.0009804 | 81.00 | 130.0 | 0.2848 |
| 40 | 0.0009992 | 165.2 | 185.2 | 0.5646 | 0.0009951 | 164.0 | 193.9 | 0.5607 | 0.0009872 | 161.86 | 211.2 | 0.5527 |
| 60 | 0.0010084 | 247.7 | 267.9 | 0.8206 | 0.0010042 | 246.1 | 276.2 | 0.8154 | 0.0009962 | 242.98 | 292.8 | 0.8052 |
| 80 | 0.0010199 | 330.4 | 350.8 | 1.0624 | 0.0010156 | 328.3 | 358.8 | 1.0561 | 0.0010073 | 324.34 | 374.7 | 1.0440 |
| 100 | 0.0010337 | 413.4 | 434.1 | 1.2917 | 0.0010290 | 410.8 | 441.7 | 1.2844 | 0.0010201 | 405.88 | 456.9 | 1.2703 |
| 120 | 0.0010496 | 496.8 | 517.8 | 1.5102 | 0.0010445 | 493.6 | 524.9 | 1.5018 | 0.0010348 | 487.65 | 539.4 | 1.4857 |
| 140 | 0.0010678 | 580.7 | 602.0 | 1.7193 | 0.0010621 | 576.9 | 608.8 | 1.7098 | 0.0010515 | 569.77 | 622.4 | 1.6915 |
| 160 | 0.0010885 | 665.4 | 687.1 | 1.9204 | 0.0010821 | 660.8 | 693.3 | 1.9096 | 0.0010703 | 652.41 | 705.9 | 1.8891 |
| 180 | 0.0011120 | 751.0 | 773.2 | 2.1147 | 0.0011047 | 745.6 | 778.7 | 2.1024 | 0.0010912 | 735.69 | 790.3 | 2.0794 |
| 200 | 0.0011388 | 837.7 | 860.5 | 2.3031 | 0.0011302 | 831.4 | 865.3 | 2.2893 | 0.0011146 | 819.70 | 875.5 | 2.2634 |
| 220 | 0.0011695 | 925.9 | 949.3 | 2.4870 | 0.0011590 | 918.3 | 953.1 | 2.4711 | 0.0011408 | 904.70 | 961.7 | 2.4419 |
| 240 | 0.0012046 | 1016.0 | 1040.0 | 2.6674 | 0.0011920 | 1006.9 | 1042.6 | 2.6490 | 0.0011702 | 990.70 | 1049.2 | 2.6158 |
| 260 | 0.0012462 | 1108.6 | 1133.5 | 2.8459 | 0.0012303 | 1097.4 | 1134.3 | 2.8243 | 0.0012034 | 1078.10 | 1138.2 | 2.7860 |
| 280 | 0.0012965 | 1204.7 | 1230.6 | 3.0248 | 0.0012755 | 1190.7 | 1229.0 | 2.9986 | 0.0012415 | 1167.20 | 1229.3 | 2.9537 |
| 300 | 0.0013596 | 1306.1 | 1333.3 | 3.2071 | 0.0013307 | 1287.9 | 1327.8 | 3.1741 | 0.0012860 | 1258.70 | 1323.0 | 3.1200 |
| 320 | 0.0014437 | 1415.7 | 1444.6 | 3.3979 | 0.0013997 | 1390.7 | 1432.7 | 3.3539 | 0.0013388 | 1353.30 | 1420.2 | 3.2868 |
| 340 | 0.0015684 | 1539.7 | 1571.0 | 3.6075 | 0.0014920 | 1501.7 | 1546.5 | 3.5426 | 0.0014032 | 1452.00 | 1522.1 | 3.4557 |
| 360 | 0.0018226 | 1702.8 | 1739.3 | 3.8772 | 0.0016265 | 1626.6 | 1675.4 | 3.7494 | 0.0014838 | 1556.00 | 1630.2 | 3.6291 |
| 380 | | | | | 0.0018691 | 1781.4 | 1837.5 | 4.0012 | 0.0015884 | 1667.20 | 1746.6 | 3.8101 |

# Appendix 9: Properties of R-134a

## Appendix 9a: Properties of Saturated R-134a (Temperature Table)

| Temp., °C | Pressure, MPa | Specific volume, m³/kg | | Internal energy, kJ/kg | | Enthalpy, kJ/kg | | Entropy, kJ/kgK | |
|---|---|---|---|---|---|---|---|---|---|
| $T_{sat}$ | $P$ | $v_f$ | $v_g$ | $u_f$ | $u_g$ | $h_f$ | $h_g$ | $s_f$ | $s_g$ |
| −24 | 0.11160 | 0.0007296 | 0.1728 | 19.21 | 213.57 | 19.29 | 232.85 | 0.0798 | 0.9370 |
| −22 | 0.12192 | 0.0007328 | 0.1590 | 21.68 | 214.70 | 21.77 | 234.08 | 0.0897 | 0.9351 |
| −20 | 0.13299 | 0.0007361 | 0.1464 | 24.17 | 215.84 | 24.26 | 235.31 | 0.0996 | 0.9332 |
| −18 | 0.14483 | 0.0007395 | 0.1350 | 26.67 | 216.97 | 26.77 | 236.53 | 0.1094 | 0.9315 |
| −16 | 0.15748 | 0.0007428 | 0.1247 | 29.18 | 218.10 | 29.30 | 237.74 | 0.1192 | 0.9298 |
| −12 | 0.18540 | 0.0007498 | 0.1068 | 34.25 | 220.36 | 34.39 | 240.15 | 0.1388 | 0.9267 |
| −8 | 0.21704 | 0.0007569 | 0.0919 | 39.38 | 222.60 | 39.54 | 242.54 | 0.1583 | 0.9239 |
| −4 | 0.25274 | 0.0007644 | 0.0794 | 44.56 | 224.84 | 44.75 | 244.90 | 0.1777 | 0.9213 |
| 0 | 0.29282 | 0.0007721 | 0.0689 | 49.79 | 227.06 | 50.02 | 247.23 | 0.1970 | 0.9190 |
| 4 | 0.33765 | 0.0007801 | 0.0600 | 55.08 | 229.27 | 55.35 | 249.53 | 0.2162 | 0.9169 |
| 8 | 0.38756 | 0.0007884 | 0.0525 | 60.43 | 231.46 | 60.73 | 251.80 | 0.2354 | 0.9150 |
| 12 | 0.44294 | 0.0007971 | 0.0460 | 65.83 | 233.63 | 66.18 | 254.03 | 0.2545 | 0.9132 |
| 16 | 0.50416 | 0.0008062 | 0.0405 | 71.29 | 235.78 | 71.69 | 256.22 | 0.2735 | 0.9116 |
| 20 | 0.57160 | 0.0008157 | 0.0358 | 76.80 | 237.91 | 77.26 | 258.35 | 0.2924 | 0.9102 |
| 24 | 0.64566 | 0.0008257 | 0.0317 | 82.37 | 240.01 | 82.90 | 260.45 | 0.3113 | 0.9089 |
| 26 | 0.68530 | 0.0008309 | 0.0298 | 85.18 | 241.05 | 85.75 | 261.48 | 0.3208 | 0.9082 |
| 28 | 0.72675 | 0.0008362 | 0.0281 | 88.00 | 242.08 | 88.61 | 262.50 | 0.3302 | 0.9076 |
| 30 | 0.77006 | 0.0008417 | 0.0265 | 90.84 | 243.10 | 91.49 | 263.50 | 0.3396 | 0.9070 |
| 32 | 0.81528 | 0.0008473 | 0.0250 | 93.70 | 244.12 | 94.39 | 264.48 | 0.3490 | 0.9064 |
| 34 | 0.86247 | 0.0008530 | 0.0236 | 96.58 | 245.12 | 97.31 | 265.45 | 0.3584 | 0.9058 |
| 36 | 0.91168 | 0.0008590 | 0.0223 | 99.47 | 246.11 | 100.25 | 266.40 | 0.3678 | 0.9053 |
| 38 | 0.96298 | 0.0008651 | 0.0210 | 102.38 | 247.09 | 103.21 | 267.33 | 0.3772 | 0.9047 |
| 40 | 1.0164 | 0.0008714 | 0.0199 | 105.30 | 248.06 | 106.19 | 268.24 | 0.3866 | 0.9041 |
| 42 | 1.0720 | 0.0008780 | 0.0188 | 108.25 | 249.02 | 109.19 | 269.14 | 0.3960 | 0.9035 |
| 44 | 1.1299 | 0.0008847 | 0.0177 | 111.22 | 249.96 | 112.22 | 270.01 | 0.4054 | 0.9030 |
| 48 | 1.2526 | 0.0008989 | 0.0159 | 117.22 | 251.79 | 118.35 | 271.68 | 0.4243 | 0.9017 |
| 52 | 1.3851 | 0.0009142 | 0.0142 | 123.31 | 253.55 | 124.58 | 273.24 | 0.4432 | 0.9004 |
| 56 | 1.5278 | 0.0009308 | 0.0127 | 129.51 | 255.23 | 130.93 | 274.68 | 0.4622 | 0.8990 |
| 60 | 1.6813 | 0.0009488 | 0.0114 | 135.82 | 256.81 | 137.42 | 275.99 | 0.4814 | 0.8973 |
| 70 | 2.1162 | 0.0010027 | 0.0086 | 152.22 | 260.15 | 154.34 | 278.43 | 0.5302 | 0.8918 |
| 80 | 2.6324 | 0.0010766 | 0.0064 | 169.88 | 262.14 | 172.71 | 279.12 | 0.5814 | 0.8827 |
| 90 | 3.2435 | 0.0011949 | 0.0046 | 189.82 | 261.34 | 193.69 | 276.32 | 0.6380 | 0.8655 |
| 100 | 3.9742 | 0.0015443 | 0.0027 | 218.60 | 248.49 | 224.74 | 259.13 | 0.7196 | 0.8117 |

## Appendix 9b: Properties of Saturated R-134a (Pressure Table)

| Pressure, MPa | Temp., °C | Specific volume, m³/kg | | Internal energy, kJ/kg | | Enthalpy, kJ/kg | | Entropy, kJ/kgK | |
|---|---|---|---|---|---|---|---|---|---|
| $P$ | $T_{sat}$ | $v_f$ | $v_g$ | $u_f$ | $u_g$ | $h_f$ | $h_g$ | $s_f$ | $s_g$ |
| 0.06 | −37.07 | 0.000710 | 0.3100 | 3.41 | 206.12 | 3.46 | 224.72 | 0.0147 | 0.9520 |
| 0.08 | −31.21 | 0.000718 | 0.2366 | 10.41 | 209.46 | 10.47 | 228.39 | 0.044 | 0.9447 |
| 0.10 | −26.43 | 0.000726 | 0.1917 | 16.22 | 212.18 | 16.29 | 231.35 | 0.0678 | 0.9395 |
| 0.12 | −22.36 | 0.000732 | 0.1614 | 21.23 | 214.50 | 21.32 | 233.86 | 0.0879 | 0.9354 |
| 0.14 | −18.80 | 0.000738 | 0.1395 | 25.66 | 216.52 | 25.77 | 236.04 | 0.1055 | 0.9322 |
| 0.16 | −15.62 | 0.000744 | 0.1229 | 29.66 | 218.32 | 29.78 | 237.97 | 0.1211 | 0.9295 |
| 0.18 | −12.73 | 0.000749 | 0.1098 | 33.31 | 219.94 | 33.45 | 239.71 | 0.1352 | 0.9273 |
| 0.20 | −10.09 | 0.000753 | 0.0993 | 36.69 | 221.43 | 36.84 | 241.30 | 0.1481 | 0.9253 |
| 0.24 | −5.37 | 0.000762 | 0.0834 | 42.77 | 224.07 | 42.95 | 244.09 | 0.171 | 0.9222 |
| 0.28 | −1.23 | 0.000770 | 0.0719 | 48.18 | 226.38 | 48.39 | 246.52 | 0.1911 | 0.9197 |
| 0.32 | 2.48 | 0.000777 | 0.0632 | 53.06 | 228.43 | 53.31 | 248.66 | 0.2089 | 0.9177 |
| 0.36 | 5.84 | 0.000784 | 0.0564 | 57.54 | 230.28 | 57.82 | 250.58 | 0.2251 | 0.9160 |
| 0.40 | 8.93 | 0.000790 | 0.0509 | 61.69 | 231.97 | 62.00 | 252.32 | 0.2399 | 0.9145 |
| 0.50 | 15.74 | 0.000806 | 0.0409 | 70.93 | 235.64 | 71.33 | 256.07 | 0.2723 | 0.9117 |
| 0.60 | 21.58 | 0.000820 | 0.0341 | 78.99 | 238.74 | 79.48 | 259.19 | 0.2999 | 0.9097 |
| 0.70 | 26.72 | 0.000833 | 0.0292 | 86.19 | 241.42 | 86.78 | 261.85 | 0.3242 | 0.9080 |
| 0.80 | 31.33 | 0.000845 | 0.0255 | 92.75 | 243.78 | 93.42 | 264.15 | 0.3459 | 0.9066 |
| 0.90 | 35.53 | 0.000858 | 0.0226 | 98.79 | 245.88 | 99.56 | 266.18 | 0.3656 | 0.9054 |
| 1.00 | 39.39 | 0.000870 | 0.0202 | 104.42 | 247.77 | 105.29 | 267.97 | 0.3838 | 0.9043 |
| 1.20 | 46.32 | 0.000893 | 0.0166 | 114.69 | 251.03 | 115.76 | 270.99 | 0.4164 | 0.9023 |
| 1.40 | 52.43 | 0.000916 | 0.0140 | 123.98 | 253.74 | 125.26 | 273.40 | 0.4453 | 0.9003 |
| 1.60 | 57.92 | 0.000939 | 0.0121 | 132.52 | 256.00 | 134.02 | 275.33 | 0.4714 | 0.8982 |
| 1.80 | 62.91 | 0.000963 | 0.0105 | 140.49 | 257.88 | 142.22 | 276.83 | 0.4954 | 0.8959 |
| 2.00 | 67.49 | 0.000988 | 0.0093 | 148.02 | 259.41 | 149.99 | 277.94 | 0.5178 | 0.8934 |
| 2.50 | 77.59 | 0.001056 | 0.0069 | 165.48 | 261.84 | 168.12 | 279.17 | 0.5687 | 0.8854 |
| 3.00 | 86.22 | 0.001142 | 0.0053 | 181.88 | 262.16 | 185.30 | 278.01 | 0.6156 | 0.8735 |

# Appendix 9c: Properties of Superheated R-134a

| T, °C | P=0.06 MPa (T_sat = −37.07 °C) v, m³/kg | u, kJ/kg | h, kJ/kg | s, kJ/kgK | P=0.10 MPa (T_sat = −26.43 °C) v, m³/kg | u, kJ/kg | h, kJ/kg | s, kJ/kgK | P=0.14 MPa (T_sat = −18.08 °C) v, m³/kg | u, kJ/kg | h, kJ/kg | s, kJ/kgK |
|---|---|---|---|---|---|---|---|---|---|---|---|---|
| T_sat | 0.31003 | 206.12 | 224.72 | 0.9520 | 0.19170 | 212.18 | 231.35 | 0.9395 | 0.13945 | 216.52 | 236.04 | 0.9322 |
| −20 | 0.33536 | 217.86 | 237.98 | 1.0062 | 0.19770 | 216.77 | 236.54 | 0.9602 | | | | |
| −10 | 0.34992 | 224.97 | 245.96 | 1.0371 | 0.20686 | 224.01 | 244.70 | 0.9918 | 0.14549 | 223.03 | 243.40 | 0.9606 |
| 0 | 0.36433 | 232.24 | 254.10 | 1.0675 | 0.21587 | 231.41 | 252.99 | 1.0227 | 0.15219 | 230.55 | 251.86 | 0.9922 |
| 10 | 0.37861 | 239.69 | 262.41 | 1.0973 | 0.22473 | 238.96 | 261.43 | 1.0531 | 0.15875 | 238.21 | 260.43 | 1.0230 |
| 20 | 0.39279 | 247.32 | 270.89 | 1.1267 | 0.23349 | 246.67 | 270.02 | 1.0829 | 0.16520 | 246.01 | 269.13 | 1.0532 |
| 30 | 0.40688 | 255.12 | 279.53 | 1.1557 | 0.24216 | 254.54 | 278.76 | 1.1122 | 0.17155 | 253.96 | 277.97 | 1.0828 |
| 40 | 0.42091 | 263.10 | 288.35 | 1.1844 | 0.25076 | 262.58 | 287.66 | 1.1411 | 0.17783 | 262.06 | 286.96 | 1.1120 |
| 50 | 0.43487 | 271.25 | 297.34 | 1.2126 | 0.25930 | 270.79 | 296.72 | 1.1696 | 0.18404 | 270.32 | 296.09 | 1.1407 |
| 60 | 0.44879 | 279.58 | 306.51 | 1.2405 | 0.26779 | 279.16 | 305.94 | 1.1977 | 0.19020 | 278.74 | 305.37 | 1.1690 |
| 70 | 0.46266 | 288.08 | 315.84 | 1.2681 | 0.27623 | 287.70 | 315.32 | 1.2254 | 0.19633 | 287.32 | 314.80 | 1.1969 |
| 80 | 0.47650 | 296.75 | 325.34 | 1.2954 | 0.28464 | 296.40 | 324.87 | 1.2528 | 0.20241 | 296.06 | 324.39 | 1.2244 |
| 90 | 0.49031 | 305.58 | 335.00 | 1.3224 | 0.29302 | 305.27 | 334.57 | 1.2799 | 0.20846 | 304.95 | 334.14 | 1.2516 |
| 100 | | | | | | | | | 0.21449 | 314.01 | 344.04 | 1.2785 |

| T, °C | P=0.18 MPa (T_sat = −12.73 °C) v, m³/kg | u, kJ/kg | h, kJ/kg | s, kJ/kgK | P=0.20 MPa (T_sat = −10.09 °C) v, m³/kg | u, kJ/kg | h, kJ/kg | s, kJ/kgK | P=0.24 MPa (T_sat = −5.37 °C) v, m³/kg | u, kJ/kg | h, kJ/kg | s, kJ/kgK |
|---|---|---|---|---|---|---|---|---|---|---|---|---|
| T_sat | 0.10983 | 219.94 | 239.71 | 0.9273 | 0.09933 | 221.43 | 241.30 | 0.9253 | 0.08343 | 224.07 | 244.09 | 0.9222 |
| −10 | 0.11135 | 222.02 | 242.06 | 0.9362 | 0.09938 | 221.50 | 241.38 | 0.9256 | | | | |
| 0 | 0.11678 | 229.67 | 250.69 | 0.9684 | 0.10438 | 229.23 | 250.10 | 0.9582 | 0.08574 | 228.31 | 248.89 | 0.9399 |
| 10 | 0.12207 | 237.44 | 259.41 | 0.9998 | 0.10922 | 237.05 | 258.89 | 0.9898 | 0.08993 | 236.26 | 257.84 | 0.9721 |
| 20 | 0.12723 | 245.33 | 268.23 | 1.0304 | 0.11394 | 244.99 | 267.78 | 1.0206 | 0.09339 | 244.30 | 266.85 | 1.0034 |
| 30 | 0.13230 | 253.36 | 277.17 | 1.0604 | 0.11856 | 253.06 | 276.77 | 1.0508 | 0.09794 | 252.45 | 275.95 | 1.0339 |
| 40 | 0.13730 | 261.53 | 286.24 | 1.0898 | 0.12311 | 261.26 | 285.88 | 1.0804 | 0.10181 | 260.72 | 285.16 | 1.0637 |
| 50 | 0.14222 | 269.85 | 295.45 | 1.1187 | 0.12758 | 269.61 | 295.12 | 1.1094 | 0.10562 | 269.12 | 294.47 | 1.0930 |

(continued)

(continued)

Top section:

| T, °C | P=0.18 MPa ($T_{sat}=-12.73\,°C$) | | | | P=0.20 MPa ($T_{sat}=-10.09\,°C$) | | | | P=0.24 MPa ($T_{sat}=-5.37\,°C$) | | | |
|---|---|---|---|---|---|---|---|---|---|---|---|---|
| | v, m³/kg | u, kJ/kg | h, kJ/kg | s, kJ/kgK | v, m³/kg | u, kJ/kg | h, kJ/kg | s, kJ/kgK | v, m³/kg | u, kJ/kg | h, kJ/kg | s, kJ/kgK |
| 60 | 0.14710 | 278.31 | 304.79 | 1.1472 | 0.13201 | 278.10 | 304.50 | 1.1380 | 0.10937 | 277.67 | 303.91 | 1.1218 |
| 70 | 0.15193 | 286.93 | 314.28 | 1.1753 | 0.13639 | 286.74 | 314.02 | 1.1661 | 0.11307 | 286.35 | 313.49 | 1.1501 |
| 80 | 0.15672 | 295.71 | 323.92 | 1.2030 | 0.14073 | 295.53 | 323.68 | 1.1939 | 0.11674 | 295.18 | 323.19 | 1.1780 |
| 90 | 0.16148 | 304.63 | 333.70 | 1.2303 | 0.14504 | 304.47 | 333.48 | 1.2212 | 0.12037 | 304.15 | 333.04 | 1.2055 |
| 100 | 0.16622 | 313.72 | 343.63 | 1.2573 | 0.14932 | 313.57 | 343.43 | 1.2483 | 0.12398 | 313.27 | 343.03 | 1.2326 |

Bottom section:

| T, °C | P=0.28 MPa ($T_{sat}=-1.23\,°C$) | | | | P=0.32 MPa ($T_{sat}=-2.48\,°C$) | | | | P=0.40 MPa ($T_{sat}=8.93\,°C$) | | | |
|---|---|---|---|---|---|---|---|---|---|---|---|---|
| | v, m³/kg | u, kJ/kg | h, kJ/kg | s, kJ/kgK | v, m³/kg | u, kJ/kg | h, kJ/kg | s, kJ/kgK | v, m³/kg | u, kJ/kg | h, kJ/kg | s, kJ/kgK |
| $T_{sat}$ | 0.07193 | 226.38 | 246.52 | 0.9197 | 0.06322 | 228.43 | 248.66 | 0.9177 | 0.05089 | 231.97 | 252.32 | 0.9145 |
| 0 | 0.07240 | 227.37 | 247.64 | 0.9238 | | | | | | | | |
| 10 | 0.07613 | 235.44 | 256.76 | 0.9566 | 0.06576 | 234.61 | 255.65 | 0.9427 | 0.05119 | 232.87 | 253.35 | 0.9182 |
| 20 | 0.07972 | 243.59 | 265.91 | 0.9883 | 0.06901 | 242.87 | 264.95 | 0.9749 | 0.05397 | 241.37 | 262.96 | 0.9515 |
| 30 | 0.08320 | 251.83 | 275.12 | 1.0192 | 0.07214 | 251.19 | 274.28 | 1.0062 | 0.05662 | 249.89 | 272.54 | 0.9837 |
| 40 | 0.08660 | 260.17 | 284.42 | 1.0494 | 0.07518 | 259.61 | 283.67 | 1.0367 | 0.05917 | 258.47 | 282.14 | 1.0148 |
| 50 | 0.08992 | 268.64 | 293.81 | 1.0789 | 0.07815 | 268.14 | 293.15 | 1.0665 | 0.06164 | 267.13 | 291.79 | 1.0452 |
| 60 | 0.09319 | 277.23 | 303.32 | 1.1079 | 0.08106 | 276.79 | 302.72 | 1.0957 | 0.06405 | 275.89 | 301.51 | 1.0748 |
| 70 | 0.09641 | 285.96 | 312.95 | 1.1364 | 0.08392 | 285.56 | 312.41 | 1.1243 | 0.06641 | 284.75 | 311.32 | 1.1038 |
| 80 | 0.09960 | 294.82 | 322.71 | 1.1644 | 0.08674 | 294.46 | 322.22 | 1.1525 | 0.06873 | 293.73 | 321.23 | 1.1322 |
| 90 | 0.10275 | 303.83 | 332.60 | 1.1920 | 0.08953 | 303.50 | 332.15 | 1.1802 | 0.07102 | 302.84 | 331.25 | 1.1602 |
| 100 | 0.10587 | 312.98 | 342.62 | 1.2193 | 0.09229 | 312.68 | 342.21 | 1.2076 | 0.07327 | 312.07 | 341.38 | 1.1878 |
| 110 | 0.10897 | 322.27 | 352.78 | 1.2461 | 0.09503 | 322.00 | 352.40 | 1.2345 | 0.07550 | 321.44 | 351.64 | 1.2149 |
| 120 | 0.11205 | 331.71 | 363.08 | 1.2727 | 0.09774 | 331.45 | 362.73 | 1.2611 | 0.07771 | 330.94 | 362.03 | 1.2417 |
| 130 | | | | | | | | | 0.07991 | 340.58 | 372.54 | 1.2681 |
| 140 | | | | | | | | | 0.08208 | 350.35 | 383.18 | 1.2941 |

## P = 0.50 MPa (T_sat = 15.74 °C)

| T, °C | v, m³/kg | u, kJ/kg | h, kJ/kg | s, kJ/kgK |
|---|---|---|---|---|
| T_sat | 0.04086 | 235.64 | 256.07 | 0.9117 |
| 20 | 0.04188 | 239.40 | 260.34 | 0.9264 |
| 30 | 0.04416 | 248.20 | 270.28 | 0.9597 |
| 40 | 0.04633 | 256.99 | 280.16 | 0.9918 |
| 50 | 0.04842 | 265.83 | 290.04 | 1.0229 |
| 60 | 0.05043 | 274.73 | 299.95 | 1.0531 |
| 70 | 0.05240 | 283.72 | 309.92 | 1.0825 |
| 80 | 0.05432 | 292.80 | 319.96 | 1.1114 |
| 90 | 0.05620 | 302.00 | 330.10 | 1.1397 |
| 100 | 0.05805 | 311.31 | 340.33 | 1.1675 |
| 110 | 0.05988 | 320.74 | 350.68 | 1.1949 |
| 120 | 0.06168 | 330.30 | 361.14 | 1.2218 |
| 130 | 0.06347 | 339.98 | 371.72 | 1.2484 |
| 140 | 0.06524 | 349.79 | 382.42 | 1.2746 |

## P = 0.60 MPa (T_sat = 21.58 °C)

| T, °C | v, m³/kg | u, kJ/kg | h, kJ/kg | s, kJ/kgK |
|---|---|---|---|---|
| T_sat | 0.03408 | 238.74 | 259.19 | 0.9097 |
| 20 | | | | |
| 30 | 0.03581 | 246.41 | 267.89 | 0.9388 |
| 40 | 0.03774 | 255.45 | 278.09 | 0.9719 |
| 50 | 0.03958 | 264.48 | 288.23 | 1.0037 |
| 60 | 0.04134 | 273.54 | 298.35 | 1.0346 |
| 70 | 0.04304 | 282.66 | 308.48 | 1.0645 |
| 80 | 0.04469 | 291.86 | 318.67 | 1.0938 |
| 90 | 0.04631 | 301.14 | 328.93 | 1.1225 |
| 100 | 0.04790 | 310.53 | 339.27 | 1.1505 |
| 110 | 0.04946 | 320.03 | 349.70 | 1.1781 |
| 120 | 0.05099 | 329.64 | 360.24 | 1.2053 |
| 130 | 0.05251 | 339.38 | 370.88 | 1.2320 |
| 140 | 0.05402 | 349.23 | 381.64 | 1.2584 |
| 150 | 0.05550 | 359.21 | 392.52 | 1.2844 |
| 160 | 0.05698 | 369.32 | 403.51 | 1.3100 |

## P = 0.70 MPa (T_sat = 26.72 °C)

| T, °C | v, m³/kg | u, kJ/kg | h, kJ/kg | s, kJ/kgK |
|---|---|---|---|---|
| T_sat | 0.02918 | 241.42 | 261.85 | 0.9080 |
| 30 | 0.02979 | 244.51 | 265.37 | 0.9197 |
| 40 | 0.03157 | 253.83 | 275.93 | 0.9539 |
| 50 | 0.03324 | 263.08 | 286.35 | 0.9867 |
| 60 | 0.03482 | 272.31 | 296.69 | 1.0182 |
| 70 | 0.03634 | 281.57 | 307.01 | 1.0487 |
| 80 | 0.03781 | 290.88 | 317.35 | 1.0784 |
| 90 | 0.03924 | 300.27 | 327.74 | 1.1074 |
| 100 | 0.04064 | 309.74 | 338.19 | 1.1358 |
| 110 | 0.04201 | 319.31 | 348.71 | 1.1637 |
| 120 | 0.04335 | 328.98 | 359.33 | 1.1910 |
| 130 | 0.04468 | 338.76 | 370.04 | 1.2179 |
| 140 | 0.04599 | 348.66 | 380.86 | 1.2444 |
| 150 | 0.04729 | 358.68 | 391.79 | 1.2706 |
| 160 | 0.04857 | 368.82 | 402.82 | 1.2963 |

## P = 0.80 MPa (T_sat = 31.33 °C)

| T, °C | v, m³/kg | u, kJ/kg | h, kJ/kg | s, kJ/kgK |
|---|---|---|---|---|
| T_sat | 0.02547 | 243.78 | 264.15 | 0.9066 |
| 40 | 0.02691 | 252.13 | 273.66 | 0.9374 |
| 50 | 0.02846 | 261.62 | 284.39 | 0.9711 |
| 60 | 0.02992 | 271.04 | 294.98 | 1.0034 |
| 70 | 0.03131 | 280.45 | 305.50 | 1.0345 |
| 80 | 0.03264 | 289.89 | 316.00 | 1.0647 |
| 90 | 0.03393 | 299.37 | 326.52 | 1.0940 |
| 100 | 0.03519 | 308.93 | 337.08 | 1.1227 |

## P = 0.90 MPa (T_sat = 35.53 °C)

| T, °C | v, m³/kg | u, kJ/kg | h, kJ/kg | s, kJ/kgK |
|---|---|---|---|---|
| T_sat | 0.02255 | 245.88 | 266.18 | 0.9054 |
| 40 | 0.02325 | 250.32 | 271.25 | 0.9217 |
| 50 | 0.02472 | 260.09 | 282.34 | 0.9566 |
| 60 | 0.02609 | 269.72 | 293.21 | 0.9897 |
| 70 | 0.02738 | 279.30 | 303.94 | 1.0214 |
| 80 | 0.02861 | 288.87 | 314.62 | 1.0521 |
| 90 | 0.02980 | 298.46 | 325.28 | 1.0819 |
| 100 | 0.03095 | 308.11 | 335.96 | 1.1109 |

## P = 1.00 MPa (T_sat = 39.33 °C)

| T, °C | v, m³/kg | u, kJ/kg | h, kJ/kg | s, kJ/kgK |
|---|---|---|---|---|
| T_sat | 0.02020 | 247.77 | 267.97 | 0.9043 |
| 40 | 0.02029 | 248.39 | 268.68 | 0.9066 |
| 50 | 0.02171 | 258.48 | 280.19 | 0.9428 |
| 60 | 0.02301 | 268.35 | 291.36 | 0.9768 |
| 70 | 0.02423 | 278.11 | 302.34 | 1.0093 |
| 80 | 0.02538 | 287.82 | 313.20 | 1.0405 |
| 90 | 0.02649 | 297.53 | 324.01 | 1.0707 |
| 100 | 0.02755 | 307.27 | 334.82 | 1.1000 |

(continued)

(continued)

### P=0.80 MPa ($T_{sat}$=31.33°C), P=0.90 MPa ($T_{sat}$=35.53°C), P=1.00 MPa ($T_{sat}$=39.33°C)

| T, °C | \|0.80\| v, m³/kg | u, kJ/kg | h, kJ/kg | s, kJ/kgK | \|0.90\| v, m³/kg | u, kJ/kg | h, kJ/kg | s, kJ/kgK | \|1.00\| v, m³/kg | u, kJ/kg | h, kJ/kg | s, kJ/kgK |
|---|---|---|---|---|---|---|---|---|---|---|---|---|
| 110 | 0.03642 | 318.57 | 347.71 | 1.1508 | 0.03207 | 317.82 | 346.68 | 1.1392 | 0.02858 | 317.06 | 345.65 | 1.1286 |
| 120 | 0.03762 | 328.31 | 358.40 | 1.1784 | 0.03316 | 327.62 | 357.47 | 1.1670 | 0.02959 | 326.93 | 356.52 | 1.1567 |
| 130 | 0.03881 | 338.14 | 369.19 | 1.2055 | 0.03423 | 337.52 | 368.33 | 1.1943 | 0.03058 | 336.88 | 367.46 | 1.1841 |
| 140 | 0.03997 | 348.09 | 380.07 | 1.2321 | 0.03529 | 347.51 | 379.27 | 1.2211 | 0.03154 | 346.92 | 378.46 | 1.2111 |
| 150 | 0.04113 | 358.15 | 391.05 | 1.2584 | 0.03633 | 357.61 | 390.31 | 1.2475 | 0.03250 | 357.06 | 389.56 | 1.2376 |
| 160 | 0.04227 | 368.32 | 402.14 | 1.2843 | 0.03736 | 367.82 | 401.44 | 1.2735 | 0.03344 | 367.31 | 400.74 | 1.2638 |
| 170 | 0.04340 | 378.61 | 413.33 | 1.3098 | 0.03838 | 378.14 | 412.68 | 1.2992 | 0.03436 | 377.66 | 412.02 | 1.2895 |
| 180 | 0.04452 | 389.02 | 424.63 | 1.3351 | 0.03939 | 388.57 | 424.02 | 1.3245 | 0.03528 | 388.12 | 423.40 | 1.3149 |

### P=1.20 MPa ($T_{sat}$=46.32°C), P=1.40 MPa ($T_{sat}$=52.43°C), P=1.60 MPa ($T_{sat}$=57.92°C)

| T, °C | \|1.20\| v, m³/kg | u, kJ/kg | h, kJ/kg | s, kJ/kgK | \|1.40\| v, m³/kg | u, kJ/kg | h, kJ/kg | s, kJ/kgK | \|1.60\| v, m³/kg | u, kJ/kg | h, kJ/kg | s, kJ/kgK |
|---|---|---|---|---|---|---|---|---|---|---|---|---|
| $T_{sat}$ | 0.01663 | 251.03 | 270.99 | 0.9023 | 0.01405 | 253.74 | 273.40 | 0.9003 | 0.01208 | 256.00 | 275.33 | 0.8982 |
| 50 | 0.01712 | 254.98 | 275.52 | 0.9164 | | | | | | | | |
| 60 | 0.01835 | 265.42 | 287.44 | 0.9527 | 0.01495 | 262.17 | 283.10 | 0.9297 | 0.01233 | 258.48 | 278.20 | 0.9069 |
| 70 | 0.01947 | 275.59 | 298.96 | 0.9868 | 0.01603 | 272.87 | 295.31 | 0.9658 | 0.01340 | 269.89 | 291.33 | 0.9457 |
| 80 | 0.02051 | 285.62 | 310.24 | 1.0192 | 0.01701 | 283.29 | 307.10 | 0.9997 | 0.01435 | 280.78 | 303.74 | 0.9813 |
| 90 | 0.02150 | 295.59 | 321.39 | 1.0503 | 0.01792 | 293.55 | 318.63 | 1.0319 | 0.01521 | 291.39 | 315.72 | 1.0148 |
| 100 | 0.02244 | 305.54 | 332.47 | 1.0804 | 0.01878 | 303.73 | 330.02 | 1.0628 | 0.01601 | 301.84 | 327.46 | 1.0467 |
| 110 | 0.02335 | 315.50 | 343.52 | 1.1096 | 0.01960 | 313.88 | 341.32 | 1.0927 | 0.01677 | 312.20 | 339.04 | 1.0773 |
| 120 | 0.02423 | 325.51 | 354.58 | 1.1381 | 0.02039 | 324.05 | 352.59 | 1.1218 | 0.01750 | 322.53 | 350.53 | 1.1069 |
| 130 | 0.02508 | 335.58 | 365.68 | 1.1660 | 0.02115 | 334.25 | 363.86 | 1.1501 | 0.01820 | 332.87 | 361.99 | 1.1357 |
| 140 | 0.02592 | 345.73 | 376.83 | 1.1933 | 0.02189 | 344.50 | 375.15 | 1.1777 | 0.01887 | 343.24 | 373.44 | 1.1638 |
| 150 | 0.02674 | 355.95 | 388.04 | 1.2201 | 0.02262 | 354.82 | 386.49 | 1.2048 | 0.01953 | 353.66 | 384.91 | 1.1912 |
| 160 | 0.02754 | 366.27 | 399.33 | 1.2465 | 0.02333 | 365.22 | 397.89 | 1.2315 | 0.02017 | 364.15 | 396.43 | 1.2181 |
| 170 | 0.02834 | 376.69 | 410.70 | 1.2724 | 0.02403 | 375.71 | 409.36 | 1.2576 | 0.02080 | 374.71 | 407.99 | 1.2445 |
| 180 | 0.02912 | 387.21 | 422.16 | 1.2980 | 0.02472 | 386.29 | 420.90 | 1.2834 | 0.02142 | 385.35 | 419.62 | 1.2704 |
| 190 | | | | | 0.02541 | 396.96 | 432.53 | 1.3088 | 0.02203 | 396.08 | 431.33 | 1.2960 |
| 200 | | | | | 0.02608 | 407.73 | 444.24 | 1.3338 | 0.02263 | 406.90 | 443.11 | 1.3212 |

# Appendix 10: Generalised Compressibility

## Appendix 10a: Generalised Compressibility Chart 0<$P_r$<1

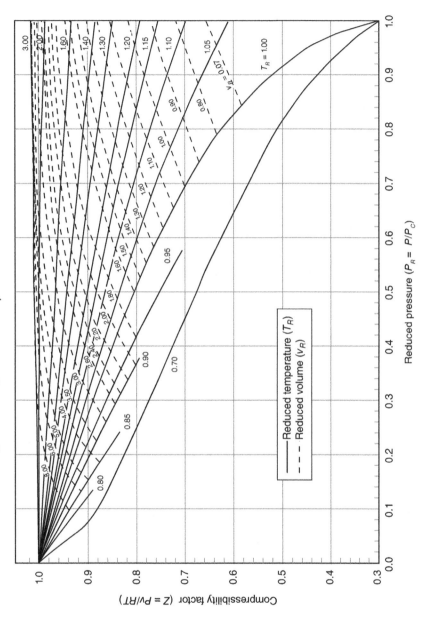

**Figure A10.1**

# Appendix 10b: Generalised Compressibility Chart $0<P_r<7$

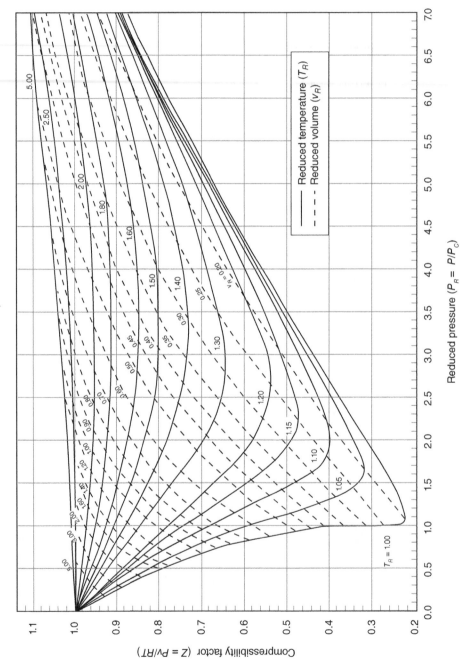

**Figure A10.2**

# Index

*Energy, Entropy and Engines: An Introduction to Thermodynamics*, First Edition. Sanjeev Chandra.
© 2016 John Wiley & Sons, Ltd. Published 2016 by John Wiley & Sons, Ltd.
Companion website: www.wiley.com/go/chandraSol16

Printed and bound by CPI Group (UK) Ltd, Croydon, CR0 4YY

27/10/2024

14580200-0002